Introduction to Nonlinear Optimization

NORTH HOLLAND SERIES IN

SYSTEM SCIENCE AND ENGINEERING

Andrew P. Sage, *Editor*

Wismer and Chattergy	Introduction to Nonlinear Optimization: A Problem Solving Approach
Sutherland	Societal Systems: Methodology, Modeling, and Management
Šiljak	Large Scale Dynamic Systems

INTRODUCTION TO NONLINEAR OPTIMIZATION

A Problem Solving Approach

David A. Wismer
Wismer Associates, Inc.

R. Chattergy
University of Hawaii

NORTH-HOLLAND
New York · Amsterdam · Oxford

Elsevier Science Publishing Co., Inc.
52 Vanderbilt Avenue, New York, New York 10017

Elsevier Science Publishers B.V.
P.O. Box 211, 1000 AE Amsterdam, The Netherlands

Third Printing, 1983.

Library of Congress Cataloging in Publication Data

Wismer, David A 1938-
Introduction to nonlinear optimization: a problem solving approach
(North-Holland series in system science and engineering; 1)

Bibliography: p.
Includes index.
1. Mathematical optimization. 2. Nonlinear theories.
I. Chattergy, R., joint author. II. Title.
QA402.5.W55 519.7 77-8213
ISBN 0-444-00234-0

Manufactured in the United States of America

To Mary Anne, Susan, David, and Jennifer
—DAW

To Virgie, Paul, and David
—RC

Contents

Preface	**xi**
Chapter 1 The Optimization Problem	**1**
Mathematical Preliminaries	2
Nonlinear Programming	14
Linear Programming	19
Solved Problems	20
Supplementary Problems	31
Chapter 2 Classical Optimization	**33**
Maxima, Minima, and Saddle Points	33
Unconstrained Problems	40
Computational Considerations	44
Solved Problems	47
Supplementary Problems	51
Chapter 3 Constraints and Lagrange Multipliers	**55**
Lagrange Multipliers	55
Economic Interpretation	60
Existence of λ	63
Inequality Constraints	65
Computational Methods	67
Solved Problems	69
Supplementary Problems	79
Chapter 4 Optimization with Inequality Constraints	**81**
The Kuhn-Tucker Theorem	81
The Constraint Qualification	86
Saddle-Point Conditions	90
Duality	91
Solved Problems	94
Supplementary Problems	114

Chapter 5 One-Dimensional Search Techniques **117**
Unimodal Functions 117
Dichotomous Search 119
Equal-Interval Search 120
Fibonacci Search 122
Golden-Section Search 127
Quadratic Interpolation 129
Exhaustive Search 132
Solved Problems 133
Supplementary Problems 138

Chapter 6 Unconstrained Gradient Techniques **139**
The Gradient Direction 139
The Gradient Algorithm 145
The Optimal Gradient 147
Convergence Considerations 148
Solved Problems 151
Supplementary Problems 161

Chapter 7 Constrained Gradient Techniques **163**
Boundary Following with Inequality Constraints 163
Boundary Following with Equality Constraints 173
Solved Problems 179
Supplementary Problems 194

Chapter 8 Penalty-Function Methods **197**
A Simple Penalty Function 197
Other Penalty Functions 203
Solved Problems 209
Supplementary Problems 217

Chapter 9 Quadratically Convergent Minimization Algorithms **219**
Quadratic Functions 220
Conjugate Directions in E^n 223
Minimization with Conjugate Directions of Descent 224
Conjugate-Direction Algorithms 226
Generation of Conjugate Directions by One-Dimensional Search 229
Variable-Metric Algorithms 232
Solved Problems 238
Supplementary Problems 242

Chapter 10 Quadratic Programming **245**
Equality Constraints 245
Feasible Solutions 245
Kuhn-Tucker Conditions 247
Influence of the Constraints on the Minimization Problem 248
Method of Theil and Van de Panne 250
Dual Problem 253
Method of Hildreth and D'Espo 254
Capacity Method of Houthakker 258
Modified Simplex Methods 260
Solved Problems 264
Supplementary Problems 268

Chapter 11 Integer Programming **269**
Truncation and Rounding 271
Branch-and-Bound Method 272
Cutting-Plane Methods 276
Solved Problems 280
Supplementary Problems 288

Chapter 12 Dynamic Programming **291**
Multistage Decision Processes 291
Method of Direct Enumeration 292
Principle of Optimality 294
Functional Equation 294
Computational Requirements with One State Variable 300
Computational Requirements with Two State Variables 304
Infinite-Stage Processes and Successive Approximations 307
Solved Problems 309
Supplementary Problems 324

Chapter 13 Large-Scale Programming **325**
Lagrangian Decomposition 326
Feasible Methods 332
Dual Feasible Methods 335
Economic Interpretation 337
Solved Problems 339
Supplementary Problems 349

Appendix Linear Programming **351**
Basic Feasible Solutions 355
Simplex Algorithm 358

Simplex Tableau 360
Dual Problem 364
Solved Problems 365
Supplementary Problems 370

References **373**

Answers to Supplementary Problems **381**

Index **387**

Preface

Over the past ten years, the authors have taught nonlinear optimization to undergraduate, graduate, and continuing education classes for engineering and business administration students. This subject matter is intrinsically multifaceted, while lacking a common theory. Some of its many facets include a diverse mathematical structure rooted in the calculus but containing many recent advances; a plethora of computational algorithms falling into many classes; and endless specialized examples of theory, computation, and application.

Without requiring extensive mathematical prerequisites, our goal has been to cover this broad spectrum of material in an introductory course. In attempting to meet this goal, we originally employed a wide variety of available texts supplemented by extensive class notes, examples, and problems. Gradually the roles of this supplementary material and the available texts were interchanged until this text emerged as the primary means to fulfill our needs.

One of the major goals of this book is to convey to the reader some intuitive insight into the formulation and solution of nonlinear optimization problems by the extensive use of examples, illustrative solved problems, and supplementary problems with answers.

In this text, we present a succinct discussion of most of the major topical areas comprising the subject of nonlinear optimization. These areas include:

- Classical unconstrained optimization techniques from the calculus
- Saddle points arising from the inclusion of constraints using Lagrange multipliers
- The Kuhn-Tucker Theory, duality, and their relation to classical methods
- One-dimensional search algorithms and computational results
- Gradient search algorithms for constrained and unconstrained problems

- Penalty functions and related computational algorithms
- Conjugate direction algorithms and quadratic convergence
- Specialized theory and computational algorithms for quadratic objective functions with linear constraints
- A review of linear programming concepts and solutions
- Formulation and solution of problems having only integer-valued answers
- Multistage decision problems and their solution by dynamic programming
- Decomposition and hierarchical methods for dealing with problems of high dimensionality

In addition, the book contains a total of 122 examples, 153 illustrative solved problems, and 136 supplementary problems with answers included at the back of the book.

This selection of topics is intended to strike a balance between the theoretical underpinnings of nonlinear optimization and the practical need to achieve computational solutions. Sometimes these two areas are indistinguishable; however throughout, the goal to convey intuitive insight is pursued through the use of numerous examples and illustrative solved problems. These examples and solved problems serve as a means to expand on the condensed material discussed in each chapter as well as to apply that material to the solution of relevant problems.

It is intended that this book should serve as an introductory text to the topic of nonlinear optimization. In addition, it can serve as a useful supplement to more advanced courses in nonlinear optimization where a large supply of examples and solved problems would be helpful to student and teacher alike.

The authors are grateful to the many students who, through the years, have helped shape this book and who have labored over its many now-solved problems and examples. In particular, the outstanding efforts of Dr. P. Bansal, a former graduate assistant, are acknowledged, with thanks.

David A. Wismer
R. Chattergy

Chapter 1

The Optimization Problem

The problem of optimization deals with how to do things in the *best* possible manner. For obvious reasons, the solution of such problems is highly desirable, and it has received an increasing amount of attention in recent years. Of course, mathematicians have worked on methods for achieving the optimum for many years, beginning perhaps with Descartes and Fermat, who worked on such problems in the seventeenth century even before the development of the calculus by Newton.

If the solution to an optimization problem does not change with time, the optimization problem is termed *static optimization* or *nonlinear programming*. Such a problem is described by a set of algebraic or transcendental equations termed the mathematical model, or just the model. The solution to this problem is of course only optimal for the model in question. Thus for physical systems the model must closely approximate the actual behavior of the systems in order for the solution obtained to be near-optimal or (often) even useful.

The equation which describes the return or benefit from any proposed solution is called the *objective function*. Other names commonly used are the cost function and the performance function. For the special case when both the system model and the objective function are linear equations, we have the *linear programming problem*.

If the solutions to an optimization are functions of time, the problem is termed *dynamic optimization*. Such a problem is generally described by a model consisting of a set of differential equations and an objective function, which may be an integral. The mathematical theory describing the solution to these problems has been developed from several mutually consistent points of view. Examples of these approaches bear the names calculus of variations, dynamic programming, and Pontryagin's maximum principle. By considering the continuous time variable only at discrete points, the dynamic optimization problem can be formulated approximately and solved by nonlinear programming.

In most optimization problems having physical significance, the solution must satisfy restrictions in the model describing the physical system as well as maximizing or minimizing the objective function. Problems of this sort are termed *constrained optimization problems* for both the static and the dynamic cases. The mathematical models represent the constraints in these problems. Constraints may be further classified as *equality constraints* or *inequality constraints* depending on whether the signs $(=)$ or $(\geqslant, \leqslant)$, respectively, are used in the equations. In some cases no constraints are present, and this problem (consisting of an objective function only) is termed an *unconstrained optimization problem*.

Mathematical Preliminaries

Before we begin, it will be helpful to define some notation and to state several mathematical preliminaries.

Throughout this book, we shall deal almost exclusively with differentiable functions of a vector variable. Thus unless we indicate to the contrary,

$$f(\mathbf{x})$$

will indicate that f is a real-valued differentiable function and $\mathbf{x}$ is an n-dimensional collection of variables. Thus

$$\mathbf{x} = \begin{bmatrix} x_1 \\ x_2 \\ \vdots \\ x_n \end{bmatrix}$$

and

$$\mathbf{x}^T = [x_1, x_2, \ldots, x_n].$$

If we wish to indicate that f is differentiable more than once, we will indicate that specifically. If we wish to indicate that $\mathbf{x}$ is evaluated at a particular point, a subscript or a superscript will be used. Thus $\mathbf{x}_0$, $\mathbf{x}_1$, or $\mathbf{x}^*$ indicates that the n variables $\mathbf{x}$ are indeed a set of specific numbers. In general, we will reserve $\mathbf{x}^*$ to denote the optimal solution.

Euclidean n-Space

The set of *all* n-tuples of real numbers, denoted by E^n, is called an n-space. It is called *Euclidean n-space* if the usual rules of vector multiplication (parallelogram rule) and scalar multiplication apply, and if a *norm* or distance between two vectors $\mathbf{x}$ and y in E^n is defined by

$$\|\mathbf{x}-\mathbf{y}\| = \sqrt{(x_1-y_1)^2+(x_2-y_2)^2+\cdots+(x_n\ \ y_n)^2}\ .$$

In case $\mathbf{y}=\mathbf{0}$, the norm or magnitude of $\mathbf{x}$ is defined by

$$\|\mathbf{x}\| = \sqrt{x_1^2+x_2^2+\cdots+x_n^2}\ .$$

Unless otherwise specified, $\mathbf{x}\in E^n$. For example, E^1 is the entire real line, E^2 is the plane of this page, E^3 is the world in which we live, etc. It is harder to imagine E^n for $n>3$.

The *inner* (or *dot*) *product* $\mathbf{x}^T\mathbf{y}$ between two vectors $\mathbf{x}\in\mathbf{E}^n$, $\mathbf{y}\in\mathbf{E}^n$ is the scalar obtained by multiplying corresponding elements and adding the products; thus

$$\mathbf{x}^T\mathbf{y} = \sum_{i=1}^{n} x_i y_i.$$

The vectors are said to be *orthogonal* if their inner product equals zero, i.e., $\mathbf{x}^T\mathbf{y}=0$.

It follows from the properties of the inner product and the Euclidean norm that for all $\mathbf{x},\mathbf{y}\in E^n$,

$$|\mathbf{x}^T\mathbf{y}| \leqslant \|\mathbf{x}\|\ \|\mathbf{y}\|,$$

where the equality holds if and only if $\mathbf{x}=\alpha\mathbf{y}$ for some scalar α. This important relation is called the *Schwarz inequality*.

The vectors $\mathbf{x}_1,\mathbf{x}_2,\ldots,\mathbf{x}_n$ are said to be *linearly independent* if any relation of the form

$$\sum_{i=1}^{n} \alpha_i\mathbf{x}_i = \mathbf{0}$$

implies that $\alpha_i=0$ for $i=1,2,\ldots,n$.

A set S of vectors of a linear vector space E^n is said to be a *basis* of E^n if (1) the vectors of S are linearly independent and (2) every vector of E^n can be expressed as a linear combination of the vectors of S.

A set of vectors is said to *span* the vector space E^n if any vector in E^n can be expressed as a linear combination of vectors of that set. For example, the vectors $\mathbf{x}_1=(1,0,0)$, $\mathbf{x}_2=(0,1,0)$, $\mathbf{x}_3=(0,0,1)$, and $\mathbf{x}_4=(1,1,0)$ span E^3, but they do not constitute a basis. However, the first three vectors above do constitute a basis as well as spanning E^3.

CONVEXITY

The concepts of convex sets and convex functions are particularly important in optimization. To develop these ideas we first define a *line*.

A *line* passing through two different points $\mathbf{x}_1$ and $\mathbf{x}_2$ in E^n is defined to be the set of points

$$X = \{\mathbf{x} | \mathbf{x} = \lambda \mathbf{x}_2 + (1-\lambda)\mathbf{x}_1, \text{ all } \lambda\},$$

where λ is a scalar. If $0 \leqslant \lambda \leqslant 1$, then X represents a *line segment* joining $\mathbf{x}_1$ and $\mathbf{x}_2$.

EXAMPLE 1.1 For $\mathbf{x} \in E^2$, the set X defining a line is shown in Fig. 1.1.

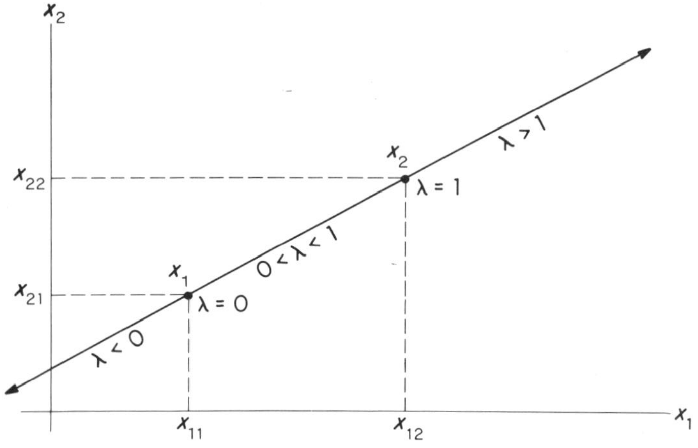

FIGURE 1.1

A set X is *convex* if for any points $\mathbf{x}_1$ and $\mathbf{x}_2$ in the set, the line segment joining these points is also in the set.

EXAMPLE 1.2 The sets shown in Fig. 1.2 (a) and (b) are convex. Those in Fig. 1.2 (c) and (d) are not.

A *hyperplane* in E^n is defined to be the set of points

$$X = \{\mathbf{x} | \mathbf{c}^T \mathbf{x} = z\},$$

with $\mathbf{c} \neq \mathbf{0}$ an n-component vector and z a fixed scalar. A hyperplane

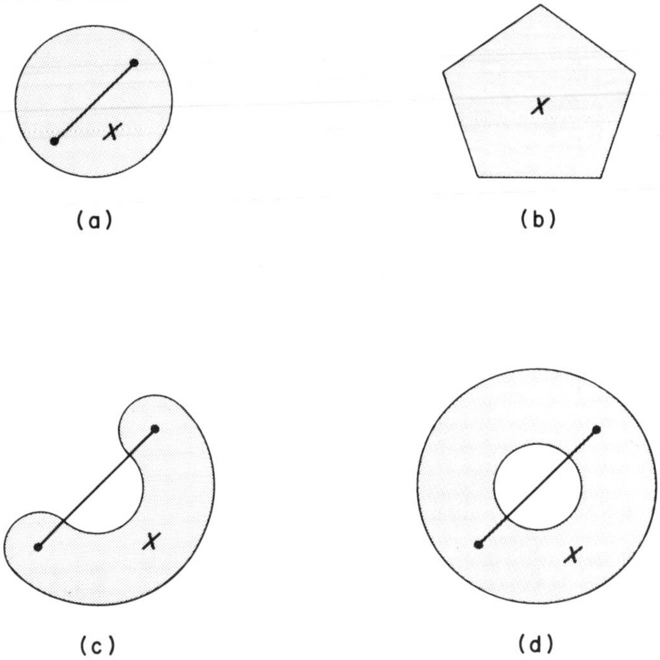

FIGURE 1.2

divides E^n into two *closed half spaces* defined by

$$X = \{\mathbf{x}|\mathbf{c}^T\mathbf{x} \geqslant z\} \quad \text{and} \quad X = \{\mathbf{x}|\mathbf{c}^T\mathbf{x} \leqslant z\}.$$

In case the strict inequality holds, X is an *open half space.*

EXAMPLE 1.3 The hyperplane

$$X = \{\mathbf{x}|2x_1 + x_2 = 3\}$$

divides E^2 into the half spaces X_1 and X_2 shown in Fig. 1.3. If $z=0$, the hyperplane passes through the origin.

Given the hyperplane $\mathbf{c}^T\mathbf{x} = z$ in E^n, the vector $\mathbf{c}$ is called the *vector normal* to the hyperplane. That is, the vector $\mathbf{c}$ is orthogonal to any vector lying in the hyperplane.

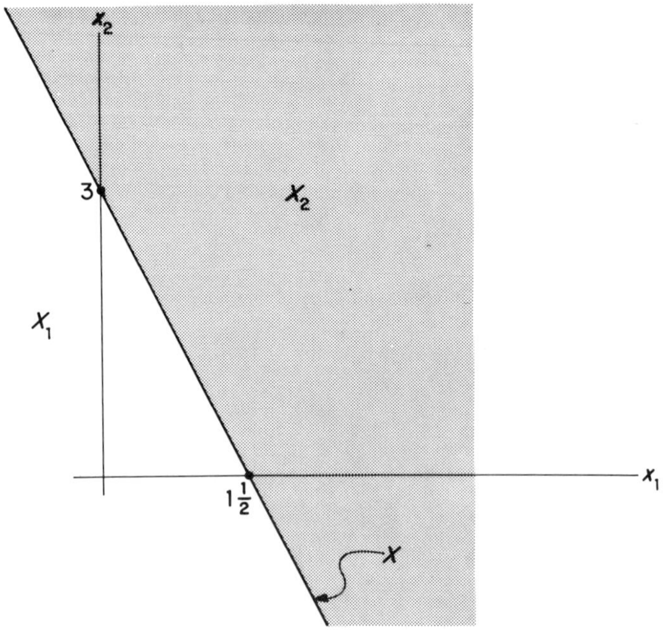

FIGURE 1.3

EXAMPLE 1.4 For the hyperplane $X=\{\mathbf{x}|-2x_1+x_2=0\}$, $\mathbf{x}=\begin{bmatrix}1\\2\end{bmatrix}$ lies in the hyperplane and $\mathbf{c}=\begin{bmatrix}-2\\1\end{bmatrix}$. Thus

$$\mathbf{c}^T\mathbf{x}=(1)(-2)+(2)(1)=0,$$

and the vectors are orthogonal (perpendicular; see Fig. 1.4).

One important type of hyperplane is called a *hyperplane of support*. Given a boundary point $\mathbf{w}$ of a convex set X, the hyperplane $\mathbf{c}^T\mathbf{x}=z$ is called a hyperplane of support at $\mathbf{w}$ if $\mathbf{c}^T\mathbf{w}=z$ and if all of X lies in one closed half space produced by the hyperplane. That is, $\mathbf{c}^T\mathbf{x}\geqslant z$ for all $\mathbf{x}\in X$ or $\mathbf{c}^T\mathbf{x}\leqslant z$ for all $\mathbf{x}\in X$.

EXAMPLE 1.5 In E^2 a hyperplane of support to a convex set is a tangent line at some boundary point of the set; in E^3 it is a tangent plane at some boundary point (see Fig. 1.5).

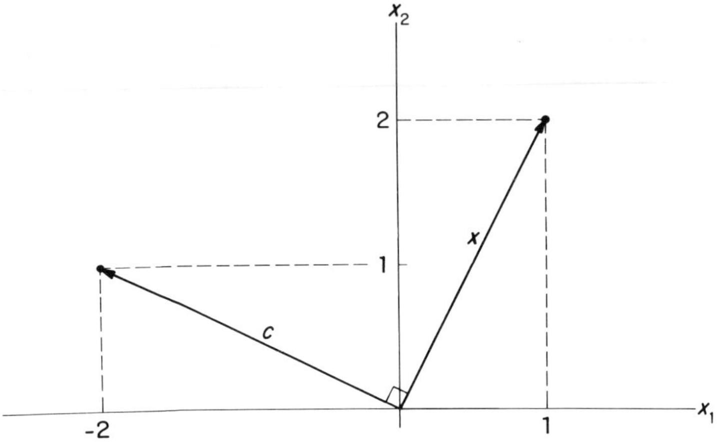

FIGURE 1.4

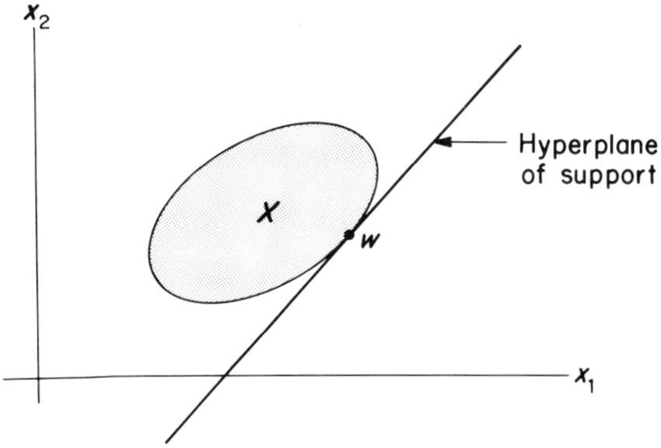

FIGURE 1.5

A function $f(\mathbf{x})$ is said to be a *convex function* over a convex set X in E^n if for any two points $\mathbf{x}_1$ and $\mathbf{x}_2$ in X and for all λ, $0 \leqslant \lambda \leqslant 1$,

$$f(\lambda \mathbf{x}_2 + (1-\lambda)\mathbf{x}_1) \leqslant \lambda f(\mathbf{x}_2) + (1-\lambda) f(\mathbf{x}_1).$$

A function is *strictly convex* if the above definition holds with the strict inequality when $0 < \lambda < 1$ and $\mathbf{x}_1 \neq \mathbf{x}_2$.

Similarly a *concave function* is defined in the same way with the sense of the inequality reversed. A linear function is both convex and concave over all of E^n. However, it is neither strictly convex nor strictly concave.

EXAMPLE 1.6 Consider the function of a single variable shown in Fig. 1.6. For all λ, $0 \leqslant \lambda \leqslant 1$, the function lies below the line segment joining $(x_1, f(x_1))$ to $(x_2, f(x_2))$, and so the function is convex; in fact, it is strictly convex.

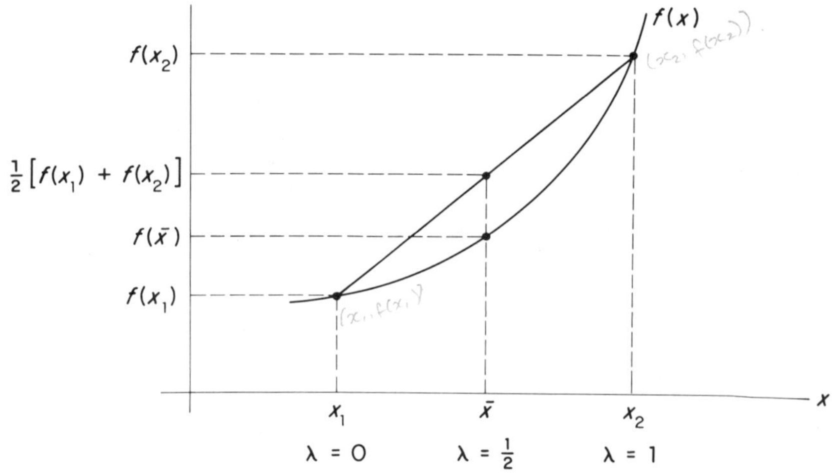

FIGURE 1.6

From the definition we see that if $f(\mathbf{x})$ is (strictly) convex, then $-f(\mathbf{x})$ is (strictly) concave, and vice versa. Thus it is sufficient to develop a theory of (say) convex functions only, since it is a simple matter to convert any concave function to a convex one.

EXAMPLE 1.7 In Fig. 1.7, the function $f(x)$ is convex, while $-f(x)$ is concave. Also the minimizing point x^* of the convex function $f(x)$ equals the maximizing point x^* of the concave function $-f(x)$. However, the two functions are not equal at that point. Rather $\min f(x) = -\max[-f(x)]$. This statement is true in general.

Note that a convex function is defined only over the domain of a convex set. Also the definition does not require that $f(\mathbf{x})$ be either continuous or differentiable.

For a differentiable function $f(\mathbf{x})$, the *gradient* is the n-dimensional vector of partial derivatives

$$\nabla f(\mathbf{x}) = \left[\frac{\partial f}{\partial x_1}(\mathbf{x}), \frac{\partial f}{\partial x_2}(\mathbf{x}), \ldots, \frac{\partial f}{\partial x_n}(\mathbf{x}) \right]^T.$$

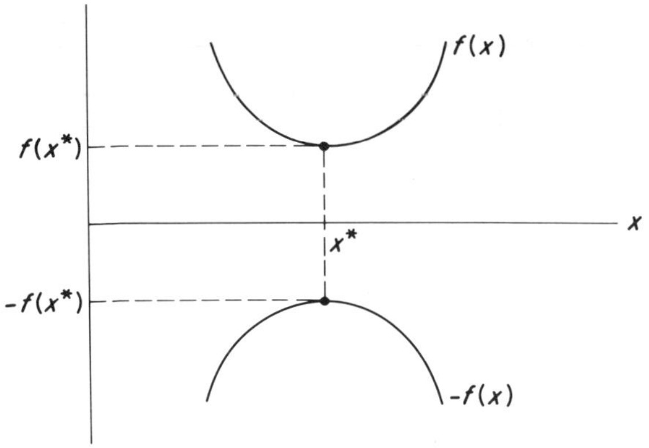

FIGURE 1.7

If in addition $f(\mathbf{x})$ possesses second partial derivatives, the *Hessian matrix* is the $n \times n$ symmetric matrix of second partial derivatives,

$$\mathbf{H}(\mathbf{x}) = \begin{bmatrix} \frac{\partial^2 f}{\partial x_1^2}(\mathbf{x}) & \frac{\partial^2 f}{\partial x_1 \partial x_2}(\mathbf{x}) & \cdots & \frac{\partial^2 f}{\partial x_1 \partial x_n}(\mathbf{x}) \\ \frac{\partial^2 f}{\partial x_2 \partial x_1}(\mathbf{x}) & \frac{\partial^2 f}{\partial x_2^2}(\mathbf{x}) & \cdots & \frac{\partial^2 f}{\partial x_2 \partial x_n}(\mathbf{x}) \\ \cdots & \cdots & \cdots & \cdots \\ \frac{\partial^2 f}{\partial x_n \partial x_1}(\mathbf{x}) & \frac{\partial^2 f}{\partial x_n \partial x_2}(\mathbf{x}) & \cdots & \frac{\partial^2 f}{\partial x_n^2}(\mathbf{x}) \end{bmatrix}.$$

The symmetry of $\mathbf{H}(\mathbf{x})$ follows from the fact that cross partial derivatives are independent of the order of differentiation. Thus

$$\frac{\partial^2 f}{\partial x_i \partial x_j} = \frac{\partial^2 f}{\partial x_j \partial x_i}, \qquad i,j = 1,2,\ldots,n.$$

If $\mathbf{u}$ is a unit vector in E^n (that is, $\|\mathbf{u}\| = 1$), then the *directional derivative* of $f(\mathbf{x})$ is defined to be

$$\lim_{\tau \to 0} \frac{f(\mathbf{x} + \tau\mathbf{u}) - f(\mathbf{x})}{\tau} = \nabla f(\mathbf{x})^T \mathbf{u}.$$

The directional derivative is the instantaneous rate of change of $f(\mathbf{x})$ in the direction $\mathbf{u}$.

EXAMPLE 1.8 Find the directional derivative of $f(x,y)=x^2+2y^2$ at $x=1$, $y=1$ in the direction of $\mathbf{d}=\begin{bmatrix}1\\3\end{bmatrix}$.

$$\nabla f(x,y)=\begin{bmatrix}2x\\4y\end{bmatrix} \quad \text{and} \quad \nabla f(1,1)=\begin{bmatrix}2\\4\end{bmatrix}.$$

We must make the direction vector into a unit vector by *normalizing*, that is, dividing $\mathbf{d}$ by its norm $\|\mathbf{d}\|$. Thus

$$\mathbf{u}=\frac{\mathbf{d}}{\|\mathbf{d}\|}=\frac{1}{\sqrt{1^2+3^2}}\begin{bmatrix}1\\3\end{bmatrix}=\begin{bmatrix}1/\sqrt{10}\\3/\sqrt{10}\end{bmatrix}.$$

Now $\|\mathbf{u}\|=\sqrt{(1/\sqrt{10})^2+(3/\sqrt{10})^2}=1.$ The directional derivative at $x=1$, $y=1$ is

$$\nabla f(1,1)^T\mathbf{u}=2(1/\sqrt{10})+4(3/\sqrt{10})=14/\sqrt{10}.$$

One important class of functions in optimization problems is the *quadratic form*. A quadratic form is a function of n variables which can be written

$$f(\mathbf{x})=\sum_{i=1}^{n}\sum_{j=1}^{n}a_{ij}x_ix_j,$$

where the a_{ij} are constants. In matrix-vector notation the quadratic form becomes

$$\mathbf{x}^T\mathbf{A}\mathbf{x},$$

where $\mathbf{A}=[a_{ij}]$ is an $n\times n$ symmetric matrix. Note that this combination of a matrix and vectors always takes on a scalar value.

EXAMPLE 1.9 Some quadratic forms are

$$\text{(a) } x_1^2+x_1x_2+3x_2^2, \quad \text{(b) } x_1^2+x_1x_2+4x_1x_3+2x_3^2, \quad \text{(c) } x_1^2+2x_2^2+3x_3^2,$$

or in matrix form, $\mathbf{x}^T\mathbf{A}\mathbf{x}$, where

$$\text{(a) } \mathbf{A}=\begin{bmatrix}1 & 0.5\\0.5 & 3\end{bmatrix}, \quad \text{(b) } \mathbf{A}=\begin{bmatrix}1 & 0.5 & 2\\0.5 & 0 & 0\\2 & 0 & 2\end{bmatrix}, \quad \text{(c) } \mathbf{A}=\begin{bmatrix}1 & 0 & 0\\0 & 2 & 0\\0 & 0 & 3\end{bmatrix}.$$

The matrix representation of **A** is not unique. However, **A** can always be taken to be symmetric without loss of generality.

EXAMPLE 1.10 In Example 1.9(a), any matrix representation of the form

$$\mathbf{A}=\begin{bmatrix} 1 & b \\ a & 3 \end{bmatrix}$$

such that $a+b=1$ will accurately represent the quadratic form. From among the infinite number of matrices satisfying this relation we can always choose the symmetric form

$$\mathbf{A}=\begin{bmatrix} 1 & c \\ c & 3 \end{bmatrix}$$

where $c=(a+b)/2=0.5$.

A quadratic form $\mathbf{x}^T\mathbf{A}\mathbf{x}$ is called *positive definite* if $\mathbf{x}^T\mathbf{A}\mathbf{x}>0$ for every $\mathbf{x}$ except $\mathbf{x}=\mathbf{0}$ (i.e., $x_1=x_2=\cdots=x_n=0$).

The quadratic form is called *positive semidefinite* if $\mathbf{x}^T\mathbf{A}\mathbf{x}\geqslant 0$ for every $\mathbf{x}$ and if there is some $\mathbf{x}\neq\mathbf{0}$ for which $\mathbf{x}^T\mathbf{A}\mathbf{x}=0$.

Similar definitions hold for *negative definite* and *negative semidefinite*, except that the sense of the inequalities is reversed.

A quadratic form is called *indefinite* if it is positive for some values of $\mathbf{x}$ and negative for others.

EXAMPLE 1.11 Identify the regions where the quadratic form $z=\mathbf{x}^T\mathbf{A}\mathbf{x}$ is positive and negative and determine the type of quadratic form represented.

From Fig. 1.8:

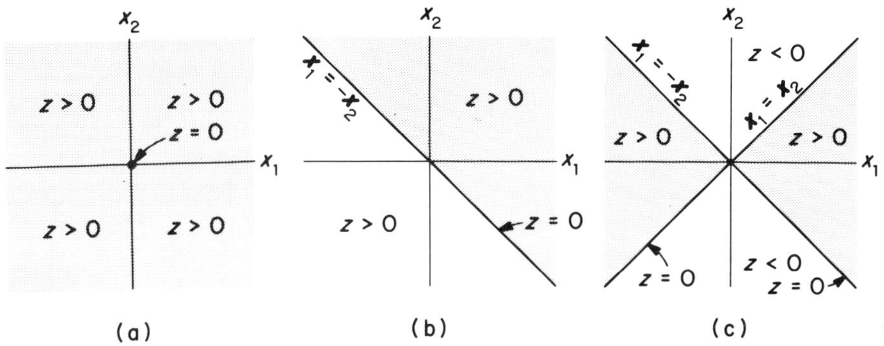

FIGURE 1.8

(a) $z = x_1^2 + x_2^2$ is positive definite, since z is positive except when $x_1 = x_2 = 0$.

(b) $z = (x_1 + x_2)^2$ is positive semidefinite, since z is positive except when $x_1 = -x_2$.

(c) $z = x_1^2 - x_2^2$ is indefinite, since z is positive for $|x_1| > |x_2|$ and negative for $|x_1| < |x_2|$.

Convexity of Differentiable Functions

For differentiable functions, the following theorems may be used to ensure the convexity of $f(\mathbf{x})$.

Theorem 1.1 *Let $f(\mathbf{x})$ be a differentiable function on the convex set $X \subset E^n$. Then $f(\mathbf{x})$ is convex on X if and only if*

$$f(\mathbf{x}^2) - f(\mathbf{x}^1) \geqslant \nabla f(\mathbf{x}^1)(\mathbf{x}^2 - \mathbf{x}^1)$$

for each $\mathbf{x}^1, \mathbf{x}^2 \in X$.

Theorem 1.2 *Let $f(\mathbf{x})$ be a twice differentiable function on the open convex set $X \subset E^n$. Then $f(\mathbf{x})$ is convex on X if and only if the Hessian matrix $\mathbf{H}(\mathbf{x})$ is positive semidefinite for every $\mathbf{x} \in \mathbf{X}$.*

Theorem 1.3 *Let $f(\mathbf{x})$ be a twice differentiable function on the open set $X \subset E^n$. If the Hessian matrix $\mathbf{H}(\mathbf{x})$ is positive definite for every $\mathbf{x} \in X$, then $f(\mathbf{x})$ is strictly convex on X. Note that this condition is sufficient but not necessary.*

Sylvester's Theorem

To determine whether or not a quadratic form is positive definite, we may use the following theorem.

Sylvester's Theorem *A quadratic form $\mathbf{x}^T\mathbf{H}\mathbf{x}$ is positive definite if and only if all the successive principal minors of the matrix $\mathbf{H}$ are positive.*

The successive principal minors are the determinants of the square submatrices obtained by successively deleting lower rows and right-hand columns. For an $n \times n$ matrix, there are n principal minors. For example, if

$$\mathbf{H} = \begin{bmatrix} h_{11} & h_{12} & h_{13} \\ h_{21} & h_{22} & h_{23} \\ h_{31} & h_{32} & h_{33} \end{bmatrix},$$

then the three principal minors are the three determinants

$$h_{11}, \qquad \begin{vmatrix} h_{11} & h_{12} \\ h_{21} & h_{22} \end{vmatrix}, \qquad \begin{vmatrix} h_{11} & h_{12} & h_{13} \\ h_{21} & h_{22} & h_{23} \\ h_{31} & h_{32} & h_{33} \end{vmatrix}.$$

A quadratic form $\mathbf{x}^T\mathbf{H}\mathbf{x}$ is negative definite if $-\mathbf{x}^T\mathbf{H}\mathbf{x}$ is positive definite. Since

$$-\mathbf{x}^T\mathbf{H}\mathbf{x}=\mathbf{x}^T(-\mathbf{H})\mathbf{x},$$

we may apply Sylvester's theorem to $-\mathbf{H}$ to determine the negative definiteness of $\mathbf{H}$.

We cannot determine whether or not a matrix is positive semidefinite by simply allowing the successive principal minors to be nonnegative ($\geqslant 0$) instead of positive (>0). Rather *all* the principal minors must be nonnegative. The matrix must be permuted in all possible combinations to determine all of the $\binom{n}{p}^2$ principal minors of order p, $p=1,2,\ldots,n$. This is seldom feasible. For a real symmetric matrix, if the successive principal minors are positive, then all the principal minors are positive. A matrix which is not positive definite, negative definite, positive semidefinite, or negative semidefinite is indefinite.

EXAMPLE 1.12 Determine the sign definiteness for each of the following matrices:

(a) $\mathbf{H}=\begin{bmatrix} 2 & 1 & 2 \\ 1 & 3 & 0 \\ 2 & 0 & 5 \end{bmatrix}$,

$$h_{11}=2, \quad \begin{vmatrix} 2 & 1 \\ 1 & 3 \end{vmatrix}=6-1=5, \quad \begin{vmatrix} 2 & 1 & 2 \\ 1 & 3 & 0 \\ 2 & 0 & 5 \end{vmatrix}=13.$$

Since all the successive principal minors are positive, $\mathbf{H}$ is positive definite.

(b) $\mathbf{H}=\begin{bmatrix} 2 & 1 & 2 \\ 1 & -3 & 0 \\ 2 & 0 & -5 \end{bmatrix}$,

$$h_{11}=2, \quad \begin{vmatrix} 2 & 1 \\ 1 & -3 \end{vmatrix}=-6-(1)=-7, \quad \begin{vmatrix} 2 & 1 & 2 \\ 1 & -3 & 0 \\ 2 & 0 & -5 \end{vmatrix}=47.$$

H is not positive definite, so form $-\mathbf{H}$:

$$-\mathbf{H}=\begin{bmatrix} -2 & -1 & -2 \\ -1 & 3 & 0 \\ -2 & 0 & 5 \end{bmatrix}.$$

Now

$$h_{11}=-2, \qquad \begin{vmatrix} -2 & -1 \\ -1 & 3 \end{vmatrix}=-7, \qquad \begin{vmatrix} -2 & -1 & -2 \\ -1 & 3 & 0 \\ -2 & 0 & 5 \end{vmatrix}=-47.$$

So **H** is negative definite.

Nonlinear Programming

We now define the nonlinear programming problem to be considered in the remaining chapters as follows:

$$\min_{\mathbf{x}} f(\mathbf{x})$$

such that

$$g_i(\mathbf{x})=0, \qquad i=1,2,\dots,m<n,$$

$$h_j(\mathbf{x})\geqslant 0, \qquad j=1,2,\dots,r,$$

where $\mathbf{x}=[x_1,x_2,\dots,x_n]^T$ and all functions f, g_i, and h_j are differentiable. In words, we wish to determine from among the set of all numbers which satisfy the m equality constraints

$$g_i(\mathbf{x})=0, \qquad i=1,2,\dots,m<n,$$

and the r inequality constraints

$$h_j(\mathbf{x})\geqslant 0, \qquad j=1,2,\dots,r,$$

that particular set (or sets, since the solution may not be unique) $[x_1^*,x_2^*,\dots,x_n^*]$ which yields a minimum value of $f(\mathbf{x})$. Note that m, the number of equality constraints, must be strictly less than n, the number of variables. If $m\geqslant n$ the problem is termed *overconstrained*, since there are no degrees of freedom left for optimizing. The number of *degrees of freedom* is given by $n-m$.

The set of all possible values which satisfy all the constraints is termed the *feasible set*. Any element in this set will be called a *feasible point*.

The nonlinear programming problem defined above deals only with differentiable functions. One class of problems where this restriction is violated is designated as *integer programming*. These problems, discussed in Chapter 11, require that the solution be chosen from the set of integers rather than the set of real numbers. At present, computational methods for solving such problems are much less highly developed than those for nonlinear programming.

We shall also designate as nonlinear programming problems those special cases where either the equality constraints, the inequality constraints, or both are not present.

EXAMPLE 1.13 For the nonlinear programming problem

$$\min_{x_1, x_2} (x_1-2)^2+(x_2-2)^2$$

such that

$$x_1+2x_2=4 \qquad x_1 \geqslant 0, \quad x_2 \geqslant 0,$$

the feasible set is shown in Fig. 1.9. The feasible region is the set of $\{x_1, x_2\}$ such that $x_1+2x_2=4$ and $x_1 \geqslant 0$, $x_2 \geqslant 0$. We can designate this set F by

$$F=\{(x_1,x_2)|x_1+2x_2=4, \ x_1 \geqslant 0, \ x_2 \geqslant 0\}.$$

Much insight can be gained into two-dimensional optimization problems by plotting the *level curves* for the objective function. Level curves are the parametric family of curves such that the objective function $f(x_1,x_2)$ is

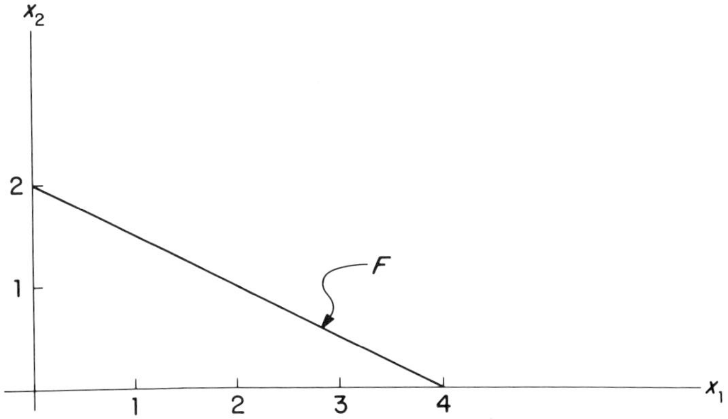

FIGURE 1.9

constant, where the constant value k is the parameter. Thus

$$f(x_1, x_2) = k \qquad \text{for} \quad k = k_1, k_2, \ldots$$

represents a family of level curves.

EXAMPLE 1.14 For Example 1.13 the level curves are shown in Fig. 1.10.

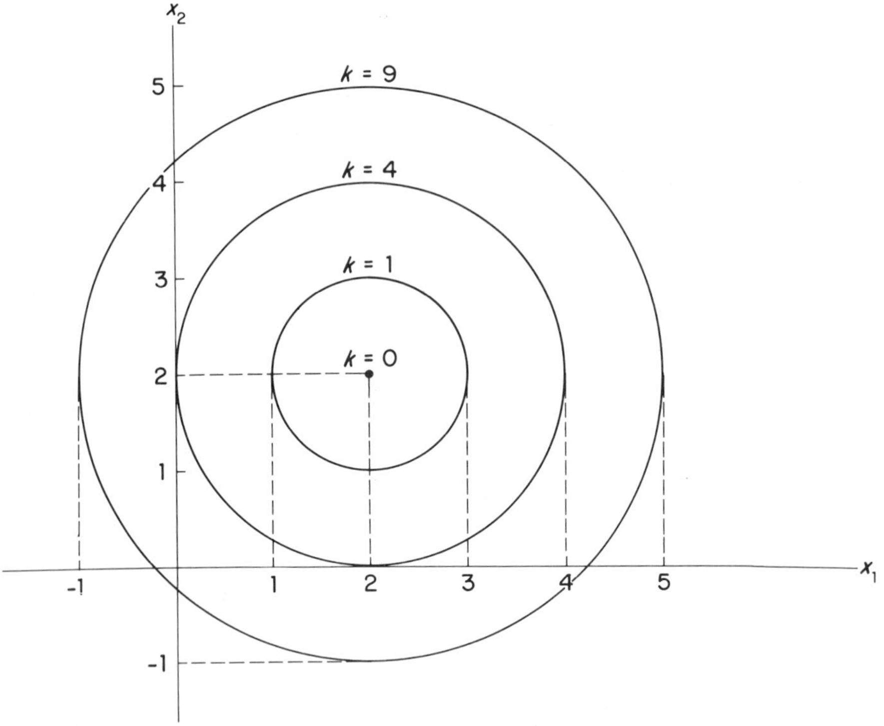

FIGURE 1.10

By superimposing the feasible region on the level curves, two-dimensional problems can usually be solved graphically.

EXAMPLE 1.15 A graphical solution for Example 1.13 is shown in Fig. 1.11. The solution is the point on the level curve having the minimum value of k which is also in the feasible region. For this problem the minimum occurs at the point of tangency between F and the level curves, namely $x_1^* = \frac{8}{5}$, $x_2^* = \frac{6}{5}$; and the objective function has the minimum value

$$f(x_1^*, x_2^*) = k^* = \left(\tfrac{8}{5} - 2\right)^2 + \left(\tfrac{6}{5} - 2\right)^2 = \tfrac{4}{5}.$$

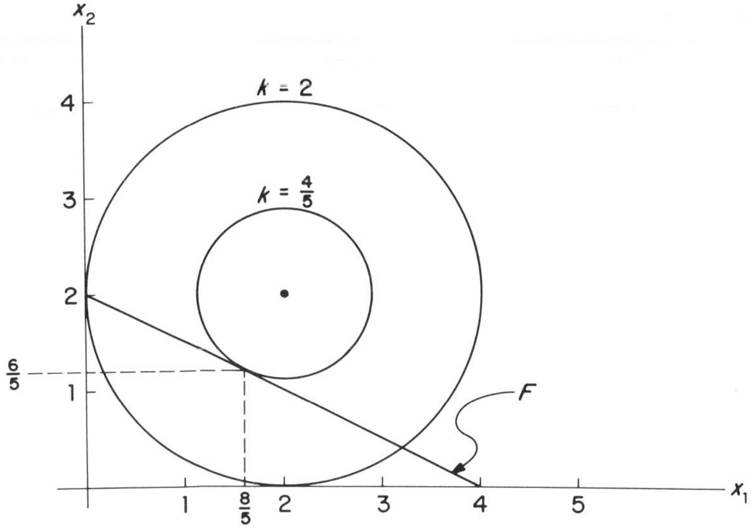

FIGURE 1.11

We have already discussed how any maximization problem can be converted into a minimization problem by multiplying $f(\mathbf{x})$ by -1. Thus there is no loss of generality in considering the nonlinear programming problem as a minimization problem. In case the inequality constraints have the form

$$h_j(\mathbf{x}) \leqslant 0, \qquad j = 1, 2, \ldots, r, \tag{1.1}$$

they can be converted to the "standard form" by multiplying by -1 and changing the sign of the inequality. Thus (1.1) is equivalent to

$$-h_j(\mathbf{x}) \geqslant 0, \qquad j = 1, 2, \ldots, r,$$

so there is no loss of generality by considering the problem as written.

In fact, the "standard form" could be simplified by writing each equality constraint as two inequalities, since the only values of $\mathbf{x}$ which satisfy

$$\mathbf{g}(\mathbf{x}) \geqslant 0 \quad \text{and} \quad \mathbf{g}(\mathbf{x}) \leqslant 0$$

are those values for which $\mathbf{g}(\mathbf{x}) = 0$. Thus the m equalities

$$g_i(\mathbf{x}) = 0, \qquad i = 1, 2, \ldots, m,$$

could be replaced by the $2m$ inequalities

$$g_i(\mathbf{x}) \geqslant 0, \qquad i = 1, 2, \ldots, m,$$

$$-g_i(\mathbf{x}) \geqslant 0 \qquad i = 1, 2, \ldots, m.$$

The nonlinear programming problem could then be written

$$\min_{\mathbf{x}} f(\mathbf{x})$$

such that

$$\begin{bmatrix} g_i(\mathbf{x}) \\ -g_i(\mathbf{x}) \\ h_j(\mathbf{x}) \end{bmatrix} \geqslant 0, \qquad \begin{array}{l} i=1,2,\ldots,m<n, \\ j=1,2,\ldots,r. \end{array}$$

Since the theoretical and the computational methods for dealing with equality and inequality constraints differ, we shall prefer to consider the nonlinear programming problem in the form shown at the beginning of this section.

EXAMPLE 1.16 The only solution to the constraint

$$e^{-x}-1=0$$

is $x=0$. This can be obtained by finding the intersection of the two feasible sets

$$F_1=\{x|e^{-x}-1\geqslant 0\},$$

$$F_2=\{x|-e^{-x}+1\geqslant 0\},$$

or

$$F=F_1\cap F_2=\{x\leqslant 0\}\cap\{x\geqslant 0\}=\{0\},$$

as shown in Fig. 1.12.

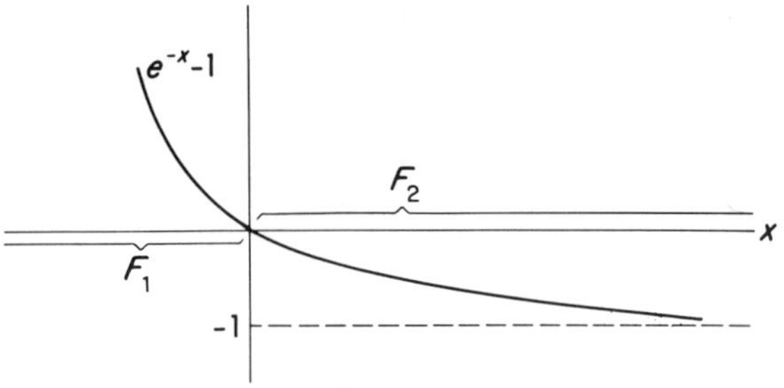

FIGURE 1.12

Linear Programming

The most important special case of the nonlinear programming problem of the previous section is the linear programming problem shown below:

$$\min_{x_j} \sum_{j=1}^{n} c_j x_j$$

such that

$$\sum_{j=1}^{n} a_{ij} x_j \leqslant b_i, \qquad i=1,2,\ldots,m,$$

$$x_j \geqslant 0, \qquad j=1,2,\ldots,n,$$

or in vector-matrix notation

$$\min_{\mathbf{x}} \mathbf{c}^T \mathbf{x}$$

such that

$$\mathbf{A}\mathbf{x} \leqslant \mathbf{b},$$

$$x_j \geqslant 0, \qquad j=1,2,\ldots,n,$$

where **c** and **x** are n-vectors, **b** is an m-vector, and **A** is an $m \times n$ matrix.

EXAMPLE 1.17 For the linear programming problem

$$\min_{x_1, x_2} \tfrac{3}{2} x_1 + x_2$$

such that

$$x_1 + 3x_2 \leqslant 15,$$

$$x_1 + x_2 \leqslant 6,$$

$$2x_1 + x_2 \leqslant 10,$$

$$x_1 \geqslant 0, \qquad x_2 \geqslant 0,$$

the feasible region, the level curves, and the graphical solution are shown in Fig. 1.13.

A review of linear programming is given in the Appendix.

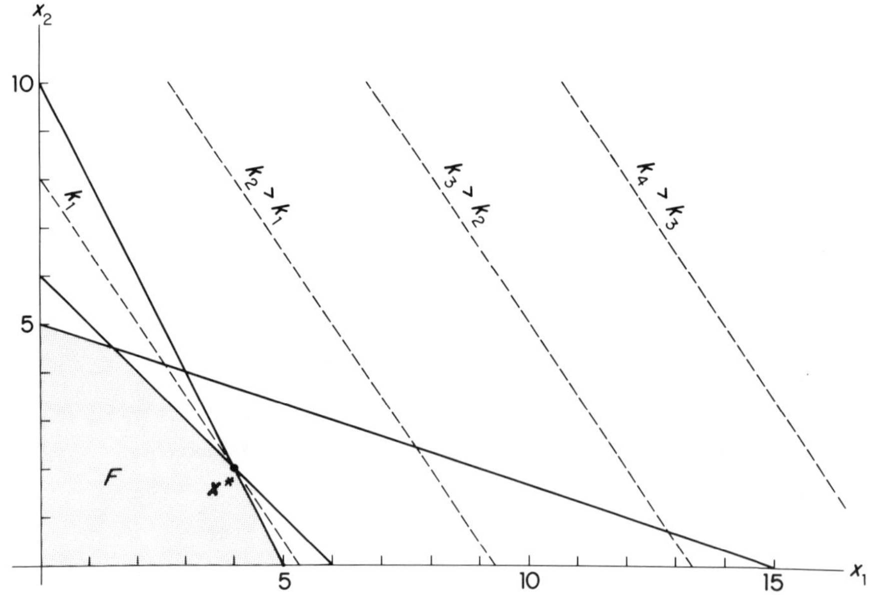

FIGURE 1.13

Solved Problems

Mathematical Preliminaries

1.1 Show directly from the definition that x^2 is a convex function.

For convexity,

$$f(\lambda x_2+(1-\lambda)x_1)\leqslant \lambda f(x_2)+(1-\lambda)f(x_1) \qquad \text{for all} \quad 0\leqslant\lambda\leqslant 1.$$

Then $f(x)=x^2$ is convex if

$$\left[\lambda x_2+(1-\lambda)x_1\right]^2\leqslant \lambda x_2^2+(1-\lambda)x_1^2 \qquad \text{for all} \quad 0\leqslant\lambda\leqslant 1.$$

Expanding and collecting terms,

$$-\lambda(1-\lambda)x_2^2+2\lambda(1-\lambda)x_1x_2-\lambda(1-\lambda)x_1^2\leqslant 0 \quad \text{or} \quad (1-\lambda)(x_1-x_2)^2\geqslant 0.$$

For $0\leqslant\lambda\leqslant 1$, $0\leqslant\lambda(1-\lambda)\leqslant\frac{1}{4}$, since $\lambda(1-\lambda)=\lambda-\lambda^2$, which is never negative for $0\leqslant\lambda\leqslant 1$. Hence

$$\lambda(1-\lambda)(x_1-x_2)^2\geqslant 0 \qquad \text{for} \quad 0\leqslant\lambda\leqslant 1.$$

1.2 Consider the quadratic form $f(x_1,x_2)=x_1^2+2x_1x_2+3x_2^2$:

(a) Write the quadratic form in matrix-vector notation.
(b) Is the matrix singular?
(c) Show that $f(x_1,x_2)$ is positive definite.
(d) Is $f(x_1,x_2)$ convex?
(e) Find the equation of the supporting hyperplane at 2, 1.

(a) $f(x_1,x_2)=[x_1,x_2]\begin{bmatrix}1 & 1\\ 1 & 3\end{bmatrix}\begin{bmatrix}x_1\\ x_2\end{bmatrix}.$

(b) $\det\begin{vmatrix}1 & 1\\ 1 & 3\end{vmatrix}=2\neq 0$, and so the matrix is nonsingular.

(c) $f(x_1,x_2)=(x_1+x_2)^2+2x_2^2$, which is $\geqslant 0$ for all $x_1,x_2\neq 0$, and so $f(x_1,x_2)$ is positive definite.

(d) Yes, all positive definite quadratic forms are strictly convex and therefore convex (see Problem 1.8).

(e) The general expression for a supporting hyperplane which is tangent to some convex set at z_0 is

$$z-z_0=\left[\nabla f(\mathbf{x}_0)\right]^T(\mathbf{x}-\mathbf{x}_0).$$

Here

$$\mathbf{x}_0=\begin{bmatrix}2\\ 1\end{bmatrix} \quad\text{and}\quad z_0=f(\mathbf{x}_0)=4+4+3=11,$$

$$\nabla f(\mathbf{x}_0)=\begin{bmatrix}2x_1+2x_2\\ 2x_1+6x_2\end{bmatrix}_{2,1}=\begin{bmatrix}6\\ 10\end{bmatrix},$$

$$z=6(x_1-2)+10(x_2-1)+11, \quad\text{or}\quad z=6x_1+10x_2-11.$$

1.3 Determine which of the following equations are quadratic forms:

(a) $z=x_1+2x_2^2$.
(b) $z=x_1x_2$.
(c) $z=x_1^2-x_2^2$.
(d) $z=x_1^2+3x_1x_2+3x_2^2$.

For z to be a quadratic form, we must be able to write it in the form

$$z=\mathbf{x}^T\mathbf{H}\mathbf{x}.$$

(a) Not a quadratic form, since it is linear in x_1.

(b) Quadratic form, $\mathbf{H}=\begin{bmatrix}0 & \frac{1}{2}\\ \frac{1}{2} & 0\end{bmatrix}$.

(c) Quadratic form, $\mathbf{H}=\begin{bmatrix} 1 & 0 \\ 0 & -1 \end{bmatrix}$.

(d) Quadratic form, $\mathbf{H}=\begin{bmatrix} 1 & \frac{3}{2} \\ \frac{3}{2} & 3 \end{bmatrix}$.

1.4 Determine the gradient for each of the following functions. Indicate singular points for both the function and the gradient.

(a) $f(x)=\ln x$.
(b) $f(x,y)=x^2-y^2-1$.
(c) $f(x)=\dfrac{\sin x}{x}$.
(d) $f(x)=\dfrac{\sin x}{xe^x}$.

Any point at which a function does not exist is called a *singular point*.
(a) $\nabla f(x)=1/x$. Both $\ln x$ and its gradient have singular points at $x=0$.

(b) $\nabla f(x,y)=\begin{bmatrix} 2x \\ -2y \end{bmatrix}$. No singular points for $f(x,y)$ or $\nabla f(x,y)$.

(c) $\nabla f(x)=\dfrac{x\cos x-\sin x}{x^2}$.

Both the function $(\sin x)/x$ and its gradient are indeterminate forms at $x=0$. By l'Hospital's rule, they are not singular points if for $f(x)=N(x)/D(x)$, the ratio $N_x(x)/D_x(x)$ exists at $x=0$.

Since

$$\frac{\partial N}{\partial x}=\cos x, \qquad \frac{\partial D}{\partial x}=1, \quad \text{and} \quad \frac{N_x}{D_x}(0)=\frac{\cos 0}{1}=-1,$$

$f(x)$ is not singular at $x=0$.

Similarly for $\nabla f(0)$,

$$\frac{N_{xx}}{D_{xx}}=-\frac{\cos 0}{2}=-\frac{1}{2},$$

and $\nabla f(x)$ is not singular at $x=0$.

(d) $\nabla f(x)=\dfrac{\cos x-\sin x}{xe^x}-\dfrac{\sin x}{x^2e^x}$.

Neither $f(x)$ nor $\nabla f(x)$ has any singular points. Check $x=0$ using l'Hospital's rule.

1.5 Determine whether or not the quadratic forms $\mathbf{x}^T\mathbf{A}\mathbf{x}$ are positive definite, where

$$\text{(a) } \mathbf{A}=\begin{bmatrix} 1 & 4 \\ 0 & 3 \end{bmatrix}, \qquad \text{(b) } \mathbf{A}=\begin{bmatrix} 1 & 1 & 0 \\ 1 & 1 & 0 \\ 0 & 0 & 2 \end{bmatrix}, \qquad \text{(c) } \mathbf{A}=\begin{bmatrix} 1 & 2 \\ 2 & 6 \end{bmatrix}.$$

We first check the principal minors using Sylvester's theorem.

(a) $|1|>0$, $\begin{vmatrix} 1 & 4 \\ 0 & 3 \end{vmatrix}=3>0$, and so $\mathbf{A}$ is positive definite.

(b) $|1|>0$, $\begin{vmatrix} 1 & 1 \\ 1 & 1 \end{vmatrix}=0$, $|\mathbf{A}|=0$, and so $\mathbf{A}$ is not positive definite.

(c) $|1|>0$, $\begin{vmatrix} 1 & 2 \\ 2 & 6 \end{vmatrix}=2$, and so $\mathbf{A}$ is positive definite.

1.6 Prove that the function $1/x$ is strictly convex for $x>0$ and strictly concave for $x<0$.

We must show that

$$f(\lambda x_2+(1-\lambda)x_1)>\lambda f(x_2)+(1-\lambda)f(x_1) \qquad \text{for} \quad (x_1,x_2)>0, \quad 0<\lambda<1$$

and

$$f(\lambda x_2+(1-\lambda)x_1)<\lambda f(x_2)+(1-\lambda)f(x_1) \qquad \text{for} \quad (x_1,x_2)<0, \quad 0<\lambda<1.$$

For $f(x)=1/x$,

$$\lambda f(x_2)+(1-\lambda)f(x_1)-f(\lambda x_2+(1-\lambda)x_1)=\frac{\lambda x_1+(1-\lambda)x_2}{x_1x_2}-\frac{1}{\lambda x_2+(1-\lambda)x_1}$$

$$=\frac{\lambda(1-\lambda)(x_1-x_2)^2}{x_1x_2[\lambda x_2+(1-\lambda)x_1]}.$$

For $0<\lambda<1$, $\lambda(1-\lambda)>0$; and for $x_1\neq x_2$, $(x_1-x_2)^2>0$. Also for $x_1,x_2>0$ or $x_1,x_2<0$, we have $x_1x_2>0$ and

$$\lambda x_2+(1-\lambda)x_1>0 \qquad \text{when} \quad x_1,x_2>0,$$

$$\lambda x_2+(1-\lambda)x_1<0 \qquad \text{when} \quad x_1,x_2<0.$$

Thus

$$\frac{\lambda(1-\lambda)(x_1-x_2)^2}{x_1x_2[\lambda x_2+(1-\lambda)x_1]}>0 \qquad \text{for all} \quad x_1,x_2>0,$$

$$\frac{\lambda(1-\lambda)(x_1-x_2)^2}{x_1x_2[\lambda x_2+(1-\lambda)x_1]}<0 \qquad \text{for all} \quad x_1,x_2<0,$$

and $1/x$ is strictly convex for $x>0$ and strictly concave for $x<0$.

1.7 Determine the gradient vector and the Hessian matrix for each function:

(a) $f(\mathbf{x})=3x_1x_2^2+4e^{x_1x_2}$.
(b) $f(\mathbf{x})=x_1^{x_2}+\ln x_1x_2$.
(c) $f(\mathbf{x})=x_1^2+x_2^2+x_3^2$.
(d) $f(\mathbf{x})=\ln(x_1^2+x_1x_2+x_2^2)$.

(a) $\nabla f(\mathbf{x})=\begin{bmatrix} 3x_2^2+4x_2e^{x_1x_2} \\ 6x_1x_2+4x_1e^{x_1x_2} \end{bmatrix}$,

$$\mathbf{H}=\begin{bmatrix} 4x_2^2e^{x_1x_2} & 6x_2+4x_1x_2e^{x_1x_2}+4e^{x_1x_2} \\ 6x_2+4e^{x_1x_2}(1+x_1x_2) & 6x_1+4x_1^2e^{x_1x_2} \end{bmatrix}.$$

(b) $\nabla f(\mathbf{x})=\begin{bmatrix} x_2x_1^{x_2-1}+1/x_1 \\ x_1x_2\ln x_1+1/x_2 \end{bmatrix}$,

$$\mathbf{H}=\begin{bmatrix} x_2(x_2-1)x_1^{x_2-2}-1/x_1^2 & x_2x_1^{x_2-1}\ln x_1+x_1^{x_2-1} \\ x_2x_1^{x_2-1}\ln x_1+x_1^{x_2-1} & x_1x_2(\ln x_1)^2-1/x_2^2 \end{bmatrix}.$$

(c) $\nabla f(\mathbf{x})=\begin{bmatrix} 2x_1 \\ 2x_2 \\ 2x_3 \end{bmatrix}$, $\mathbf{H}=\begin{bmatrix} 2 & 0 & 0 \\ 0 & 2 & 0 \\ 0 & 0 & 2 \end{bmatrix}$.

(d) $\nabla f(\mathbf{x})=\begin{bmatrix} \dfrac{2x_1+x_2}{x_1^2+x_1x_2+x_2^2} \\ \dfrac{x_1+2x_2}{x_1^2+x_1x_2+x_2^2} \end{bmatrix}$,

$$\mathbf{H}=\frac{1}{(x_1^2+x_1x_2+x_2^2)^2}\begin{bmatrix} -2x_1^2+x_2^2-2x_1x_2 & -x_1^2-x_2^2-4x_1x_2 \\ -x_1^2-x_2^2-4x_1x_2 & x_1^2-2x_2^2-2x_1x_2 \end{bmatrix}.$$

1.8 Prove that $f(\mathbf{x})=\mathbf{q}^T\mathbf{x}+\frac{1}{2}\mathbf{x}^T\mathbf{Q}\mathbf{x}$ is strictly convex if and only if $\mathbf{Q}$ is positive definite.

From the definition,

$$f(\lambda\mathbf{x}_2+(1-\lambda)\mathbf{x}_1)<\lambda f(\mathbf{x}_2)+(1-\lambda)f(\mathbf{x}_1) \qquad \text{for} \quad 0<\lambda<1, \quad \mathbf{x}_1\neq\mathbf{x},$$

and so

$$\lambda f(\mathbf{x}_2)+(1-\lambda)f(\mathbf{x}_1)-f(\lambda\mathbf{x}_2+(1-\lambda)\mathbf{x}_1)>0.$$

Substituting $f(\mathbf{x})=\mathbf{q}^T\mathbf{x}+\frac{1}{2}\mathbf{x}^T\mathbf{Q}\mathbf{x}$,

$$\lambda\left[\mathbf{q}^T\mathbf{x}_2+\tfrac{1}{2}\mathbf{x}^T\mathbf{Q}\mathbf{x}_2\right]+(1-\lambda)\left[\mathbf{q}^T\mathbf{x}_1+\tfrac{1}{2}\mathbf{x}_1^T\mathbf{Q}\mathbf{x}_1\right]-\mathbf{q}^T\left[\lambda\mathbf{x}_2+(1-\lambda)\mathbf{x}_1\right]$$

$$-\tfrac{1}{2}\left[\lambda\mathbf{x}_2+(1-\lambda)\mathbf{x}_1\right]^T\mathbf{Q}\left[\lambda\mathbf{x}_2+(1-\lambda)\mathbf{x}_1\right]>0,$$

from which

$$\lambda(1-\lambda)\left[(\mathbf{x}_1-\mathbf{x}_2)^T\mathbf{Q}(\mathbf{x}_1-\mathbf{x}_2)\right]>0.$$

Now if $\mathbf{x}_1-\mathbf{x}_2=\mathbf{y}\neq\mathbf{0}$ and $0<\lambda<1$, then

$$\lambda(1-\lambda)\mathbf{y}^T\mathbf{Q}\mathbf{y}>0$$

if and only if $\mathbf{Q}$ is positive definite.

1.9 Prove that the sum of convex functions is convex.

Let the functions $f_i(\mathbf{x})$, $i=1,2,\ldots,m$ be convex over some convex set $X\subset E^n$. Define

$$F(\mathbf{x})=\sum_{i=1}^{m} f_i(\mathbf{x}).$$

Since $f_i(\mathbf{x})$, $i=1,2,\ldots,m$, are convex,

$$f_i(\lambda\mathbf{x}_2+(1-\lambda)\mathbf{x}_1)\leqslant\lambda f_i(\mathbf{x}_2)+(1-\lambda)f_i(\mathbf{x}_1),$$

where $i=1,2,\ldots,m$, $0\leqslant\lambda\leqslant 1$, $\mathbf{x}_1,\mathbf{x}_2\in X$.
Summing over all i,

$$\sum_{i=1}^{m} f_i(\lambda\mathbf{x}_2+(1-\lambda)\mathbf{x}_1)\leqslant\sum_{i=1}^{m}\left[\lambda f_i(\mathbf{x}_2)+(1-\lambda)f_i(\mathbf{x}_1)\right],$$

or

$$F(\lambda\mathbf{x}_2+(1-\lambda)\mathbf{x}_1)\leqslant\lambda F(\mathbf{x}_2)+(1-\lambda)F(\mathbf{x}_1),$$

and therefore $F(\mathbf{x})$ is a convex function. Of course, the sum of concave functions is also a concave function.

1.10 Is the function $f(x_1,x_2)=x_1x_2$ convex over E^2? Is it convex over $X=\{\mathbf{x}|x_1\geqslant 0,\ \mathbf{x}_2\geqslant 0\}$?

$f(x_1,x_2)=x_1x_2$ is not convex over E^2, as the following counterexample shows. For convexity,

$$f(\lambda\mathbf{x}_2+(1-\lambda)\mathbf{x}_1)\leqslant\lambda f(\mathbf{x}_2)+(1-\lambda)f(\mathbf{x}_1)$$

for all $\mathbf{x}_1,\mathbf{x}_2\in E^2$ and all $0\leqslant\lambda\leqslant 1$.*

*Note to Reader. Throughout this text x_1 and x_2 (in italic type) refer to the components of $\mathbf{x}$, while $\mathbf{x}_1$ and $\mathbf{x}_2$ (in bold type) are distinct vectors (with components $\mathbf{x}_{11}$, $\mathbf{x}_{12}$ etc.).

Take

$$\mathbf{x}_1=\begin{bmatrix} 1 \\ -1 \end{bmatrix}, \mathbf{x}_2=\begin{bmatrix} -2 \\ 2 \end{bmatrix}.$$

Then $f(\mathbf{x}_1)=1(-1)=-1$, $f(\mathbf{x}_2)=(-2)2=-4$,

$$\lambda\mathbf{x}_2+(1-\lambda)\mathbf{x}_1=\lambda\begin{bmatrix} 1 \\ -1 \end{bmatrix}+(1-\lambda)\begin{bmatrix} -2 \\ 2 \end{bmatrix}=\begin{bmatrix} 3\lambda-2 \\ -3\lambda+2 \end{bmatrix},$$

and so

$$f((3\lambda-2),(-3\lambda+2))=(3\lambda-2)(-3\lambda+2)=-9\lambda^2+12\lambda-4,$$

$$\lambda f(\mathbf{x}_2)+(1-\lambda)f(\mathbf{x}_1)=\lambda(-4)+(1-\lambda)(-1)=-3\lambda-1.$$

Question: Is $-9\lambda^2+12\lambda-4\leqslant-3\lambda-1$ for all $0\leqslant\lambda\leqslant1$? Letting $\lambda=1$, we see that the inequality is violated. Hence x_1x_2 is not convex over E^2.

Now consider two points in X, say

$$\mathbf{x}_1=\begin{bmatrix} 2 \\ 2 \end{bmatrix}, \qquad \mathbf{x}_2=\begin{bmatrix} 4 \\ 1 \end{bmatrix}.$$

Then $f(\mathbf{x}_1)=(2)(2)=4$, $f(\mathbf{x}_2)=(4)(1)=4$,

$$\lambda\mathbf{x}_2+(1-\lambda)\mathbf{x}_1=\lambda\begin{bmatrix} 4 \\ 1 \end{bmatrix}+(1-\lambda)\begin{bmatrix} 2 \\ 2 \end{bmatrix}=\begin{bmatrix} 2\lambda+2 \\ -\lambda+2 \end{bmatrix},$$

and so

$$f[2\lambda+2,-\lambda+2]=(2\lambda+2)(-\lambda+2)=-2\lambda^2+2\lambda+4,$$

$$\lambda f(\mathbf{x}_2)+(1-\lambda)f(\mathbf{x}_1)=\lambda(4)+(1-\lambda)4=4.$$

Question: Is $-2\lambda^2+2\lambda+4\leqslant4$ for all $0\leqslant\lambda\leqslant1$? Letting $\lambda=\frac{1}{2}$, we see that the inequality does not hold. Hence x_1x_2 is not convex even over X. However, we note that x_1x_2 is convex in X along any straight line through the origin.

Let $x_2=\alpha x_1$, $0\leqslant\alpha<\infty$; then $f(\mathbf{x})=\alpha x_1^2$, which is convex for all $0\leqslant\alpha<\infty$. Also, along any line parallel to the coordinate axes, $f(\mathbf{x})$ is linear and therefore convex; e.g., if $x_2=4$ then $f(\mathbf{x})=x_1x_2=4x_1$. This completes the proof.

1.11 In each of the following cases write the objective function in the form $y=\mathbf{x}^T\mathbf{H}\mathbf{x}+\mathbf{q}^T\mathbf{x}$:

(a) $y=x_1^2+2x_1x_2+4x_1x_3+3x_2^2+2x_2x_3+5x_3^2+4x_1-2x_2+3x_3$.

(b) $y=5x_1^2+12x_1x_2-16x_1x_3+10x_2^2-26x_2x_3+17x_3^2-2x_1-4x_2-6x_3$.

(c) $y=x_1^2-4x_1x_2+6x_1x_3+5x_2^2-10x_2x_3+8x_3^2$.

(a) $\mathbf{H}=\begin{bmatrix} 1 & 1 & 2 \\ 1 & 3 & 1 \\ 2 & 1 & 5 \end{bmatrix}, \quad \mathbf{q}=\begin{bmatrix} 4 \\ -2 \\ 3 \end{bmatrix}.$

(b) $\mathbf{H}=\begin{bmatrix} 5 & 6 & -8 \\ 6 & 10 & -13 \\ -8 & -13 & 17 \end{bmatrix}, \quad \mathbf{q}=\begin{bmatrix} -2 \\ -4 \\ -6 \end{bmatrix}.$

(c) $\mathbf{H}=\begin{bmatrix} 1 & -2 & 3 \\ -2 & 5 & -5 \\ 3 & -5 & 8 \end{bmatrix}, \quad \mathbf{q}=\begin{bmatrix} 0 \\ 0 \\ 0 \end{bmatrix}.$

1.12 Determine properties of sign definiteness for the following quadratic form:

$$y=x_1^2-4x_1x_2+6x_1x_3+5x_2^2-10x_2x_3+8x_3^2,$$

or $y=\mathbf{x}^T\mathbf{H}\mathbf{x}$, where

$$\mathbf{H}=\begin{bmatrix} 1 & -2 & 3 \\ -2 & 5 & -5 \\ 3 & -5 & 8 \end{bmatrix}.$$

The successive principal minors are

$$|1|=1, \qquad \begin{vmatrix} 1 & -2 \\ -2 & 5 \end{vmatrix}=5-4=1, \qquad |\mathbf{H}|=-2$$

Thus by Sylvester's theorem, **H** is not positive definite. Similarly for $-\mathbf{H}$, $|-1|=-1$, so **H** is not negative definite. Thus **H** is either positive semidefinite, negative semidefinite, or indefinite. We see that **H** is indeed indefinite by displaying two points which make y positive and negative, respectively.

Let $x_2=0$; then $y=x_1^2+6x_1x_3+8x_3^2=(x_1+4x_3)(x_1+2x_3)$.
If $x_1=3$, $x_3=-1$, then $y=(3-4)(3-2)=-1<0$.
If $x_1=0$, $x_3=1$, then $y=4(2)=8>0$.

Nonlinear Programming

1.13 For the equations in Problem 1.3, sketch the level curves in E^2 and indicate regions where z is positive and negative.

(a) $z=x_1+2x_2^2$. For $z=0$, $x_1=-2x_2^2$. See Fig. 1.14(a).

(b) $z=x_1x_2$. For $z=0$, either $x_1=0$ or $x_2=0$. See Fig. 1.14(b).

(c) $z=x_1^2-x_2^2$. For $z=0$, $x_1=\pm x_2$. See Fig. 1.14(c).

(d) $z=x_1^2+3x_1x_2+3x_2^2=(x_1+\frac{3}{2}x_2)^2+\frac{3}{4}x_2^2$. For $z=0$, $x_1=-\frac{3}{2}x_2=0$. See Fig. 1.14(d).

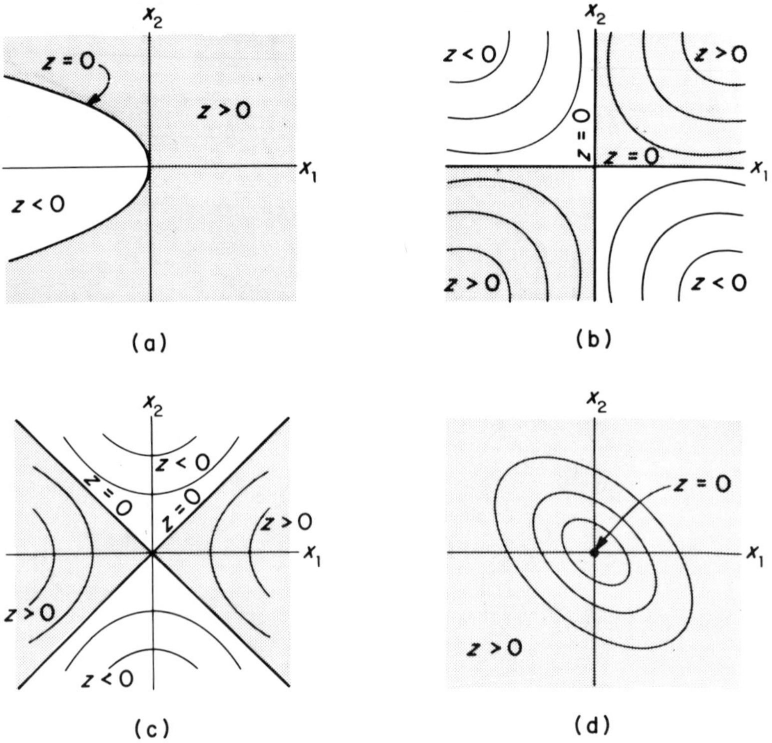

FIGURE 1.14

1.14 For each of the following, construct a diagram indicating the set of feasible points. Is the set convex? Assume $x_1 \geqslant 0$, $x_2 \geqslant 0$.

(a) $x_1^2 + (x_2 - 1)^2 - 1 \leqslant 0,$
$(x_1 - 1)^2 + x_2^2 - 1 \leqslant 0.$

(b) $x_1^2 + (x_2 - 1)^2 - 1 \leqslant 0,$
$x_1^2 + x_2^2 - 1 \leqslant 0.$

(c) $x_1^2 + x_2^2 - 1 \leqslant 0,$
$x_1 + x_2 - \frac{1}{2} \leqslant 0.$

(a) F is strictly convex, therefore convex. See Fig. 1.15(a).

(b) F is convex but not strictly convex. See Fig. 1.15(b).

(c) F is convex but not strictly convex. See Fig. 1.15(c).

1.15 A manufacturing process produces x_i items in a given time period at a cost $f_i(x_i)$. In each period there is a demand d_i for the product and a cost $c_i(y_i)$ for holding inventory in amount y_i. The production cannot exceed X in each time

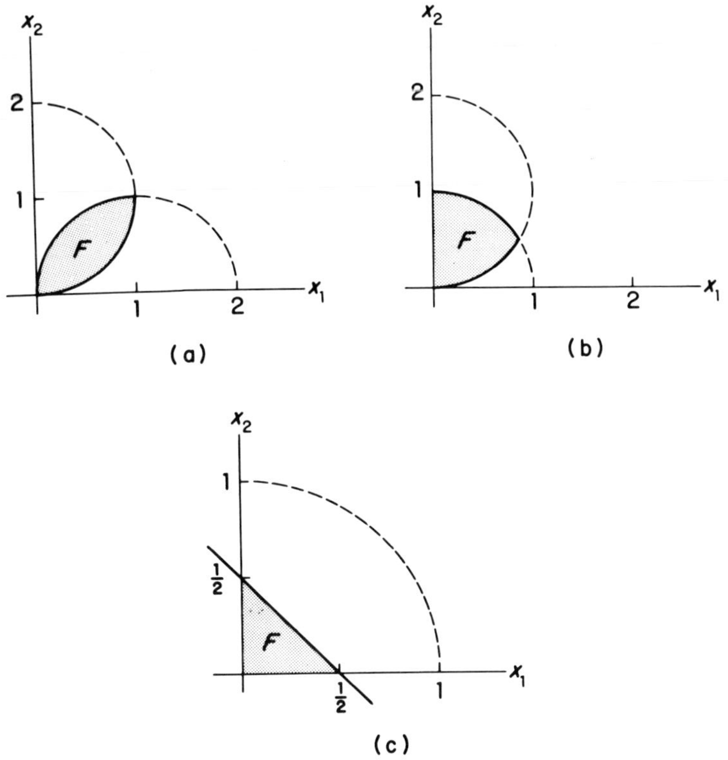

FIGURE 1.15

period and the inventory cannot exceed Y. If the initial inventory is zero, formulate a nonlinear programming problem for minimizing total cost over n periods.

The total cost is given by

$$F(\mathbf{x},\mathbf{y}) = \sum_{i=1}^{n} \left[f_i(x_i) + c_i(y_i) \right].$$

The inventory in any period is given by

$$y_i = y_{i-1} + x_i - d_i, \qquad \text{where} \quad y_0 = 0, \quad i = 1, 2, \ldots, n.$$

The nonlinear programming problem is

$$\min_{x_i, y_i} \sum_{i=1}^{n} \left[f_i(x_i) + c_i(y_i) \right]$$

such that

$$y_i = y_{i-1} + x_i - d_i,$$

$$y_0 = 0, \qquad x_i \leqslant X, \qquad y_i \leqslant Y.$$

Note that the x_i variables can be easily eliminated, so the problem becomes

$$\min_{y_i} \sum_{i=1}^{n} \left[f_i(y_i - y_{i-1} + d_i) + c_i(y_i) \right]$$

such that

$$y_i - y_{i-1} + d_i \leqslant X, \qquad y_0 = 0, \quad y_i \leqslant Y, \qquad i = 1, 2, \ldots, n.$$

Linear Programming

1.16 Under what conditions would Problem 1.15 be a linear programming problem?

If $f_i(x_i) = a_i x_i$, $c_i(y_i) = b_i y_i$, and $y_i \geqslant 0$, then the linear program is

$$\min_{y_i} \sum_{i=1}^{n} [a_i(y_i - y_{i-1} + d_i) + b_i y_i]$$

such that

$$y_i - y_{i-1} + d_i \leqslant X, \qquad y_i \leqslant Y, \quad y_i \geqslant 0, \quad y_0 = 0, \qquad i = 1, 2, \ldots, n.$$

1.17 A large bakery has m plants located throughout the state. Daily bread production at the ith plant is limited to S_i loaves, $i = 1, 2, \ldots, m$. Each day the bakery must supply to its n distributing warehouses D_j, $j = 1, 2, \ldots, n$, loaves to meet demand requirements. The linear programming problem is to determine which plants are to supply which warehouses so that total transportation costs are minimum.

Let x_{ij} be the number of loaves shipped from plant i to warehouse j, and let c_{ij} be the associated shipping cost per loaf. Then the problem can be stated as follows:

$$\min_{x_{ij}} \sum_{i=1}^{m} \sum_{j=1}^{n} c_{ij} x_{ij}$$

such that

$$\sum_{j=1}^{n} x_{ij} \leqslant S_i, \qquad \sum_{i=1}^{m} x_{ij} \geqslant D_j, \qquad x_{ij} \geqslant 0$$

$$i = 1, 2, \ldots, m, \qquad j = 1, 2, \ldots, n.$$

Supplementary Problems

1.18 Consider the quadratic form $f(\mathbf{x})=x_1^2-2x_1x_2+4x_2^2$.

(a) Write the quadratic form in matrix vector notation.
(b) Write the Hessian matrix.
(c) Show that $f(\mathbf{x})$ is positive definite.
(d) Is $f(\mathbf{x})$ convex? Why?
(e) Find the equation of the tangent hyperplane at 2,2.

1.19 Sketch the level curves in E^2 and show regions where z is positive and negative:

(a) $z=x_1^2+2x_2^2$. (c) $z=x_1/x_2$.
(b) $z=x_1^2-x_2^2+4$. (d) $z=x_1^2-2x_1x_2+x_2^2+4x_1$.

1.20 Show that e^{-x} is strictly convex.

1.21 Which of the functions z in Problem 1.19 are quadratic forms?

1.22 Determine the gradient: (a) $z=3x_1/x_2+4e^{x_1/x_2}$, (b) $z=(x_1x_2)^{-1}$.

1.23 Determine the Hessian for each function in Problem 1.22.

1.24 Determine the sign definiteness of each of the quadratic forms $\mathbf{x}^T\mathbf{A}\mathbf{x}$:

$$\text{(a) } \mathbf{A}=\begin{bmatrix} 2 & 1 & 4 \\ 6 & 0 & 1 \\ 1 & -1 & 2 \end{bmatrix}, \quad \text{(b) } \mathbf{A}=\begin{bmatrix} 1 & 1 & 0 \\ 3 & 2 & 1 \\ 1 & 2 & 4 \end{bmatrix},$$

$$\text{(c) } \mathbf{A}=\begin{bmatrix} 1 & -2 & 1 \\ -4 & 2 & -1 \\ 1 & -1 & 0 \end{bmatrix}.$$

1.25 Write each objective function in the form $y=\mathbf{x}^T\mathbf{H}\mathbf{x}+q^T\mathbf{x}$:

(a) $y=2x_1^2+x_1x_2+9x_1x_3+3x_2^2+x_2x_3+2x_2$.
(b) $y=x_1^2-6x_1x_3+x_3^2+9x_3$.

1.26 Solve graphically:

$$\min_{x_1,x_2} (x_1-6)^2+(x_i-2)^2$$

such that

$$0.5x_1+x_2\leqslant 4, \quad 3x_1+x_2\leqslant 15, \quad x_1+x_2\geqslant 1, \quad x_1\geqslant 0, \quad x_2\geqslant 0.$$

1.27 Solve graphically:

$$\max_{x_1,x_2} 2x_1+3x_2+4x_1^2+2x_1x_2+x_2^2$$

such that

$$x_1 - x_2 \geqslant 0, \qquad x_1 + x_2 \leqslant 4, \qquad x_1 \leqslant 3.$$

1.28 Solve graphically:

$$\max_{x_1, x_2} x_1 x_2$$

such that

$$3x_1 + 2x_2 \leqslant 9, \quad 0.5x_1 + x_2 \leqslant 4, \quad x_1 \geqslant 0, \quad x_2 \geqslant 0.$$

1.29 Solve graphically:

$$\max_{x_1, x_2} 4x_1 + 3x_2$$

such that

$$2x_1 - x_2 \geqslant 0, \quad 0.5x_1 - x_2 \leqslant 0, \quad x_1 + x_2 \leqslant 5, \quad x_1 \geqslant 0, \quad x_2 \geqslant 0.$$

1.30 Solve Problem 1.29 where x_1 and x_2 are integers.

Chapter 2

Classical Optimization

Maxima, Minima, and Saddle Points

In this chapter we treat the unconstrained static optimization problem as posed below:

$$\min_{\mathbf{x}} f(\mathbf{x}).$$

The problem is unconstrained because there are no restrictions on the values which $\mathbf{x}$ may assume. There are however other questions which must be answered. For example, when does $f(\mathbf{x})$ possess a minimum? Is it unique? Are these relative minima? We begin to answer these questions with some definitions.

> *Absolute or global minimum:* The function $f(\mathbf{x})$ defined over a closed set X in E^n is said to take on its *absolute* or *global minimum* over X at the point $\mathbf{x}^*$ if $f(\mathbf{x}) \geqslant f(\mathbf{x}^*)$ for every point $\mathbf{x} \in X$.
>
> *Strong relative or local minimum:* Let $f(\mathbf{x})$ be defined at all points in some δ-neighborhood of $\mathbf{x}^*$ in E^n. The function $f(\mathbf{x})$ is said to have a *strong relative* or *local minimum* at $\mathbf{x}^*$ if there exists an ε, $0 < \varepsilon < \delta$, such that for all $\mathbf{x}$, $0 < \|\mathbf{x} - \mathbf{x}^*\| < \varepsilon$, $f(\mathbf{x}) > f(\mathbf{x}^*)$.

A global minimum at $\mathbf{x}^*$ implies that the objective function takes on its lowest value $f(\mathbf{x}^*)$ at that point no matter where else we may search in the set X. For most practical purposes (excluding minima at $\|\mathbf{x}^*\| = \infty$) the set X may be all of E^n. A strong local minimum, on the other hand, only guarantees that the value of $f(\mathbf{x}^*)$ is a minimum with respect to other points nearby, specifically in an ε-region about $\mathbf{x}^*$ (see Fig. 2.1). Thus a function may have many local minima, each with a different value of the objective function, say, $f(\mathbf{x}_i^0)$, $i = 1, \ldots, p$. The global minimum can always be chosen from among these local minima by comparing their values and choosing one such that

$$f(\mathbf{x}^*) \leqslant f(\mathbf{x}_i^0), \qquad i = 1, 2, \ldots, p,$$

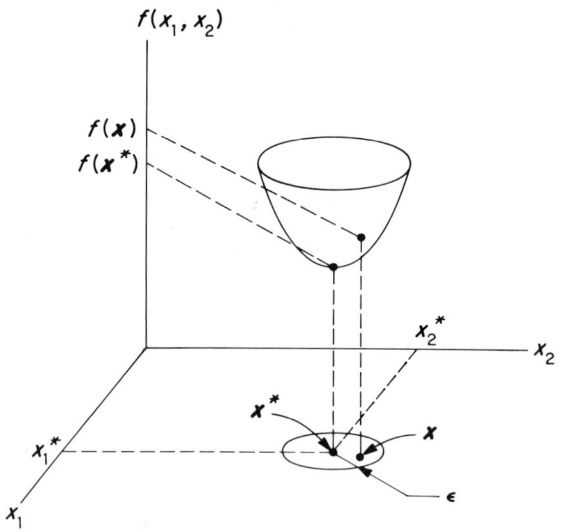

FIGURE 2.1

where

$$\mathbf{x}^* \in \left\{\mathbf{x}_i^0 \,|\, i = 1, 2, \ldots, p\right\}.$$

Every global minimum is also a local minimum; however, the converse is not true.

If $f(\mathbf{x}) > f(\mathbf{x}^*)$ for a global minimum, the minimum is unique and is called a *proper global minimum*. In the case of a local minimum, if $f(\mathbf{x}) \geqslant f(\mathbf{x}^*)$, then the local minimum is not unique and is called a *weak local minimum*. Similar definitions can be obtained for maximization problems by simply reversing the sense of the inequality.

EXAMPLES

2.1 (a) Refer to Fig. 2.2(a). $\sin x$ has an infinite number of strong local minima at $x_k^* = -\pi/2 + 2k\pi$, $k = \ldots, -2, -1, 0, 1, 2, \ldots$. Since

$$\sin x_k^* = -1 \qquad \text{for all } k,$$

the x_k^* are also global minima for all k, but there is no proper global minimum.

(b) Refer to Fig. 2.2(b). x^2 has a strong local minimum at $x^* = 0$ which is also a proper global minimum.

(c) Refer to Fig. 2.2(c). $e^{-0.1x}\cos x$ for $x \geqslant 0$ has an infinite number of

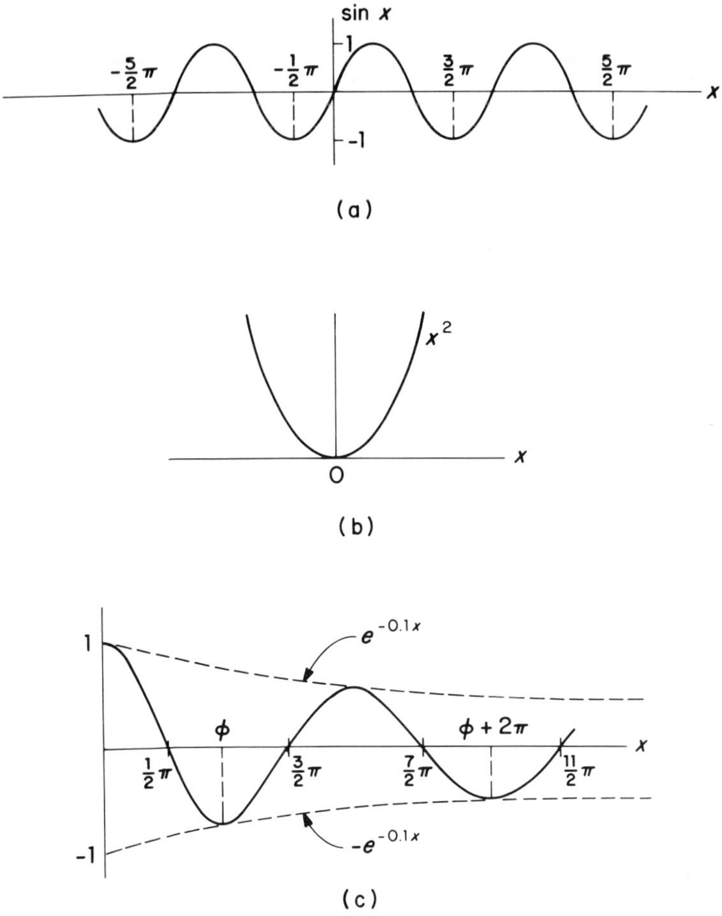

FIGURE 2.2

strong local minima at $x_k^* = \phi + 2k\pi$, $k = 0, 1, 2, \ldots$, where $\phi = \arctan(-0.1) \approx 174.3°$. There is a proper global minimum at x_0^*, since

$$e^{-0.1x_0^*}\cos x_0^* < e^{-0.1x_k^*}\cos x_k^*, \qquad k = 1, 2, \ldots.$$

2.2 (a) $\mathbf{x}^T\mathbf{A}\mathbf{x}$ has a proper global minimum at $\mathbf{x}^* = \mathbf{0}$ if $\mathbf{A}$ is positive definite. This follows from the definition of positive definiteness.

(b) $\mathbf{x}^T\mathbf{A}\mathbf{x}$ has a global minimum which is not proper if $\mathbf{A}$ is nonnegative definite. This global minimum occurs at $\mathbf{x}^*$ where $(\mathbf{x}^*)^T\mathbf{A}\mathbf{x}^* = 0$. By the definition of nonnegative definiteness, there is some $\mathbf{x}^* \neq \mathbf{0}$ in addition to $\mathbf{x}^* = \mathbf{0}$ which produces the global minimum.

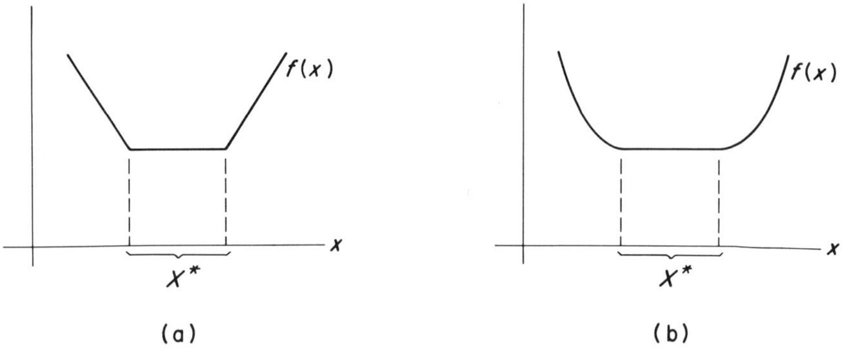

FIGURE 2.3

2.3 (a) Refer to Fig. 2.3(a). Nondifferentiable function with weak local minimum in the region X^*.

(b) Refer to Fig. 2.3(b). Differentiable function with weak local minimum in the region X^*.

It is intuitively obvious that in optimization problems having physical significance, we will usually want to determine the global optimum. However, if the objective function has more than one minimum, this will require determining all the minima, then evaluating the objective function at each local minimum and choosing the global minimum. What is more, if we should fail to determine *all* the local minima, the global minimum could be missed despite all our efforts. One convenient way to avoid this difficulty is to ensure that our objective function is convex or strictly convex. For such functions a local minimum is also a global minimum. If the function is strictly convex the global minimum is also unique, that is, a proper global minimum. Thus for convex functions we need find only one local minimum and the problem is solved. As we shall see, most of the theory and computational algorithms available are for determining a local minimum. For computational simplicity we would therefore most often like to restrict our attention to convex objective functions. For many problems having physical significance, however, this is not always possible.

EXAMPLES

2.4 Let $f(\mathbf{x})$ be a twice differentiable convex function on $X \subset E^n$. Then its Hessian $\mathbf{H}(\mathbf{x})$ is nonnegative definite for all $\mathbf{x} \in X$. The definition of nonnegative definiteness implies that $f(\mathbf{x})$ has a global minimum at $\mathbf{x}^*$ where $(\mathbf{x}^*)^T\mathbf{H}(\mathbf{x}^*)\mathbf{x}^* = 0$.

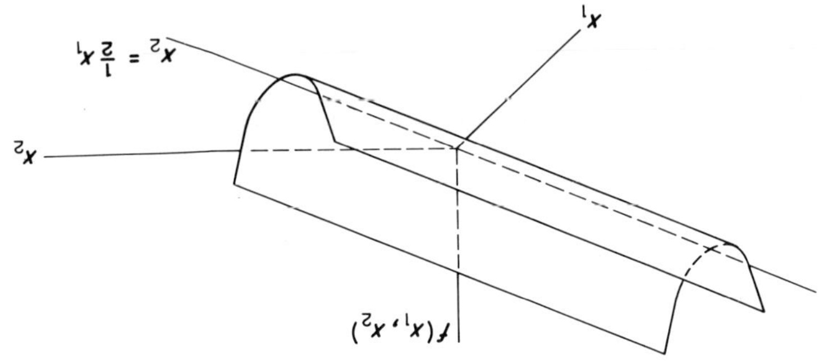

FIGURE 2.4

2.5 Refer to Fig. 2.4.

$$\min_{x_1, x_2} (x_1 - 2x_2)^2.$$

This function is positive except along $x_2 = \frac{1}{2}x_1$, where it is zero. Thus the global minimum occurs at X^* where

$$X^* = \{x_1, x_2 | x_1 - 2x_2 = 0\}.$$

2.6 $f(\mathbf{x}) = \mathbf{x}^T\mathbf{A}\mathbf{x} + b\mathbf{x}$, where

$$\mathbf{A} = \begin{bmatrix} 6 & 1 & 1 \\ 1 & 2 & 0 \\ 1 & 0 & 4 \end{bmatrix}.$$

The Hessian of $f(\mathbf{x})$ is $\mathbf{A}$, and by Sylvester's theorem $\mathbf{A}$ is positive definite for all $\mathbf{x}$, since it is not a function of $\mathbf{x}$ and its principal minors are all positive

$$A_1 = 6, \qquad A_2 = \begin{vmatrix} 6 & 1 \\ 1 & 2 \end{vmatrix} = 11, \qquad A_3 = \begin{vmatrix} 6 & 1 & 1 \\ 1 & 2 & 0 \\ 1 & 0 & 4 \end{vmatrix} = 42.$$

Thus by Theorem 1.3, p. 12, $f(\mathbf{x})$ is strictly convex and $f(\mathbf{x})$ possesses a proper global minimum.

Convex and concave functions are often called *unimodal* because they have the property of having only one optimum point (or region). Functions with many maxima and/or minima are often called multimodal.

A function $f(\mathbf{x}, \mathbf{y})$ is said to have a *saddle point* at $(\mathbf{x}_0, \mathbf{y}_0)$ if there exists an $\varepsilon > 0$ such that for all $\mathbf{x}$, $\|\mathbf{x}-\mathbf{x}_0\| < \varepsilon$, and all $\mathbf{y}$, $\|\mathbf{y}-\mathbf{y}_0\| < \varepsilon$,

$$f(\mathbf{x}, \mathbf{y}_0) \leqslant f(\mathbf{x}_0, \mathbf{y}_0) \leqslant f(\mathbf{x}_0, \mathbf{y}).$$

Thus if $f(\mathbf{x}, \mathbf{y})$ possesses a saddle point at $(\mathbf{x}_0, \mathbf{y}_0)$, this point is a maximum of the function with respect to the $\mathbf{x}$-variables and a minimum with respect to the $\mathbf{y}$-variables. For twice differentiable functions of two variables $f(x,y)$, the following discriminant (Δ) is useful for identifying saddle points as well as maxima and minima. Define (x_0,y_0) by the necessary conditions

$$\frac{\partial f}{\partial x}(x_0,y_0)=0, \qquad \frac{\partial f}{\partial y}(x_0,y_0)=0,$$

and

$$\Delta = \left\{ \frac{\partial^2 f}{\partial x^2}\frac{\partial^2 f}{\partial y^2} - \left(\frac{\partial^2 f}{\partial x\, \partial y} \right)^2 \right\}\Bigg|_{x_0,y_0}$$

Then

(1) (x_0,y_0) is a local maximum if $\Delta > 0$ and

$$\left.\frac{\partial^2 f}{\partial x^2}\right|_{x_0,y_0} < 0 \qquad \left[\text{or } \left.\frac{\partial^2 f}{\partial y^2}\right|_{x_0,y_0} < 0\right].$$

(2) (x_0,y_0) is a relative minimum if $\Delta > 0$ and

$$\left.\frac{\partial^2 f}{\partial x^2}\right|_{x_0,y_0} > 0 \qquad \left[\text{or } \left.\frac{\partial^2 f}{\partial y^2}\right|_{x_0,y_0} > 0\right].$$

(3) (x_0,y_0) is a saddle point if $\Delta < 0$.

(4) If $\Delta = 0$, no information can be obtained by this test.

Unfortunately, similar simple conditions for identifying saddle points in higher-dimensional spaces are not available. This question, and what to do when $\Delta = 0$, are considered in the next section.

EXAMPLE 2.7 Consider $f(x,y)=x^2-y^2$.

$$\frac{\partial f}{\partial x}=2x, \qquad \frac{\partial^2 f}{\partial x^2}=2, \qquad \frac{\partial f}{\partial y}=-2y, \qquad \frac{\partial^2 f}{\partial y^2}=-2, \qquad \frac{\partial^2 f}{\partial x\,\partial y}=0,$$

$$\Delta=2(-2)-0=-4<0.$$

Thus $f(x,y)$ has a saddle point at $x_0=0$, $y_0=0$. This is seen clearly from the level curves shown in Fig. 2.5.

The analogy to a saddle point for functions of only one variable is called an *inflection point*. If

$$\frac{df}{dx}(x_0)=0,$$

then x_0 is an inflection point if the first nonvanishing derivative is of odd order.

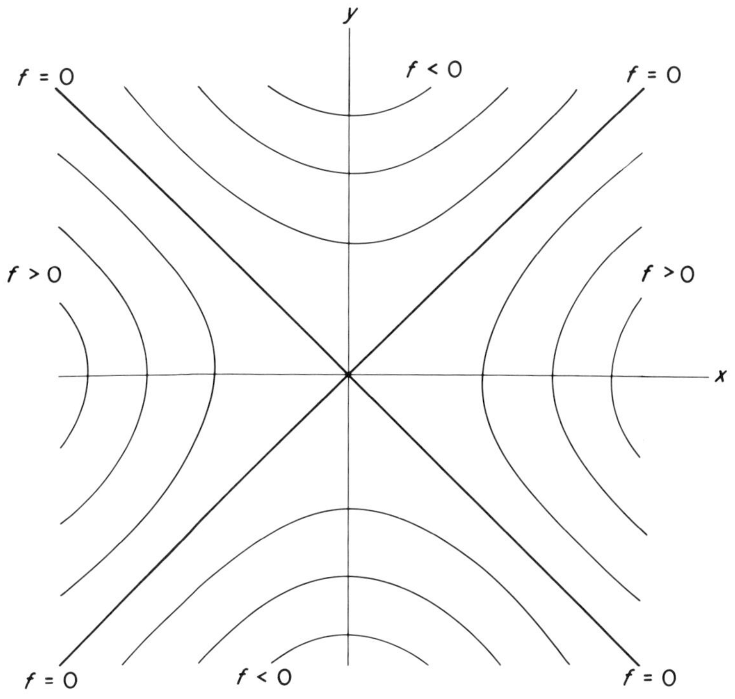

FIGURE 2.5

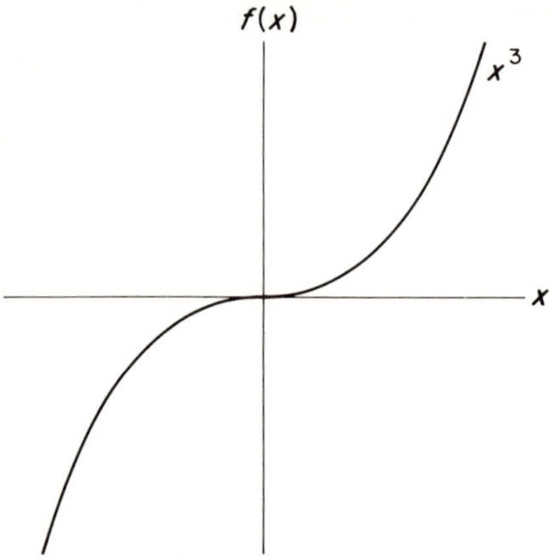

FIGURE 2.6

EXAMPLE 2.8 See Fig. 2.6. $f(x)=x^3$; then

$$\frac{df}{dx}=3x^2=0,$$

which implies $x_0=0$. But

$$\frac{d^2f}{dx^2}=6x\bigg|_{x_0=0}=0 \quad \text{and} \quad \frac{d^3f}{dx^3}=6\neq 0.$$

Hence $x_0=0$ is an inflection point.

Unconstrained Problems

Necessary and sufficient conditions for the unconstrained problem

$$\min_{\mathbf{x}} f(\mathbf{x})$$

to have a strong local minimum will now be derived. We assume that $f(\mathbf{x})$ is twice differentiable. For convenience of notation the derivation is carried out for a two-dimensional argument $\mathbf{x}=[x,y]^T$. These results are easily generalized.

Let x_0,y_0 be a candidate point for a strong local minimum. Then expand $f(x,y)$ about x_0,y_0 in a Taylor series retaining only first- and second-order terms. Letting $x = x_0 + h$, $y = y_0 + k$,

$$f(x_0+h, y_0+k) = f(x_0,y_0) + hf_x(x_0,y_0) + kf_y(x_0,y_0) + \tfrac{1}{2}h^2 f_{xx}(x_0,y_0) + hkf_{xy}(x_0,y_0) + \tfrac{1}{2}k^2 f_{yy}(x_0,y_0) + O(h^3,k^3),$$

where

$$f_x = \frac{\partial f}{\partial x}, \qquad f_y = \frac{\partial f}{\partial y},$$

$$f_{xx} = \frac{\partial^2 f}{\partial x^2}, \qquad f_{yy} = \frac{\partial^2 f}{\partial y^2}, \qquad f_{xy} = \frac{\partial^2 f}{\partial x\,\partial y}.$$

In matrix notation,

$$f(x_0+h, y_0+k) = f(x_0,y_0) + hf_x(x_0,y_0) + kf_y(x_0,y_0) + \tfrac{1}{2}[h,k]\mathbf{H}(x_0,y_0)\begin{bmatrix} h \\ k \end{bmatrix} + O(h^3,k^3)$$

where

$$\mathbf{H}(x_0,y_0) = \begin{bmatrix} f_{xx} & f_{xy} \\ f_{yx} & f_{yy} \end{bmatrix}_{x_0,y_0}$$

is the Hessian matrix evaluated at x_0, y_0, and the expression $\tfrac{1}{2}[h,k]\mathbf{H}(x_0,y_0)\begin{bmatrix} h \\ k \end{bmatrix}$ is a quadratic form.

By definition, a strong local minimum exists at x_0, y_0 if

$$f(x_0+h, y_0+k) > f(x_0,y_0)$$

for all h and k such that $0 < \sqrt{h^2+k^2} < \varepsilon$.

In terms of the Taylor series, this expression becomes

$$hf_x(x_0,y_0) + kf_y(x_0,y_0) + \tfrac{1}{2}[h,k]\mathbf{H}(x_0,y_0)\begin{bmatrix} h \\ k \end{bmatrix} + O(h^3,k^3) > 0.$$

We must now determine conditions on $f(x,y)$ such that this inequality holds. As h and k range over positive and negative values such that $0 < \|h,k\| < \varepsilon$, they describe a circular region about x_0, y_0. Since h and k may take on either sign, there is no way to ensure that the first two terms above are positive. The best we can do is to make

$$f_x(x_0,y_0)=0, \qquad f_y(x_0,y_0)=0.$$

These are *necessary conditions* for the above inequality to hold.

Now by reducing ε suitably, we can shrink the circle about x_0,y_0 until the cubic and higher terms are smaller than the value of the quadratic form. Hence there remains

$$\tfrac{1}{2}[h,k]\mathbf{H}(x_0,y_0)\begin{bmatrix} h \\ k \end{bmatrix}>0,$$

which is the definition of a positive definite quadratic form. Thus if $f_x(x_0,y_0)=0$, $f_y(x_0,y_0)=0$, and $\mathbf{H}(x_0,y_0)$ is positive definite, then $f(x,y)$ possesses a strong local minimum at x_0,y_0. These conditions are *sufficient.*

In case $\mathbf{H}(x_0,y_0)$ is identically zero, the function $f(x,y)$ could still have a minimum at x_0,y_0 if the first nonzero terms were of even order in h and k, i.e., if n were even and

$$\frac{1}{n!}\sum_{j_1=1}^{2}\cdots\sum_{j_n=1}^{2}\frac{\partial^n f}{\partial x_{j_1}\cdots\partial x_{j_n}}(x_0,y_0)\,\Delta x_{j_1}\Delta x_{j_2}\cdots\Delta x_{j_n}>0.$$

If the first nonzero terms are of odd order, then $f(x,y)$ has a saddle point at x_0,y_0.

For a function of n variables, $f(\mathbf{x})$, the necessary conditions for a *minimum* generalize to

$$f_{x_1}(\mathbf{x}_0)=f_{x_2}(\mathbf{x}_0)=\cdots=f_{x_n}(\mathbf{x}_0)=0.$$

The sufficient conditions for a strong relative minimum require in addition that the Hessian $\mathbf{H}(\mathbf{x}_0)$ be positive definite.

For a function of n variables $f(\mathbf{x})$ to have a strong relative *maximum* it is necessary that

$$f_{x_1}(\mathbf{x}_0)=f_{x_2}(\mathbf{x}_0)=\cdots=f_{x_n}(\mathbf{x}_0)=0.$$

This is the same as the necessary condition for a minimum. However, the sufficient conditions for a maximum require that in addition $H(\mathbf{x}_0)$ be negative definite.

These necessary conditions are conveniently written in terms of the gradient vector

$$\nabla f(\mathbf{x}_0)=\mathbf{0}.$$

Since this condition is necessary for both maxima and minima, the values of x_0 which satisfy this expression are collectively called *stationary points*.

EXAMPLES

2.9

$$\min_{\mathbf{x}} \mathbf{x}^T\mathbf{A}\mathbf{x} + \mathbf{B}^T\mathbf{x}$$

Necessary conditions: $\mathbf{A}\mathbf{x} + \mathbf{B} = \mathbf{0}$ or $\mathbf{x}^* = -\frac{1}{2}\mathbf{A}^{-1}\mathbf{B}$.

Sufficient conditions: $\mathbf{x}^* = -\frac{1}{2}\mathbf{A}^{-1}\mathbf{B}$ and $\mathbf{A}$ positive definite.

The positive definiteness of **A** also guarantees that **A** is nonsingular, i.e., that the inverse of **A** exists.

For the problem

$$\max_{\mathbf{x}} \mathbf{x}^T\mathbf{A}\mathbf{x} + \mathbf{B}^T\mathbf{x}$$

the same results are obtained except that **A** must be negative definite to guarantee a maximum.

2.10 Find the extreme points of $x^3 - 4x^2$. See Fig. 2.7.

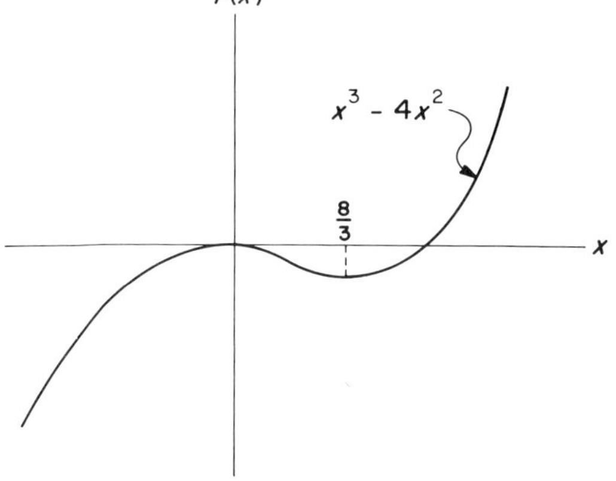

FIGURE 2.7

Necessary conditions: $3x^2-8x=0$ or $x(3x-8)=0$ which is satisfied by either $x^*=0$ or $x^*=\frac{8}{3}$.

Sufficient conditions: $H(x^*)=6x^*-8$. When $x^*=0$, $H(x^*)=8$, which is sufficient for a minimum.

When $x^*=0$, $f(0)=0$. When $x^*=\frac{8}{3}$, $f(\frac{8}{3})=-\frac{256}{27}$.

2.11

$$\min x_1^4+6x_1^2x_2^2+x_1x_2^3-6x_1x_3^3-x_2x_3+4x_3^3.$$

Necessary conditions:

$$\frac{\partial f}{\partial x_1}=4x_1^3+12x_1x_2^2+x_2^3-6x_3^3=0,$$

$$\frac{\partial f}{\partial x_2}=12x_1^2x_2+3x_1x_2^2-x_3=0,$$

$$\frac{\partial f}{\partial x_3}=-18x_1x_3^2-x_2+12x_3^2=0.$$

Sufficient conditions: **H** positive definite, where

$$\mathbf{H}=\begin{bmatrix} f_{11} & f_{12} & f_{13} \\ f_{21} & f_{22} & f_{23} \\ f_{31} & f_{32} & f_{33} \end{bmatrix}$$

and

$$\begin{aligned} f_{11}&=12x_1^2+12x_2^2, & f_{12}&=24x_1x_2+3x_2^2=f_{21}, \\ f_{22}&=12x_1^2+6x_1x_2, & f_{13}&=-18x_3^2=f_{31}, \\ f_{33}&=-36x_1x_3+24x_3, & f_{23}&=-1=f_{32}. \end{aligned}$$

Computational Considerations

Application of the necessary condition from the previous section for determining maxima and minima is referred to as an indirect method for optimization because it leads to a computational problem which must be solved in order to obtain numerical results. The solution of the nonlinear set of equations described by

$$\nabla f(\mathbf{x}_0)=\mathbf{0}$$

is in general not a trivial problem. For all but very simple problems this set of simultaneous equations has many solutions, each of which must be determined in order to obtain global results. We must evaluate the Hessian at each stationary point to determine which points are maxima and which are minima. Then each value of the objective function must be compared for all the minima in order to determine the global minimum. In cases where $f(\mathbf{x})$ is highly nonlinear, even the task of writing down analytical expressions for the gradient may be formidable.

In Chapter 6 we will discuss *direct* optimizing methods which obtain numerical results by treating the minimization problem itself rather than by formulating intermediate conditions (i.e., necessary and sufficient conditions) which must then be solved to obtain numerical answers.

Many numerical methods are available for solving sets of simultaneous nonlinear equations, and these methods are described in numerous books on numerical analysis. For completeness, one of the techniques, the Newton-Raphson method, is described here. This method is a generalization of the well-known Newton's method for finding the roots of a function

$$f(x)=0$$

To solve this scalar problem, we approximate the function at some point x_0 by its tangent and then solve for the value x_1, where the tangent crosses the x-axis. This process is repeated from the new point x_1 until the algorithm converges.

This method defines a sequence of points according to the recurrence relation

$$x_{n+1}=x_n-\frac{f(x_n)}{f'(x_n)}.$$

This sequence of points is illustrated in Fig. 2.8. The method converges rapidly if it converges at all. The convergence rate clearly depends on the initial estimate of the solution x_0, with faster convergence resulting as x_0 approaches the actual solution.

EXAMPLE 2.12 Newton's method is useful for computing square roots. To solve

$$x=\pm\sqrt{a}\ ,$$

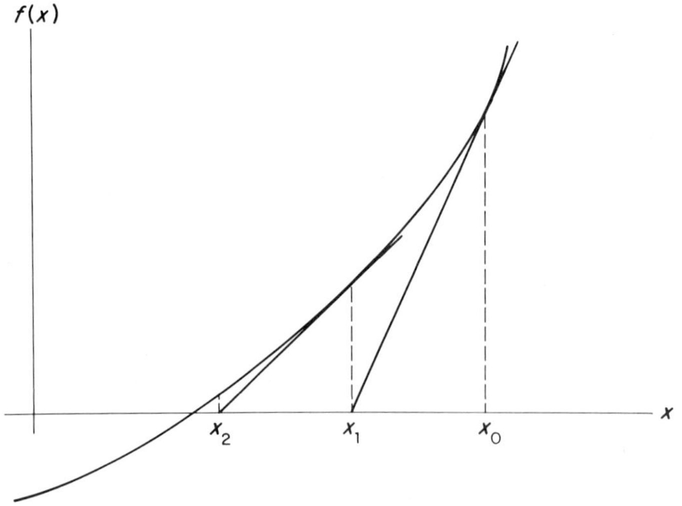

FIGURE 2.8

we must find the roots of

$$f(x)=x^2-a=0.$$

The Newton algorithm is

$$x_{n+1}=x_n-\frac{x_n^2-a}{2x_n}.$$

For $a=8$ and $x_0=3.0$, the first three iterations are

Iteration	x_n	x_n^2
1	3.000	9.000
2	2.833	8.026
3	2.829	8.003

Our motivation now is to extend the idea of Newton's method to solve the sets of simultaneous equations of the form $\nabla f(\mathbf{x}_0)=\mathbf{0}$ resulting from the necessary conditions for optimality. We first expand the set in a first-order Taylor series and set this result equal to zero:

$$\nabla f(\mathbf{x}_{n+1})=\nabla f(\mathbf{x}_n)+\nabla\nabla f(\mathbf{x}_n)(\mathbf{x}_{n+1}-\mathbf{x}_n)=\mathbf{0}.$$

The matrix $\nabla\nabla f(\mathbf{x}_n)$ is just the Hessian of $f(\mathbf{x}_n)$, which we now denote by $\mathbf{H}(\mathbf{x}_n)$. Solving for $\mathbf{x}_{n+1}$ yields

$$\mathbf{x}_{n+1} = \mathbf{x}_n - \mathbf{H}^{-1}(\mathbf{x}_n)\nabla f(\mathbf{x}_n)$$

If $\mathbf{H}(\mathbf{x}_n)$ is positive definite (or negative definite) for all $\mathbf{x}_n$, the algorithm converges rapidly. Thus convergence is guaranteed independent of the value of $\mathbf{x}_0$ only if $f(\mathbf{x})$ is strictly convex (concave). In reality, however, convergence is often attained for more general classes of functions if a suitable starting point $\mathbf{x}_0$ can be found. One disadvantage of this method is that it requires computation of the second partial derivatives of $f(\mathbf{x})$ at each iteration. If $f(\mathbf{x})$ is quadratic, the Hessian is not a function of $\mathbf{x}_n$, and therefore $\mathbf{H}^{-1}$ is a constant matrix for all iterations. Computationally it is very desirable not to recalculate $\mathbf{H}^{-1}$ at each iteration. For this reason, nonlinear functions are often approximated by quadratic forms for the purpose of determining stationary points.

For an example where $\mathbf{H}^{-1}(\mathbf{x}_n)$ approaches singularity, see Problem 2.10.

Solved Problems

Maxima, Minima, and Saddle Points

2.1 Let $f(x,y) = (y - x^2)(y - 2x^2)$. Show that the origin is a saddle point although on any line through the origin, $f(x,y)$ has a local minimum at $(0,0)$.

The origin satisfies the necessary conditions for stationarity:

$$f_x = -6xy + 8x^3 = 0,$$

$$f_y = -3x^2 + 2y = 0.$$

Computing higher-order partial derivatives,

$$\begin{aligned} f_{xx} &= (-6y + 24x^2)\big|_{0,0} = 0, & f_{xxx} &= 48x|_{0,0} = 0,\\ f_{xy} &= f_{yx} = -6x|_{0,0} = 0, & f_{xxy} &= -6,\\ f_{yy} &= 2, & f_{xyy} &= 0,\\ & & f_{yyy} &= 0, \end{aligned}$$

and expanding $f(x,y)$ in a Taylor series about the origin through third-order terms,

$$f(dx,dy) = f_x\,dx + f_y\,dy + \frac{1}{2!}[dx,dy]\begin{bmatrix} f_{xx} & f_{xy} \\ f_{yx} & f_{yy} \end{bmatrix}\begin{bmatrix} dx \\ dy \end{bmatrix}$$

$$+\frac{1}{3!}\left[f_{xxx}(dx)^3 + 3f_{xxy}(dx)^2dy + 3f_{yyx}dx(dy)^2 + f_{yyy}(dy)^3\right],$$

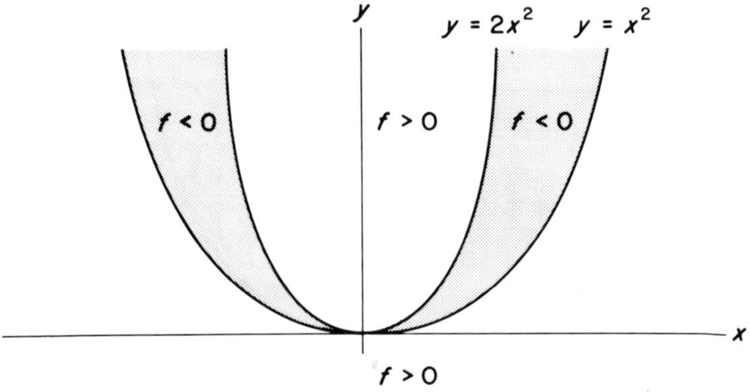

FIGURE 2.9

we see that

$$f(dx,dy)=(dy)^2-3(dx)^2dy.$$

For small dx and dy the quadratic terms are dominant. However, the quadratic term is independent of dx and therefore does not give sufficient information about $f(x,y)$ at the origin. By examining the cubic terms, we see that changing the sign of dy will change the sign of $f(x,y)$ (see Fig. 2.9). Hence the origin is a saddle point.

Along any line through the origin, $y=mx$, $m\neq 0$, and

$$g(x)=m^2x^2-3mx^3+2x^4,$$

$$\frac{dg}{dx}=2m^2x-9mx^2+8x^3=0, \qquad \frac{d^2g}{dx^2}=2m^2-18mx+24x^2.$$

At $x=0$, $dg/dx=0$, and $d^2g/dx^2=2m^2>0$, which is sufficient for a local minimum at the origin.

2.2 Let $f(\mathbf{x})$ be strictly convex over the convex set X. Prove that a strong local minimum is also a global minimum.

Let $\mathbf{x}^0$ be a local minimum and $\mathbf{x}^*$ be a global minimum, with $\mathbf{x}^0, \mathbf{x}^* \in X$. Assume $\mathbf{x}^0 \neq \mathbf{x}^*$. Then the convex combination of $\mathbf{x}^0$ and $\mathbf{x}^*$ is in X, i.e.,

$$\hat{\mathbf{x}}=\lambda\mathbf{x}^*+(1-\lambda)\mathbf{x}^0\in\mathbf{X}, \qquad 0<\lambda<1.$$

Since $f(\mathbf{x})$ is strictly convex,

$$f(\hat{\mathbf{x}})<\lambda f(\mathbf{x}^*)+(1-\lambda)f(\mathbf{x}^0).$$

But $f(\mathbf{x}^*)\leqslant f(\mathbf{x}^0)$, so $f(\hat{\mathbf{x}})<f(\mathbf{x}^0)$.

Now consider some ε-neighborhood of $\mathbf{x}^0$, N_ε, such that $0<\varepsilon<\|\mathbf{x}^*-\mathbf{x}_0\|$. Then

$$0<\lambda=\frac{\varepsilon}{\|\mathbf{x}^*-\mathbf{x}_0\|}<1 \quad \text{and} \quad \hat{x}\in N_\varepsilon.$$

Since $f(\mathbf{x}^0)$ is a strong local minimum, $f(\mathbf{x}^0)<f(\hat{\mathbf{x}})$, which leads to a contradiction. Thus $\mathbf{x}^*=\mathbf{x}^0$.

UNCONSTRAINED PROBLEMS

2.3 Find the stationary points for $f(x)=x^2-x^3$.

$\partial f/\partial x=2x-3x^2=0$. Then $2x(1-\frac{3}{2}x)=0$, and so $x^*=0$ and $x^*=\frac{2}{3}$.

$$\frac{\partial^2 f}{\partial x^2}=2-6x\Big|_{x^*=0}=2>0, \qquad \frac{\partial^2 f}{\partial x^2}=2-6x\Big|_{x^*=2/3}=-2<0.$$

Hence $x^*=0$ is a local minimum and $x^*=\frac{2}{3}$ is a local maximum. Of course, the global maximum and minimum are at $x^*=-\infty$ and $x^*=\infty$, respectively.

2.4 Give sufficient conditions for the various types of optima discussed in this chapter.

From the Taylor series expansion the dominant terms are

$$\mathbf{x}^T\mathbf{H}\mathbf{x}$$

once the necessary conditions $\nabla f(\mathbf{x})=0$ are satisfied. The following conditions on $\mathbf{H}$ are sufficient for the optima indicated.

H	$\mathbf{x}^*$
Positive definite	Strong local minimum
Negative definite	Strong local maximum
Positive semidefinite	Weak local minimum
Negative semidefinite	Weak local maximum
Indefinite	Saddle point
Identically zero	Not determined by **H**; examine higher-order terms

2.5 Determine sufficient conditions for a strong local maximum when the first k derivatives of $f(\mathbf{x})$ vanish identically at $\mathbf{x}^*$.

If the first k derivatives vanish, then the dominant terms in the Taylor series are

$$f(\mathbf{x}^*+\Delta\mathbf{x})-f(\mathbf{x}^*)=\frac{1}{(k+1)!}\sum_{j_1}\sum_{j_2}\cdots\sum_{j_{k+1}}\frac{\partial^{k+1}f}{\partial x_{j_1}\cdots\partial x_{j_{k+1}}}(\mathbf{x}^*)\cdot\Delta x_{j_1}\Delta x_{j_2}\cdots\Delta x_{j_{k+1}}.$$

For a strong local maximum, $f(\mathbf{x}^*+\Delta\mathbf{x})<f(\mathbf{x}^*)$.

Case 1. If k is odd, $k+1$ will be even, and the condition for a local maximum is

$$\sum_{j_1}\sum_{j_2}\cdots\sum_{j_{k+1}} \frac{\partial^{k+1} f}{\partial x_{j_1}\cdots\partial x_{j_{k+1}}}(\mathbf{x}^*)<0.$$

Case 2. If k is even, $k+1$ will be odd and $\Delta x_{j_1}\Delta x_{j_2}\cdots\Delta x_{j_{k+1}}$ could have either sign; hence the $(k+1)$st derivative must also be zero and

$$\sum_{j_1}\sum_{j_2}\cdots\sum_{j_{k+2}} \frac{\partial^{k+2} f}{\partial x_{j_1}\cdots\partial x_{j_{k+2}}}(\mathbf{x}^*)<0.$$

2.6 Determine $\mathbf{x}^*$ for the quadratic forms in Problem 1.11. Is $\mathbf{x}^*$ a maximum, a minimum, or a saddle point?

Since $y=\mathbf{x}^T\mathbf{H}\mathbf{x}+\mathbf{q}^T\mathbf{x}$, we have $\partial\,\nabla y(\mathbf{x})/\partial\mathbf{x}=2\mathbf{H}\mathbf{x}+\mathbf{q}=\mathbf{0}$ and $\mathbf{x}^*=-\frac{1}{2}\mathbf{H}^{-1}\mathbf{q}$.

(a) $\mathbf{H}^{-1}=\begin{bmatrix}14 & -3 & -5\\ -3 & 1 & 1\\ -5 & 1 & 2\end{bmatrix},\quad \mathbf{x}^*=-\frac{1}{2}H^{-1}\begin{bmatrix}4\\ -2\\ 3\end{bmatrix}=\begin{bmatrix}-23.5\\ 5.5\\ 8\end{bmatrix},$

$h_{11}=1>0$, $h_{11}h_{22}-(h_{12})^2=2>0$, $|\mathbf{H}|=1>0$.
Thus $\mathbf{H}$ is positive definite and $\mathbf{x}^*$ is a minimum.

(b) $\mathbf{H}^{-1}=\begin{bmatrix}1 & 2 & 2\\ 2 & 21 & 17\\ 2 & 17 & 14\end{bmatrix},\quad \mathbf{x}^*=-\frac{1}{2}\mathbf{H}^{-1}\begin{bmatrix}-2\\ -4\\ -6\end{bmatrix}=\begin{bmatrix}11\\ 95\\ 78\end{bmatrix},$

$h_{11}=5$, $h_{11}h_{22}-(h_{12})^2=2>0$, $|\mathbf{H}|=1>0$.
Hence $\mathbf{H}$ is positive definite and $\mathbf{x}^*$ is a minimum.

(c) $\mathbf{x}^*=-\frac{1}{2}\mathbf{H}^{-1}\begin{bmatrix}0\\ 0\\ 0\end{bmatrix}=\begin{bmatrix}0\\ 0\\ 0\end{bmatrix},$

$h_{11}=1>0$, $h_{11}h_{22}-(h_{12})^2=1>0$, $|\mathbf{H}|=-2<0$.
$\mathbf{H}$ is indefinite, so $\mathbf{x}^*$ is a saddle point.

2.7 Find the extreme points of $1-e^{-x^2}$.

$\frac{df}{dx}=-e^{-x^2}(-2x)=0$. Since $e^{-x^2}>0$, $x^*=0$.

$\frac{d^2f}{dx^2}=-4x^2e^{-x^2}+2e^{-x^2}\Big|_{x^*=0}=2>0$. Therefore, $x^*=0$ is a minimum.

2.8 For $f(\mathbf{x})=x_1x_2$, determine the stationary point and show that it is a saddle point.

$$\frac{\partial f}{\partial x_1}=x_2^*=0,\qquad \frac{\partial f}{\partial x_2}=x_1^*=0.$$

Since the function has only two arguments, we can test the discriminant

$$\Delta=\left(\frac{\partial^2 f}{\partial x_1^2}\right)\left(\frac{\partial^2 f}{\partial x_2^2}\right)-\left(\frac{\partial^2 f}{\partial x_1\,\partial x_2}\right)^2, \qquad \Delta=(0)(0)-1^2=-1.$$

Since $\Delta<0$, $\mathbf{x}^*=\mathbf{0}$ is a saddle point.

COMPUTATIONAL CONSIDERATIONS

2.9 Use the Newton-Raphson method to solve

$$\min_{x,y} 4x_1^2+2x_1x_2+2x_2^2+x_1+x_2.$$

$$\nabla f(\mathbf{x})=\begin{bmatrix} 8x_1+2x_2+1 \\ 2x_1+4x_2+1 \end{bmatrix}=\mathbf{0}, \qquad \mathbf{H}(\mathbf{x})=\begin{bmatrix} 8 & 2 \\ 8 & 4 \end{bmatrix}, \qquad \mathbf{H}^{-1}(\mathbf{x})=\frac{1}{14}\begin{bmatrix} 2 & -1 \\ -1 & 4 \end{bmatrix}.$$

$$\mathbf{x}^{n+1}=\mathbf{x}^n-\mathbf{H}^{-1}\nabla f(\mathbf{x}^n).$$

Substituting,

$$x_1^{n+1}=x_1^n-\tfrac{1}{14}[16x_1^n+4x_2^n+2-2x_1^n-4x_2^n-1],$$

$$x_2^{n+1}=x_2^n-\tfrac{1}{14}[-8x_1^n-2x_2^n-1+8x_1^n+16x_2^n+4],$$

or

$$x_1^{n+1}=x_1^n-x_1^n-\tfrac{1}{14}=-\tfrac{1}{14},$$

$$x_2^{n+1}=x_2^n-x_2^n-\tfrac{3}{14}=-\tfrac{3}{14}.$$

Thus the Newton-Raphson method solves linear equations in one iteration regardless of the initial estimate $\mathbf{x}^0$.

2.10 Consider the function shown in Fig. 2.10(a). Apply the Newton-Raphson method graphically, beginning at $x_0=4$. Note that the solution diverges because $f'(x_0)$ and $f'(x_1)$ are increasingly flat (singular).

See Fig. 2.10(b) for the solution.

Supplementary Problems

2.11 Show that $(0,0)$ is the saddle point for the functions (a) $y=x_1^3+x_2^3$ and (b) $y=x_1x_2$.

2.12 Write the following objective functions in the form $y=\mathbf{x}^T\mathbf{H}\mathbf{x}+\mathbf{B}\mathbf{x}$ and solve for $\mathbf{x}^*$:

(a) $y=x_1^2+2x_1x_2+4x_1x_3+3x_2^2+2x_2x_3+5x_3^2+4x_1-2x_2+3x_3$.
(b) $y=5x_1^2+12x_1x_2-16x_1x_3+10x_2^2-26x_2x_3+17x_3^2-2x_1-4x_2-6x_3$.
(c) $y=x_1^2-4x_1x_2+6x_1x_3+5x_2^2-10x_2x_3+8x_3^2$.

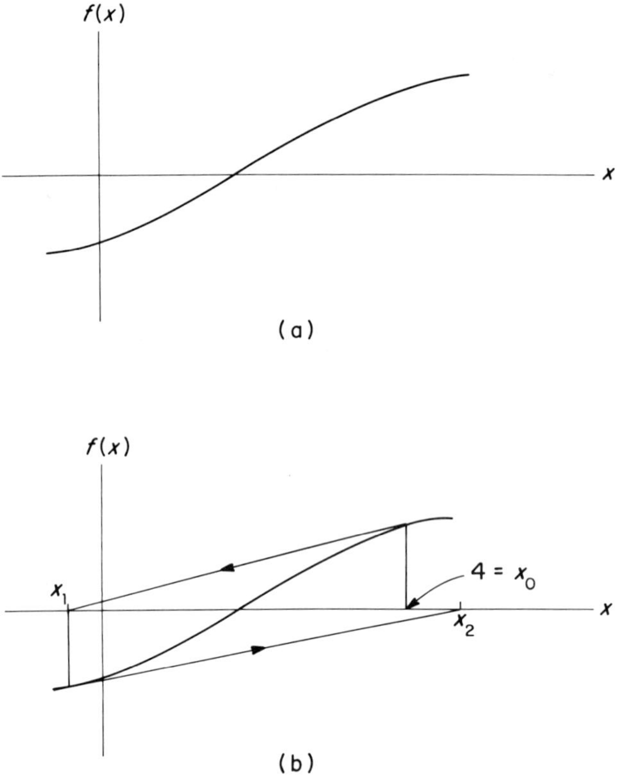

FIGURE 2.10

2.13 For Problem 2.12, determine if $\mathbf{x}^*$ is a maximum or a minimum.

2.14 Find the maximum of $f(x_1,x_2)=(x_1-a_1^2x_2)(x_1-a_2^2x_2^2)$, where a_1 and a_2 are constants and $a_1^2 \neq a_2^2$.

2.15 A sufficient condition for $f(x_1,x_2)$ to have a relative maximum is that the quadratic form

$$\mathbf{x}^T\mathbf{H}\mathbf{x}, \qquad \mathbf{x}^T=[x_1,x_2],$$

be negative definite. Derive two inequalities which constitute an equivalent sufficiency condition.

2.16 Determine the location and type of all stationary points for

(a) $y=-3x^4+9x^3-20$,
(b) $y=x^4-8x^3+24x^2-35x+16$.

2.17 Find the minimum of $y=x^x$ for $x>0$.

2.18 Find the global minimum of $y = x^2 - e^{-x^2}$.

2.19 Find the maximum and minimum of $y = e^{-x}/(1 + e^{-x})$.

2.20 A length of wire is cut into two parts. One part is to be bent into the form of a circle, and the other into the form of a square. In what ratio must the wire be cut if the sum of the areas enclosed by the circle and the square is a minimum? (*Hint:* Use direct substitution.)

2.21 Find the specifications of an open-topped rectangular tank whose total wall and base area is to be 108 m^2 if the maximum volume is required.

2.22 min $x_1^2 + 4x_2^2 - 4x_1$ such that $2x_2 - x_1 = 12$.

2.23 Expand $\sqrt{1+2x}$ in a Taylor series about $x = 0$. Determine the error at $x = 0.1$ if terms higher than first order are neglected.

2.24 Solve Problem 2.23 if terms higher than second order are neglected.

2.25 Expand $\ln x$ into a Taylor series about $x = 1$. Determine the error at $x = 1.05$ if terms higher than second order are neglected.

2.26 Using the Newton-Raphson method, find the real root of $xe^x - 2 = 0$.

2.27 Use the Newton-Raphson method to find the root of $\sin x - (x+1)/(x-1) = 0$.

2.28 Employ the Newton-Raphson method to find the root of $x^3 - 5x^2 - 6x + 4 = 0$.

2.29 Perform two iterations of the Newton-Raphson method to find the positive solution of

$$1 - 3000x^{-2}y^{-1} = 0,$$

$$2 - 4000x^{-1}y^{-2} = 0,$$

starting at (a) $(x,y) = (8,3)$, (b) $(x,y) = (10,5)$, (c) $(x,y) = (1,2)$.

Chapter 3

Constraints and Lagrange Multipliers

Lagrange Multipliers

Returning to the constrained optimization problem stated in Chapter 1, we consider first the case having equality constraints only. This problem is posed as

$$\min_{\mathbf{x}} f(\mathbf{x})$$

such that

$$g_i(\mathbf{x})=0, \qquad i=1,2,\ldots,m<n.$$

The problem is to determine a point $\mathbf{x}^*$ which yields a strong relative minimum for $f(\mathbf{x})$ and also satisfies the constraints. Obviously the feasible region is greatly reduced by the presence of the equality constraints. The solution first proposed by Lagrange in 1760 was to form a new unconstrained problem by appending the constraints to the objective function with Lagrange multipliers λ_i, $i=1,2,\ldots,m$. The new objective function $L(\mathbf{x},\boldsymbol{\lambda})$ is called the Lagrangian. It is defined on E^{m+n}, which is a higher-dimensional problem than the original, since it has $m+n$ unknowns. The price paid for disposing of the constraints is this higher problem dimensionality. Since the problem defined by

$$L(\mathbf{x},\boldsymbol{\lambda})=f(\mathbf{x})+\sum_{i=1}^{m}\lambda_i g_i(\mathbf{x})$$

is now unconstrained, we can apply the necessary conditions for stationarity defined in Chapter 2. These are given by

$$\frac{\partial L}{\partial x_j}=\frac{\partial f}{\partial x_j}+\sum_{i=1}^{m}\lambda_i\frac{\partial g_i}{\partial x_j}=0, \qquad j=1,2,\ldots,n,$$

$$\frac{\partial L}{\partial \lambda_i}=g_i(\mathbf{x})=0, \qquad i=1,2,\ldots,m,$$

which yield a set of $m+n$ equations in $m+n$ unknowns $(\mathbf{x},\boldsymbol{\lambda})$ to be solved for the optimal values $(\mathbf{x}^*,\boldsymbol{\lambda}^*)$. Note that the conditions

$$\frac{\partial L}{\partial \lambda_i}=0, \qquad i=1,2,\ldots,m,$$

guarantee that the constraints are satisfied at the optimal solution. In this case $g_i(\mathbf{x})=0$, $i=1,2,\ldots,m$, and the optimal value of the Lagrangian is equal to the optimal return for the original problem, i.e.,

$$L(\mathbf{x}^*,\boldsymbol{\lambda}^*)=f(\mathbf{x}^*).$$

We note however that even if $f(\mathbf{x})$ is strictly convex and possesses a minimum at $\mathbf{x}^*$, the Lagrangian surface is not strictly convex at $(\mathbf{x}^*,\boldsymbol{\lambda}^*)$. More will be said of this in Chapter 4. Since the above stationarity conditions are necessary conditions only, they may be applied for both minimization and maximization problems without change. They will attain correct solutions, however, only if a solution to the unconstrained problem exists. Generally speaking, this implies that we are minimizing convex-type functions and maximizing concave-type functions.

EXAMPLE 3.1

$$\max x_1^2+4x_2^2$$

such that

$$x_1+2x_2=6.$$

We have

$$L=x_1^2+4x_2^2+\lambda(x_1+2x_2-6),$$

$$\frac{\partial L}{\partial x_1}=2x_1+\lambda=0, \qquad \frac{\partial L}{\partial x_2}=8x_2+2\lambda=0, \qquad \frac{\partial L}{\partial \lambda}=x_1+2x_2-6=0.$$

So $x_1^*=3$, $x_2^*=\frac{3}{2}$, $\lambda^*=-6$. However, this is a solution to the problem

$$\min x_1^2+4x_2^2$$

such that

$$x_1+2x_2=6,$$

since the solution to the maximization problem is unbounded, i.e., $x_1^*=\pm\infty$, $x_2^*=+\infty$, and $f(x_1^*,x_2^*)=\infty$.

As a general rule, if any of the equality constraints can be solved explicitly for one or more variables, it is best to use this relation to eliminate these variables from the problem. This reduces both the problem dimensionality and the number of constraints. Once the reduced problem is solved, the explicit equality constraints can be used to recover the optimal values of the eliminated variables.

EXAMPLES

3.2

$$\min x_1^2+2x_2^2+x_3^2$$

such that

$$x_1^2+3x_1x_2+x_2^2=1, \qquad x_2+x_3=4.$$

First eliminate x_3 by $x_3=4-x_2$. Then

$$L(x_1,x_2,\lambda)=x_1^2+2x_2^2+(4-x_2)^2+\lambda[x_1^2+3x_1x_2+x_2^2-1],$$

which yields the three necessary conditions

$$\frac{\partial L}{\partial x_1}=2x_1+2\lambda x_1+3\lambda x_2=0,$$

$$\frac{\partial L}{\partial x_2}=4x_2-2(4-x_2)+3\lambda x_1+2\lambda x_2=0,$$

$$\frac{\partial L}{\partial \lambda}=x_1^2+3x_1x_2+x_2^2-1=0,$$

instead of the five equations in five unknowns which would result from the Lagrangian if written

$$L(x_1,x_2,x_3,\lambda_1,\lambda_2)$$
$$=x_1^2+2x_2^2+x_3^2+\lambda_1[x_1^2+3x_1x_2+x_2^2-1]+\lambda_2[x_2+x_3-4].$$

In this case,

$$\frac{\partial L}{\partial x_1}=2x_1+2\lambda_1x_1+3\lambda_1x_2=0,$$

$$\frac{\partial L}{\partial x_2}=4x_2+3\lambda_1x_1+2\lambda_1x_2+\lambda_2=0,$$

$$\frac{\partial L}{\partial x_3}=2x_3+\lambda_2=0,$$

$$\frac{\partial L}{\partial \lambda_1}=x_1^2+3x_1x_2+x_2^2-1=0,$$

$$\frac{\partial L}{\partial \lambda_2}=x_2+x_3-4=0.$$

3.3

$$\min\ (x_1-2)^2+(x_2-2)^2$$

such that

$$x_1+x_2=6.$$

The unconstrained solution is given by

$$\frac{\partial f}{\partial x_1}=2(x_1-2)=0, \qquad \frac{\partial f}{\partial x_2}=2(x_2-2)=0,$$

from which $x_1^*=2$, $x_2^*=2$, and $f(x_1^*,x_2^*)=0$.
Using the constraint to eliminate x_2, we get

$$\min\ (x_1-2)^2+(6-x_1-2)^2,$$

$$\frac{\partial f}{\partial x_1}=2(x_1-2)-2(4-x_1)=0,$$

or $x_1^*=3$, $x_2^*=6-3=3$, and $f(x_1^*,x_2^*)=(3-2)^2+(3-2)^2=2$.
The constrained and unconstrained solutions are shown in Fig. 3.1.

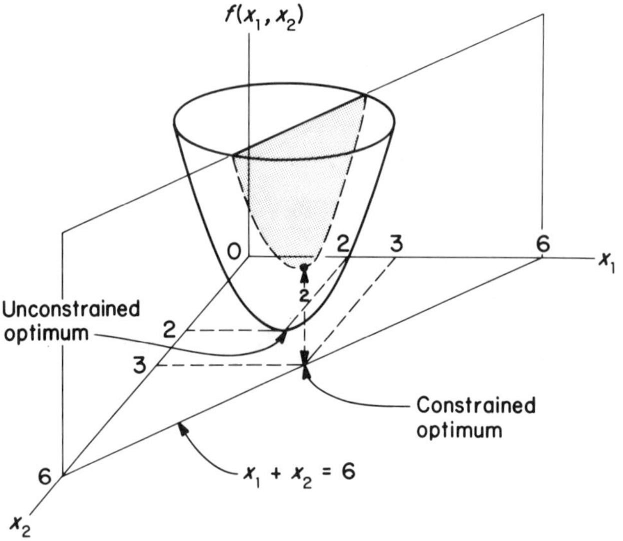

FIGURE 3.1

In general, the presence of a constraint cannot improve the value of the objective function over the unconstrained case. This follows because the presence of constraints merely restricts the feasible region. Thus an optimum found by searching over all of E^n cannot be improved by searching over only a subset of E^n.

EXAMPLE 3.4

$$\min_{\mathbf{x}} \tfrac{1}{2}\mathbf{x}^T\mathbf{A}\mathbf{x}+\mathbf{b}^T\mathbf{x}$$

such that

$$\mathbf{C}\mathbf{x}=\mathbf{d},$$

where $\mathbf{A}$ is an $n\times n$ positive definite matrix and $\mathbf{C}$ is an $m\times n$ matrix, $m<n$.

Now

$$L(\mathbf{x},\boldsymbol{\lambda})=\tfrac{1}{2}\mathbf{x}^T\mathbf{A}\mathbf{x}+\mathbf{b}^T\mathbf{x}+\boldsymbol{\lambda}^T[\mathbf{C}\mathbf{x}-\mathbf{d}]$$

and

$$\nabla_{\mathbf{x}}L(\mathbf{x}^*,\boldsymbol{\lambda}^*)=\mathbf{A}\mathbf{x}^*+\mathbf{b}+\mathbf{C}^T\boldsymbol{\lambda}^*=\mathbf{0},$$

$$\nabla_{\boldsymbol{\lambda}}L(\mathbf{x}^*,\boldsymbol{\lambda}^*)=\mathbf{C}\mathbf{x}^*-\mathbf{d}=\mathbf{0},$$

or

$$\left[\begin{array}{c|c}\mathbf{A} & \mathbf{C}^T\\ \hline \mathbf{C} & \mathbf{0}\end{array}\right]\begin{bmatrix}\mathbf{x}^*\\ \boldsymbol{\lambda}^*\end{bmatrix}=\begin{bmatrix}-\mathbf{b}\\ \mathbf{d}\end{bmatrix}.$$

The solution is

$$\begin{bmatrix}\mathbf{x}^*\\ \boldsymbol{\lambda}^*\end{bmatrix}=\mathbf{M}^{-1}\begin{bmatrix}-\mathbf{b}\\ \mathbf{d}\end{bmatrix},$$

where

$$\mathbf{M}=\left[\begin{array}{c|c}\mathbf{A} & \mathbf{C}^T\\ \hline \mathbf{C} & \mathbf{0}\end{array}\right]$$

is a square matrix of dimension $(m+n)\times(m+n)$. If M is nonsingular, then it can be inverted, and the solution $(\mathbf{x}^*,\boldsymbol{\lambda}^*)$ exists and can be determined. The question of the existence of Lagrange multipliers will be treated in a later section.

Economic Interpretation

Previously the Lagrange multipliers were treated simply as an algebraic device for solving the constrained optimization problem, and no importance was attached to their numerical value. However, these multipliers have an interesting economic interpretation, and their numerical values are often of interest. To indicate this interpretation, it is convenient to pose the problem

$$\min_{\mathbf{x}} f(\mathbf{x})$$

such that

$$g_i(\mathbf{x}) = b_i, \qquad i = 1, 2, \ldots, m < n,$$

where the numerical values b_i are regarded as the amounts of some scarce resources which are limited according to the constraints. It is intuitively clear that as the amount of resource b_i varies, the optimal solution $(\mathbf{x}^*, \boldsymbol{\lambda}^*)$ will change. Thus we can write

$$L(\mathbf{b}) = f(\mathbf{x}(\mathbf{b})) + \boldsymbol{\lambda}^T(\mathbf{b})\left[\mathbf{g}(\mathbf{x}(\mathbf{b})) - \mathbf{b}\right].$$

Differentiating with respect to b_i yields

$$\frac{\partial L}{\partial b_i} = \sum_{j=1}^{n} \frac{\partial f}{\partial x_j}\frac{\partial x_j}{\partial b_i} + \sum_{j=1}^{n}\sum_{k=1}^{m} \lambda_k \frac{\partial g_k}{\partial x_j}\frac{\partial x_j}{\partial b_i} - \lambda_i + \sum_{k=1}^{m}\left[g_k(\mathbf{x}) - b_k\right]\frac{\partial \lambda_k}{\partial b_i},$$

or in matrix form,

$$\nabla_{\mathbf{b}} L = \left[\frac{\partial \mathbf{x}}{\partial \mathbf{b}}\right]\nabla_{\mathbf{x}} f + \left[\frac{\partial \mathbf{x}}{\partial \mathbf{b}}\right]\left[\frac{\partial \mathbf{g}}{\partial \mathbf{x}}\right]\boldsymbol{\lambda} + \left[\frac{\partial \boldsymbol{\lambda}}{\partial \mathbf{b}}\right]\left[\mathbf{g}(\mathbf{x}) - \mathbf{b}\right] - \boldsymbol{\lambda},$$

where $[\partial \mathbf{x}/\partial \mathbf{b}]$, $[\partial \mathbf{g}/\partial \mathbf{x}]$, and $[\partial \boldsymbol{\lambda}/\partial \mathbf{b}]$ are appropriately defined matrices of dimension $m \times n$, $n \times m$, and $m \times m$, respectively.

Collecting terms,

$$\nabla_{\mathbf{b}} L = \left[\frac{\partial \mathbf{x}}{\partial \mathbf{b}}\right]\left[\nabla_{\mathbf{x}} f + \frac{\partial \mathbf{g}}{\partial \mathbf{x}}\boldsymbol{\lambda}\right] + \left[\frac{\partial \boldsymbol{\lambda}}{\partial \mathbf{b}}\right]\left[\mathbf{g}(\mathbf{x}) - \mathbf{b}\right] - \boldsymbol{\lambda}.$$

Now at the optimum the first two terms vanish because of the necessary conditions $\nabla_{\mathbf{x}}L=\nabla_{\boldsymbol{\lambda}}L=\mathbf{0}$. So we are left with

$$\nabla_{\mathbf{b}}L(\mathbf{x}^*,\boldsymbol{\lambda}^*)=-\boldsymbol{\lambda}^*,$$

or

$$\frac{\partial L}{\partial b_i}(\mathbf{x}^*,\boldsymbol{\lambda}^*)=\frac{\partial f}{\partial b_i}(\mathbf{x}^*)=-\lambda_i^*, \qquad i=1,2,\ldots,m,$$

since $L(\mathbf{x}^*,\boldsymbol{\lambda}^*)=f(\mathbf{x}^*)$ at the optimum.

If we think of the objective function as dollars of return, then λ_i^* may be interpreted as dollars per unit of the ith resource. Since this has the dimensions of a price, the Lagrange multipliers are sometimes called *shadow prices*.

Another interpretation which follows directly from the above result is that the Lagrange multipliers represent *sensitivity coefficients*. This means that the marginal value of additional amounts of the ith resource is given by $-\lambda_i^*$ when all resources are allocated optimally. If we convert a maximization problem to a minimization problem by changing the sign of $f(\mathbf{x})$, we also are changing the signs of all λ_i^*, $i=1,2,\ldots,m$. To avoid confusion in such cases, it is wise to convert back to the original problem when interpreting the multipliers as sensitivities.

EXAMPLE 3.5 Consider the optimal allocation of a scarce resource between two processes where the total amount of resource available is b (see Fig. 3.2). We wish to maximize the return from both processes. Thus we have

$$\max_{x_1,x_2} f(x_1)+f(x_2)$$

such that

$$x_1+x_2=b.$$

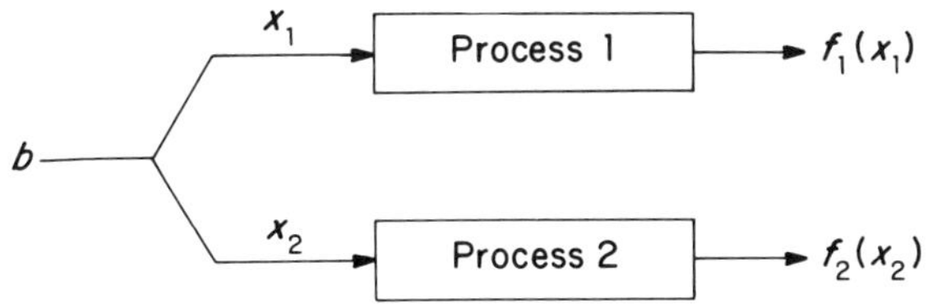

FIGURE 3.2

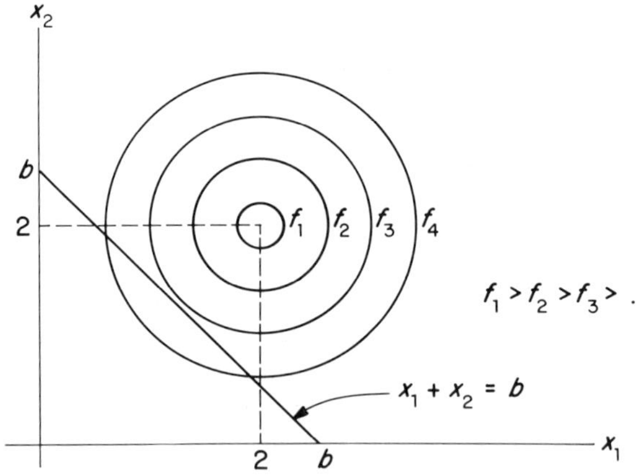

FIGURE 3.3

If $f(x_1) = 50 - (x_1 - 2)^2$ and $f(x_2) = 50 - (x_2 - 2)^2$, then (see Fig. 3.3)

$$L = 100 - (x_1 - 2)^2 - (x_2 - 2)^2 + \lambda[x_1 + x_2 - b],$$

$$\frac{\partial L}{\partial x_1} = -2(x_1 - 2) + \lambda = 0,$$

$$\frac{\partial L}{\partial x_2} = -2(x_2 - 2) + \lambda = 0,$$

$$\frac{\partial L}{\partial \lambda} = x_1 + x_2 - b = 0,$$

and

$$x_1^* = x_2^* = b/2, \qquad \lambda^* = b - 4.$$

The sensitivity

$$-\lambda^* = \frac{\partial f}{\partial b}(\mathbf{x}^*) = 4 - b$$

implies that for $b < 4$ an increase in b would have a positive return, while for $b > 4$ an increase in b would have a negative return or loss. If $b = 4$, then $\lambda^* = 0$, and the solutions to the constrained and the unconstrained problems are identical.

The Lagrange multipliers themselves can be used to solve certain types of decision-making problems. For example, if an additional amount of some ith resource can be purchased for y dollars per unit, then this purchase should be made if

$$-\lambda_i^* > y.$$

That is, if the marginal return $(-\lambda_i^*)$ from the optimal allocation of an additional unit of the ith resource is greater than its marginal cost y, then the purchase of additional amounts of the ith resource is economical.

EXAMPLE 3.6 In Example 3.5, suppose additional amounts of resource cost \$1 per unit and $b=2$. Then the marginal return, $-\lambda^*=2$, is greater than the marginal cost, $y=1$, and the purchase should be made up to the point where $b=3$. Thus $-\lambda^*=1=y$ and further increases in resource are uneconomical at that cost. If $b>4$, no additional resource should be purchased at any price or even accepted if free.

Existence of λ

If we regard the Lagrange multiplier as an algebraic contrivance to make possible the solution of constrained optimization problems, then its existence cannot be argued on the basis of a purely physical interpretation. However, since the multipliers are associated with the constraints only, it is intuitively plausible that their existence depends upon the nature of the constraints. To show this, consider the Lagrangian

$$L = f(\mathbf{x}) + \sum_{i=1}^{m} \lambda_i g_i(\mathbf{x})$$

and the resulting necessary conditions

$$\nabla_{\mathbf{x}} f(\mathbf{x}) + \left[\frac{\partial \mathbf{g}}{\partial \mathbf{x}}\right]\boldsymbol{\lambda} = \mathbf{0}, \qquad \mathbf{g}(\mathbf{x}) = \mathbf{0},$$

where $\partial \mathbf{g}/\partial \mathbf{x}$ is the $n \times m$ matrix

$$\frac{\partial \mathbf{g}}{\partial \mathbf{x}} = \begin{bmatrix} \dfrac{\partial g_1}{\partial x_1} & \dfrac{\partial g_2}{\partial x_1} & \cdots & \dfrac{\partial g_m}{\partial x_1} \\ \cdots & \cdots & \cdots & \cdots \\ \dfrac{\partial g_1}{\partial x_n} & \dfrac{\partial g_2}{\partial x_n} & \cdots & \dfrac{\partial g_m}{\partial x_n} \end{bmatrix}.$$

Suppose we have found some candidate point for the optimum, $\mathbf{x}^0$, which satisfies the constraints $\mathbf{g}(\mathbf{x}^0)=\mathbf{0}$. If this point is the optimum, there must be some $\boldsymbol{\lambda}$ for which

$$\left[\frac{\partial \mathbf{g}}{\partial \mathbf{x}}\right]_{\mathbf{x}^0}\boldsymbol{\lambda} = -\nabla_{\mathbf{x}} f(\mathbf{x}^0).$$

Since the $n\times m$ matrix $[\partial \mathbf{g}/\partial \mathbf{x}]_{\mathbf{x}^0}$ is well defined when evaluated at $\mathbf{x}^0$, we can calculate $\boldsymbol{\lambda}$ if there is some submatrix $\mathbf{G}_m \subset [\partial \mathbf{g}/\partial \mathbf{x}]_{\mathbf{x}^0}$ such that $\mathbf{G}_m$ is nonsingular. This is precisely the requirement for $[\partial \mathbf{g}/\partial \mathbf{x}]_{\mathbf{x}^0}$ to be of rank m. Thus if $[\partial \mathbf{g}/\partial \mathbf{x}]_{\mathbf{x}^0}$ is of rank m at the optimum $\mathbf{x}^0$, the optimal multipliers exist. Since the evaluation of the rank of $[\partial \mathbf{g}/\partial \mathbf{x}]_{\mathbf{x}^0}$ requires that the optimum $\mathbf{x}^0$ be known, this existence result is of little practical value.

EXAMPLE 3.7 Minimize the distance from the origin to the curve $y^2-(x-1)^3=0$, or

$$\min\sqrt{x^2+y^2}$$

such that

$$y=+\sqrt{(x-1)^3}\,, \qquad y=-\sqrt{(x-1)^3}\,.$$

The solution to this problem is easily seen graphically to be at $x^*=1$, $y^*=0$ (see Fig. 3.4).

Since $g(x,y)=y^2-(x-1)^3$,

$$\left[\frac{\partial g}{\partial \mathbf{x}}\right]=\begin{bmatrix}-3(x-1)^3\\ 2y\end{bmatrix}.$$

Evaluating at $1,0$,

$$\left[\frac{\partial g}{\partial \mathbf{x}}\right]_{1,0}=\begin{bmatrix}0\\0\end{bmatrix},$$

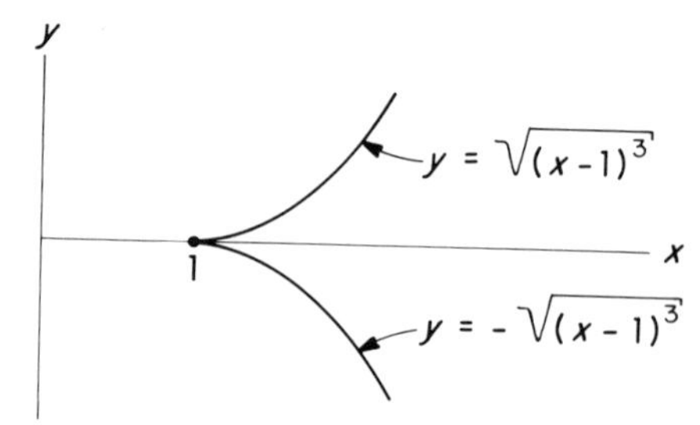

FIGURE 3.4

which contains no nonsingular submatrix of rank 1. Hence the multiplier does not exist at the optimum. Writing the Lagrangian

$$L=(x^2+y^2)^{1/2}+\lambda\left[y^2-(x-1)^3\right],$$

we have

$$\frac{\partial L}{\partial x}=(x^2+y^2)^{-1/2}x-3\lambda(x-1)^2=0,$$

$$\frac{\partial L}{\partial y}=(x^2+y^2)^{-1/2}y+2\lambda y=0,$$

$$\frac{\partial L}{\partial \lambda}=y^2-(x-1)^3=0.$$

Substituting 1,0 into these equations yields

$$1-\lambda(0)=0,$$

$$0+\lambda(0)=0,$$

$$0+0=0.$$

Clearly there is no finite value of λ which satisfies the first equation.

Inequality Constraints

In Chapter 1 we posed the general nonlinear programming problem which included inequality constraints of the form

$$h_j(\mathbf{x})\geqslant 0, \qquad j=1,2,\ldots,r.$$

In Chapter 4 we will treat this problem directly; however, to aid the transition we will first obtain results by converting the inequalities to equality constraints and applying the Lagrangian theory. This method of *slack variables* is often used for obtaining results for simple inequality-constrained problems.

First define a real-valued slack variable θ_j for each inequality $h_j(x)$. Then if

$$\theta_j^2=h_j(\mathbf{x})\geqslant 0, \qquad j=1,2,\ldots,r,$$

the inequality is satisfied by satisfying the equality for all real-valued θ_j.

Applying the Lagrangian necessary conditions,

$$L(\mathbf{x},\boldsymbol{\lambda},\boldsymbol{\theta})=f(\mathbf{x})+\sum_{j=1}^{m}\lambda_j\left(h_j(\mathbf{x})-\theta_j^2\right),$$

$$\frac{\partial L}{\partial x_i}=\frac{\partial f}{\partial x_i}+\sum_{j=1}^{m}\lambda_j\frac{\partial h_j}{\partial x_i}=0, \qquad i=1,2,\ldots,n,$$

$$\frac{\partial L}{\partial \lambda_j}=h_j(\mathbf{x})-\theta_j^2=0, \qquad j=1,2,\ldots,m,$$

$$\frac{\partial L}{\partial \theta_j}=-2\lambda_j\theta_j=0, \qquad j=1,2,\ldots,m.$$

From the last equation, either $\lambda_j^*=0$ or $\theta_j^*=0$ or both.

Case 1. If $\lambda_j^*=0$ and $\theta_j^*\neq 0$, then the constraint $h_j(\mathbf{x})\geqslant 0$ is ignored, since $h_j(\mathbf{x}^*)=(\theta_j^*)^2>0$ and the optimal solution is not changed by the presence of that constraint. For instance, if all $\lambda_j^*=0, j=1,2,\ldots,m$, the first equation yields

$$\nabla f(\mathbf{x})=\mathbf{0},$$

which is the necessary condition for unconstrained problems.

Case 2. If $\theta_j^*=0$ and $\lambda_j^*\neq 0$, then $h_j(\mathbf{x}^*)=0$ and the optimal solution is on the boundary of the jth constraint. Since $\lambda_j^*\neq 0$, the solution does not satisfy $\nabla f(\mathbf{x}^*)=\mathbf{0}$.

Case 3. If $\theta_j^*=0$ and $\lambda_j^*=0$ for all j, then $h_j(\mathbf{x}^*)=0$, and the solution satisfies $\nabla f(\mathbf{x}^*)=\mathbf{0}$. That is, the boundary passes through the unconstrained optimum.

EXAMPLE 3.8

$$\min_x (x-a)^2+b$$

such that

$$x\geqslant c.$$

Letting $\theta^2=x-c\geqslant 0$, we write

$$L=(x-a)^2+b+\lambda\left[x-c-\theta^2\right],$$

$$\frac{\partial L}{\partial x} = 2(x-a) + \lambda = 0,$$

$$\frac{\partial L}{\partial \lambda} = x - c - \theta^2 = 0,$$

$$\frac{\partial L}{\partial \theta} = -2\lambda\theta = 0.$$

Case 1: $\lambda = 0$; then $x^* = a$ and $\theta^2 = a - c$. If $c < a$, θ is real and $\theta \neq 0$, so the optimal solution is $x^* = a$ for $c < a$ [see Fig. 3.5(a)].

Case 2: $\theta = 0$, so $x^* = c$ and the optimal solution is on the constraint boundary. In this case $\lambda^* = 2(a-c)$. Since $\lambda^* \neq 0$, this case yields the optimal solution when $c > a$ [see Fig. 3.5(b)].

Case 3: If $\theta^* = 0$ and $\lambda^* = 0$, then $x^* = c$ and $c = a$ [see Fig. 3.5(c)].

Computational Methods

As with the indirect methods for unconstrained problems, the Lagrangian methods mentioned in this chapter lead to a significant computational problem, namely, the solution of $m+n$ simultaneous equations in $m+n$ unknowns resulting from the necessary conditions

$$\nabla_{\mathbf{x}} L(\mathbf{x}^*, \boldsymbol{\lambda}^*) = \mathbf{0},$$

$$\nabla_{\boldsymbol{\lambda}} L(\mathbf{x}^*, \boldsymbol{\lambda}^*) = \mathbf{0}.$$

This is the same type of problem which arose from applying indirect methods to the unconstrained case, although it is now of even higher dimensionality and increased complexity. Many numerical methods, such as the Newton-Raphson method discussed in Chapter 2, can be used to solve this problem. These methods are straightforward, but are often inadequate, especially as the dimensionality of $m+n$ becomes large.

EXAMPLE 3.9 Examine the convergence of the Newton-Raphson method for solving the system of equations

$$\nabla_{\mathbf{z}} L(\mathbf{z}^*) = \mathbf{0},$$

where

$$\mathbf{z} = \begin{bmatrix} \mathbf{x} \\ \boldsymbol{\lambda} \end{bmatrix}.$$

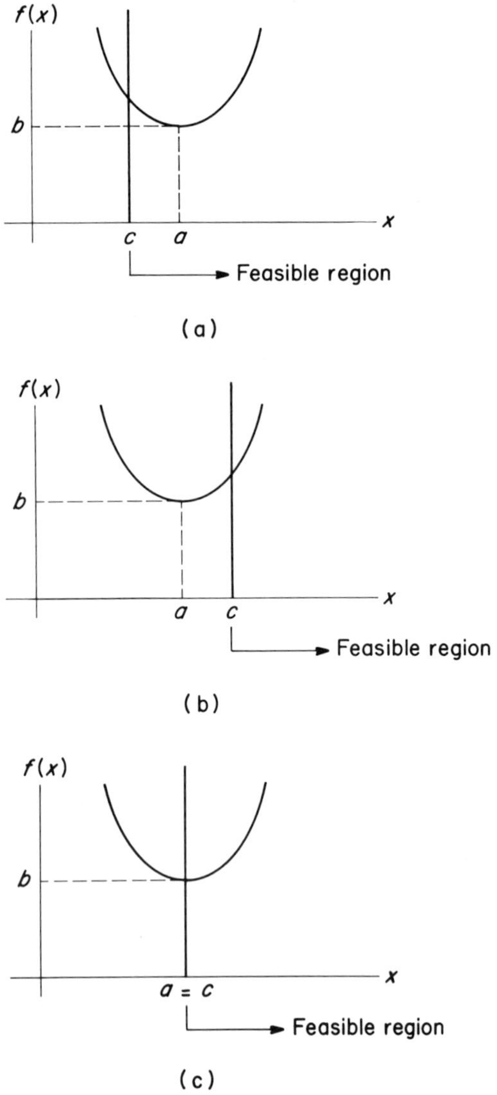
f(x)
b
x
c
a
Feasible region
(a)
f(x)
b
x
a
c
Feasible region
(b)
f(x)
b
x
a = c
Feasible region
(c)

FIGURE 3.5

The Newton-Raphson algorithm yields

$$\mathbf{z}_{n+1} = \mathbf{z}_n - \mathbf{H}^{-1}(\mathbf{z}_n)\nabla_{\mathbf{z}} L(\mathbf{z}_n),$$

where

$$\mathbf{H}(\mathbf{z}_n) = \begin{bmatrix} \nabla_{\mathbf{xx}} L & \nabla_{\mathbf{x}\lambda} L \\ \nabla_{\lambda\mathbf{x}} L & \nabla_{\lambda\lambda} L \end{bmatrix}_{\mathbf{x}_n, \lambda_n} \quad \text{and} \quad \nabla_{\mathbf{z}} L(\mathbf{z}_n) = \begin{bmatrix} \nabla_{\mathbf{x}} L \\ \nabla_{\lambda} L \end{bmatrix}_{\mathbf{x}_n, \lambda_n}.$$

Convergence is guaranteed if H is positive definite, where

$$\mathbf{H}(\mathbf{z}_n) = \left[\begin{array}{c|c} \nabla_{\mathbf{xx}} L & \nabla_{\mathbf{x}} g \\ \hline (\nabla_{\mathbf{x}} g)^T & \mathbf{0} \end{array}\right]_{\mathbf{x}_n, \lambda_n}.$$

It is necessary to examine the major minors of $\mathbf{H}(\mathbf{z}_n)$ and apply Sylvester's theorem for each case to determine the positive definiteness of $\mathbf{H}(\mathbf{z}_n)$. As a minimum requirement the matrices

$$\nabla_{xx} L \quad \text{and} \quad (\nabla_{\mathbf{x}} g)^T (\nabla_{\mathbf{xx}} L)^{-1} (\nabla_{\mathbf{x}} g)$$

must be nonsingular to ensure that H is nonsingular.

Solved Problems

Lagrange Multipliers

3.1 For the scalar problem

$$\min_x f(x)$$

such that

$$g(x) = 0,$$

show that $L(x,\lambda)$ possesses a saddle point at x^*, λ^*.

Forming the Hessian of $L(x,\lambda)$ yields

$$\begin{bmatrix} L_{xx} & g_x \\ g_x & 0 \end{bmatrix}_{x^*, \lambda^*}$$

whose discriminant is

$$-\left[g_x(x^*)\right]^2<0$$

if $g_x(x^*)\neq 0$. By Chapter 2, page 38, this negative discriminant corresponds to a saddle point of $L(x,\lambda)$ at (x^*,λ^*).

3.2 Show that in general $L(\mathbf{x},\boldsymbol{\lambda})=f(\mathbf{x})+(\boldsymbol{\lambda}^T,\mathbf{g}(\mathbf{x}))$ has a saddle point at $\mathbf{x}^*,\boldsymbol{\lambda}^*$.

Let $\mathbf{x}$ have dimension n, and $\boldsymbol{\lambda}$ have dimension $m<n$. Assume that a finite local minimum exists at $\mathbf{x}^0$, i.e.,

$$\min f(\mathbf{x})=f(\mathbf{x}^0) \quad \text{and} \quad \mathbf{g}(\mathbf{x}^0)=\mathbf{0}.$$

Further assume that there exists a $\boldsymbol{\lambda}^0$ associated with $\mathbf{x}^0$, and that within some region of $\boldsymbol{\lambda}^0$, a finite minimum of $L(\mathbf{x},\boldsymbol{\lambda})$ exists. Now define

$$L(\mathbf{x}^*,\boldsymbol{\lambda})=\min_{\mathbf{x}} L(\mathbf{x},\boldsymbol{\lambda})=H(\boldsymbol{\lambda}),$$

where $\mathbf{x}^*$ corresponds to the minimum with respect to $\mathbf{x}$ with arbitrary $\boldsymbol{\lambda}$; if $\boldsymbol{\lambda}=\boldsymbol{\lambda}^0$, then $\mathbf{x}^*=\mathbf{x}^0$. Let X be the set of all $\mathbf{x}$ such that $\mathbf{g}(\mathbf{x})=\mathbf{0}$. Then

$$\min_{\mathbf{x}\in X} L(\mathbf{x},\boldsymbol{\lambda})=\min_{\mathbf{x}\in X} f(\mathbf{x})=f(\mathbf{x}^0)=H(\boldsymbol{\lambda}^0).$$

But

$$H(\boldsymbol{\lambda})=\min_{\text{all } \mathbf{x}} L(\mathbf{x},\boldsymbol{\lambda})\leqslant \min_{\mathbf{x}\in X} L(\mathbf{x},\boldsymbol{\lambda})=H(\boldsymbol{\lambda}^0).$$

Thus

$$H(\boldsymbol{\lambda})\leqslant H(\boldsymbol{\lambda}^0),$$

i.e., $\mathbf{H}(\boldsymbol{\lambda})$ takes on its maximum at $\boldsymbol{\lambda}^0$. By our assumption that $L(\mathbf{x},\boldsymbol{\lambda})$ has a relative minimum with respect to $\mathbf{x}$ for any $\boldsymbol{\lambda}$ in the neighborhood of $\boldsymbol{\lambda}^0$, it must then be true that

$$L(\mathbf{x}^0,\boldsymbol{\lambda}^0)\leqslant L(\mathbf{x},\boldsymbol{\lambda}^0).$$

Also, since $\mathbf{x}^0$ satisfies $\mathbf{g}(\mathbf{x}^0)=\mathbf{0}$,

$$L(\mathbf{x}^0,\boldsymbol{\lambda}^0)=f(\mathbf{x}^0)=L(\mathbf{x}^0,\boldsymbol{\lambda})$$

and

$$L(\mathbf{x}^0,\boldsymbol{\lambda})\leqslant L(\mathbf{x}^0,\boldsymbol{\lambda}^0)\leqslant L(\mathbf{x},\boldsymbol{\lambda}^0),$$

which is the definition of a saddle point. The Lagrangian is actually a degenerate form of saddle point, since the left-hand inequality is actually satisfied by the equality.

3.3 Determine the volume of the largest rectangular box with sides parallel to the coordinate planes which can be inscribed in the ellipsoid

$$\left(\frac{x}{a}\right)^2+\left(\frac{y}{b}\right)^2+\left(\frac{z}{c}\right)^2=1.$$

The objective function is max xyz, and the Lagrangian is

$$L(x,y,z,\lambda)=xyz+\lambda\left[\left(\frac{x}{a}\right)^2+\left(\frac{y}{b}\right)^2+\left(\frac{z}{c}\right)^2-1\right],$$

so that

$$\frac{\partial L}{\partial x}=yz+\frac{2\lambda x}{a^2}=0,$$

$$\frac{\partial L}{\partial y}=xz+\frac{2\lambda y}{b^2}=0,$$

$$\frac{\partial L}{\partial z}=xy+\frac{2\lambda z}{c^2}=0,$$

$$\frac{\partial L}{\partial \lambda}=\frac{x^2}{a^2}+\frac{y^2}{b^2}+\frac{z^2}{c^2}-1=0.$$

The solutions are $x^*=y^*=z^*=\lambda^*=0$, which yields the minimum volume $v^*=0$; and $x^*=a/\sqrt{3}$, $y^*=b/\sqrt{3}$, $z^*=c/\sqrt{3}$, $\lambda^*=-abc/2\sqrt{3}$, which yields the maximum volume $v^*=8\sqrt{3}\,abc/9$.

3.4 Find the maxima and minima of the quadratic form

$$f(x)=x_1x_2$$

such that

$$x_1^2+x_2^2=1.$$

Illustrate geometrically.

Form the Lagrangian

$$L(x_1,x_2,\lambda)=x_1x_2+\lambda(x_1^2+x_2^2-1).$$

Then

$$\frac{\partial L}{\partial x_1}=x_2+2\lambda x_1=0, \tag{3.1}$$

$$\frac{\partial L}{\partial x_2}=x_1+2\lambda x_2=0, \tag{3.2}$$

$$\frac{\partial L}{\partial \lambda}=x_1^2+x_2^2-1=0. \tag{3.3}$$

Solving (3.1), (3.2), and (3.3) yields minima at $(-1/\sqrt{2}, 1/\sqrt{2})$ and $(1/\sqrt{2}, -1/\sqrt{2})$ with $f(x^*) = -\frac{1}{2}$, and maxima at $(1/\sqrt{2}, 1/\sqrt{2})$ and $(-1/\sqrt{2}, -1/\sqrt{2})$ with $f(x^*) = \frac{1}{2}$. The level curves are shown in Fig. 3.6.

3.5 Determine the dimensions of the solid cylinder of maximum total surface area which can be cast from a solid metallic sphere of radius r_0.

This problem can be posed as

$$\min_{r,h} 2\pi rh + 2\pi r^2$$

such that

$$\pi r^2 h = \tfrac{4}{3}\pi r_0^3,$$

where r is the radius and h is the height of the cylinder.
Form the Lagrangian

$$L(r,h,\lambda) = 2\pi rh + 2\pi r^2 + \lambda\left(\pi r^2 h - \tfrac{4}{3}\pi r_0^3\right).$$

Then

$$\frac{\partial L}{\partial r} = 2\pi h + 4\pi r + \lambda 2\pi rh = 0, \tag{3.4}$$

$$\frac{\partial L}{\partial h} = 2\pi r + \lambda\pi r^2 = 0, \tag{3.5}$$

$$\frac{\partial L}{\partial \lambda} = \pi r^2 h - \tfrac{4}{3}\pi r_0^3 = 0. \tag{3.6}$$

Solving (3.4) and (3.5) yields $h^* = 2r^*$, and combining with (3.6) gives

$$r^* = r_0\left(\tfrac{2}{3}\right)^{1/3},$$

$$h^* = 2r_0\left(\tfrac{2}{3}\right)^{1/3},$$

$$\lambda^* = -\frac{2}{r_0}\left(\tfrac{3}{2}\right)^{1/3},$$

and the maximum surface area is $6\pi r_0(\frac{2}{3})^{2/3}$.

3.6 Solve graphically:

$$\min x_1^2 + x_2^2$$

such that

$$(x_1 - 1)^3 - x_2^2 = 0.$$

The solution is shown in Fig. 3.7.

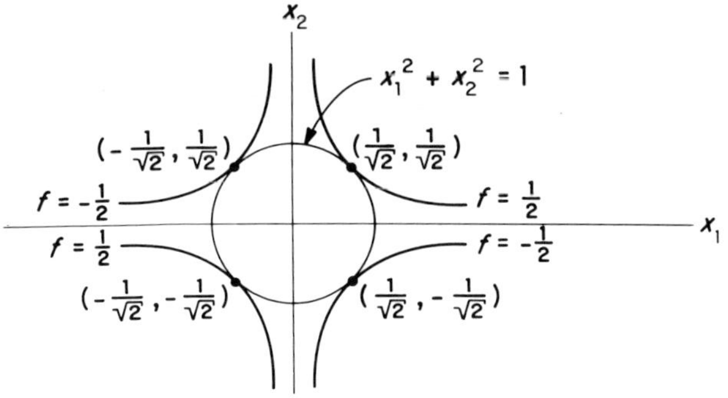

FIGURE 3.6

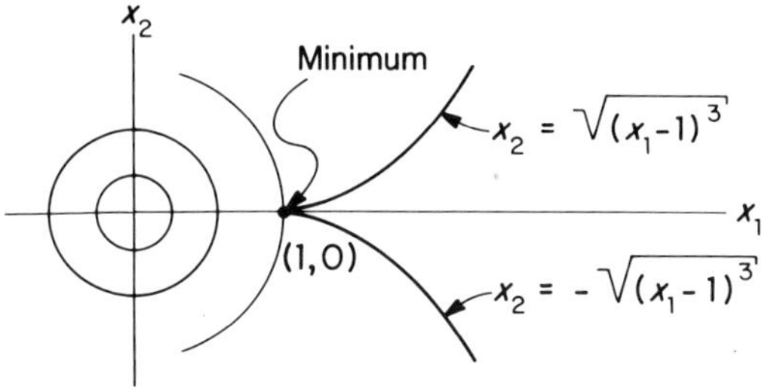

FIGURE 3.7

3.7 Show that the Lagrange multipliers do not exist for Problem 3.6.

The Lagrangian is

$$L(x_1, x_2, \lambda) = x_1^2 + x_2^2 + \lambda\left[(x_1 - 1)^3 - x_2^2\right].$$

Thus

$$\frac{\partial L}{\partial x_1} = 2x_1 + 3\lambda(x_1 - 1)^2 = 0,$$

$$\frac{\partial L}{\partial x_2} = 2x_2(1 - \lambda) = 0,$$

$$\frac{\partial L}{\partial \lambda} = (x_1 - 1)^3 - x_2^2.$$

Clearly at 1,0 there is no λ which satisfies these three equations.

Also, the Jacobian

$$\begin{bmatrix} \dfrac{\partial g}{\partial x_1} \\ \dfrac{\partial g}{\partial x_2} \end{bmatrix}_{1,0} = \begin{bmatrix} 3(x_1-1)^2 \\ -2x_2 \end{bmatrix}_{1,0} = \begin{bmatrix} 0 \\ 0 \end{bmatrix}$$

is not of rank 1 as required for the existence of λ.

3.8 Consider the problem of economically dispatching power from n power plants to m loads over a transmission network. We wish to determine the most economical way to generate power so that the demand is satisfied.

We let

$$P_i = \text{power produced by } i\text{th plant},$$

$$F_i(P_i) = \text{cost of generating power } P_i \text{ in the } i\text{th plant},$$

$$L(P_1,\ldots,P_n) = \text{power loss due to transmission},$$

$$D = \text{total demand}.$$

The objective is now

$$\min_{P_i} \sum_{i=1}^{n} F_i(P_i)$$

such that the demand is satisfied, i.e.,

$$\sum_{i=1}^{n} P_i - L(\mathbf{P}) = D.$$

The Lagrangian is

$$L(\mathbf{P},\lambda) = \sum_{i=1}^{n} F_i(P_i) + \lambda\left[\sum_{i=1}^{n} P_i - L(\mathbf{P}) - D\right].$$

Thus

$$\frac{\partial L}{\partial P_i} = \frac{dF_i}{dP_i} + \lambda - \lambda\frac{\partial L}{\partial P_i} = 0, \qquad i = 1,2,\ldots,n, \tag{3.7}$$

$$\frac{\partial L}{\partial \lambda} = \sum_{i=1}^{n} P_i - L(\mathbf{P}) - D = 0. \tag{3.8}$$

From (3.7),

$$\frac{dF_i}{dP_i} = \lambda L_i \tag{3.9}$$

where $L_i = \partial L/\partial P_i - 1$.

Thus the most economical dispatch rule is to operate all plants so that the ratio of the marginal cost of generation (dF_i/dP_i) to the line loss factor L_i is equal to λ. The value of λ is chosen to ensure that the demand (3.8) is satisfied.

Inequality Constraints

3.9 Find the maximum and the minimum values of $2x^2-3y^2-2x$ for $x^2+y^2 \leqslant 1$.

Define

$$\theta^2 = 1 - x^2 - y^2, \qquad \theta \text{ real}.$$

Then

$$L(x,y,\lambda,\theta) = 2x^2 - 3y^2 - 2x + \lambda(\theta^2 - 1 + x^2 + y^2),$$

and

$$\frac{\partial L}{\partial x} = 4x - 2 + 2\lambda x = 0,$$

$$\frac{\partial L}{\partial y} = -6y + 2\lambda y = 0,$$

$$\frac{\partial L}{\partial \lambda} = \theta^2 - 1 + x^2 + y^2,$$

$$\frac{\partial L}{\partial \theta} = 2\lambda\theta = 0.$$

Case 1: $\theta = 0$ (constrained solution). Then

$$4x - 2 + 2\lambda x = 0,$$

$$y(2\lambda - 6) = 0,$$

$$x^2 + y^2 = 1,$$

which have three solutions:

$$x^* = 0.2, \quad y^* = \pm\sqrt{0.96}, \quad \lambda^* = 3, \quad f(x^*,y^*) = -3.2, \qquad (3.10)$$

$$x^* = 1, \quad y^* = 0, \quad \lambda^* = -1, \quad f(x^*,y^*) = 0, \qquad (3.11)$$

$$x^* = -1, \quad y^* = 0, \quad \lambda^* = -3, \quad f(x^*,y^*) = x. \qquad (3.12)$$

Case 2: $\lambda = 0$ (unconstrained solution). Then

$$4x - 2 = 0,$$
$$-6y = 0,$$
$$1 + x^2 + y^2 = \theta^2,$$

and

$$x^*=0.5, \qquad y^*=0, \qquad \theta^2=1.25, \qquad f(x^*,y^*)=-0.5.$$

By inspection the minimum value of $f(x,y)$ occurs at $0.2, \pm\sqrt{0.96}$ and the maximum occurs at $-1,0$.

3.10 It is desired to control the concentration c in the reactor output stream by manipulating the reactor temperature T and the input concentration c_{in} (see Fig. 3.8). It is desired to choose c_{in} and T to minimize z, where

$$z=(c-c_{\text{d}})^2+T^2,$$

such that

$$c=c_{\text{in}}+e^T,$$

$$c_{\text{in}}\leqslant K.$$

Here c_{d} is the desired output concentration and is fixed. Find the optimal solution for

(a) $K=c_{\text{d}}$,
(b) $K=c_{\text{d}}-2$.

By eliminating c using the equality constraint, we can write the Lagrangian as

$$L(c_{\text{in}},T,\lambda,\theta)=(c_{\text{in}}+e^T-c_{\text{d}})^2+T^2+\lambda(\theta^2+c_{\text{in}}-K),$$

so that

$$\frac{\partial L}{\partial c_{\text{in}}}=2(c_{\text{in}}+e^T-c_{\text{d}})+\lambda=0, \tag{3.13}$$

$$\frac{\partial L}{\partial T}=2(c_{\text{in}}+e^T-c_{\text{d}})e^T+2T=0, \tag{3.14}$$

$$\frac{\partial L}{\partial \lambda}=\theta^2+c_{\text{in}}-K=0, \tag{3.15}$$

$$\frac{\partial L}{\partial \theta}=2\theta\lambda=0. \tag{3.16}$$

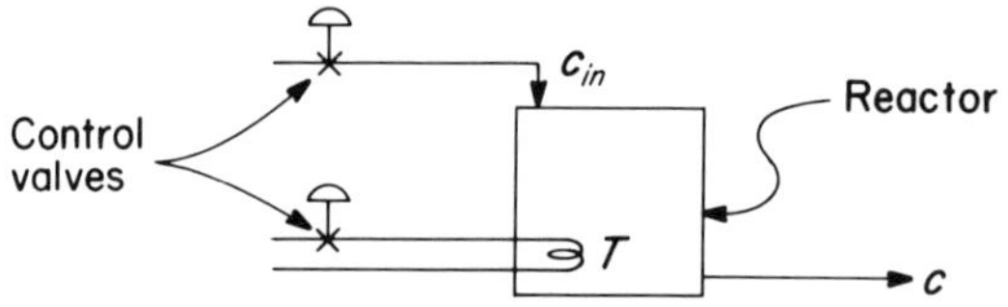

FIGURE 3.8

Case 1: $\lambda=0$ (unconstrained).

$$T^*=0, \qquad c_{\text{in}}^*=c_{\text{d}}-1, \qquad \theta^2=K \quad c_{\text{in}}^*.$$

(a) For $K=c_{\text{d}}$, $\theta^2=c_{\text{d}}-c_{\text{d}}+1=1$. Since θ is real, the constraint is satisfied and this is the optimal solution.
(b) For $K=c_{\text{d}}-2$, $\theta^2=c_{\text{d}}-2-c_{\text{d}}+1=-1$. Since θ is imaginary, the optimal solution must lie on the boundary when $K=c_{\text{d}}-2$.

Case 2: $\theta=0$ (constrained). In this case $c_{\text{in}}^*=K$ or $c_{\text{in}}^*=c_{\text{d}}-2$ for the problem in question. Substituting into (3.13) and (3.14) yields

$$e^{T^*}(e^{T^*}-2)+T^*=0 \tag{3.17}$$

and $\lambda^*=2(2-e^{T^*})$, where (3.17) must be solved numerically for T^*.

3.11 Maximize and minimize the distance from the origin under the following constraints:

$$(x_1-2)^2+(x_2-3)^2\leqslant 4,$$

$$x_1^2=4x_2.$$

The objective function is

$$\min_{\mathbf{x}} x_1^2+x_2^2,$$

and defining $\theta^2=4-(x_1-2)^2-(x_2-3)^2\geqslant 0$, the Lagrangian is

$$L(x_1,x_2,\lambda_1,\lambda_2,\theta)=x_1^2+x_2^2+\lambda_1\left[\theta^2+(x_1-2)^2+(x_2-3)^2-4\right]+\lambda_2\left[x_1^2-4x_2\right].$$

Thus

$$\frac{\partial L}{\partial x_1}=2x_1+2\lambda_1(x_1-2)+2\lambda_2x_1=0, \tag{3.18}$$

$$\frac{\partial L}{\partial x_2}=2x_2+2\lambda_1(x_2-3)-4\lambda_2=0, \tag{3.19}$$

$$\frac{\partial L}{\partial \lambda_1}=\theta^2+(x_1-2)^2+(x_2-3)^2-4=0, \tag{3.20}$$

$$\frac{\partial L}{\partial \lambda_2}=x_1^2-4x_2=0, \tag{3.21}$$

$$\frac{\partial L}{\partial \theta}=2\lambda_1\theta=0. \tag{3.22}$$

Case 1: $\lambda_1 = 0$.

$$2x_1(1+\lambda_2)=0,$$

$$2x_2-4\lambda_2=0,$$

$$\theta^2=4-(x_1-2)^2-(x_2-3)^2,$$

$$4x_2=x_1^2.$$

There are no solutions which satisfy all these equations for θ real. Hence the solution must correspond to Case 2.

Case 2: $\theta=0$. Now from (3.20) and (3.21), the two solutions are: (2, 1), which corresponds to a minimum with $f(\mathbf{x})=5$; and (3.86, 3.72), which corresponds (approximately) to a maximum with $f(x)=28.73$.

3.12 Sketch the feasible region satisfying the constraints

$$x_1^2+x_2^2 \leqslant 8,$$

$$-2x_1+x_2^2 \leqslant 0,$$

$$x_1x_2 \leqslant 4. \tag{3.22}$$

The solution is shown in Fig. 3.9. Note that the constraint (3.22) is redundant.

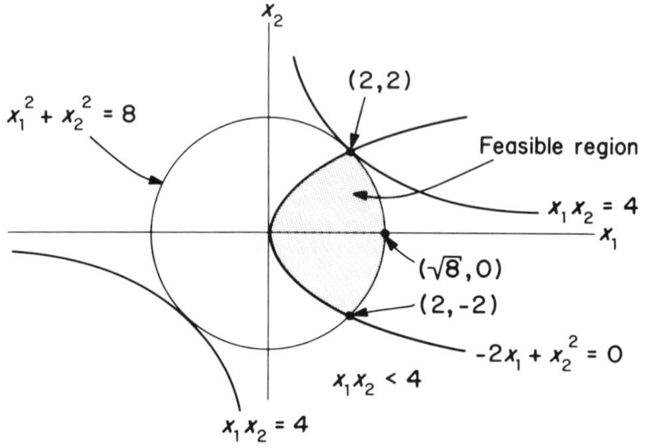

FIGURE 3.9

Supplementary Problems

3.13 Sketch the level curves for Problem 3.9.

3.14 Solve for T^* in Problem 3.10 using the Newton-Raphson method.

3.15 Solve the problem

$$\min |x_1 - 2| + |x_2 - 2|$$

such that

$$g_1(\mathbf{x}) = x_1 - x_2^2 \geqslant 0,$$

$$g_2(\mathbf{x}) = x_1^2 + x_2^2 - 1 = 0.$$

3.16 Solve Problem 3.15 ignoring the equality constraint.

3.17 Solve Problem 3.15 ignoring both constraints.

3.18 What method is the most effective in solving Problem 3.15 and why?

3.19 For a nonideal solid solution of two elements A and B, the heat of mixing is given by

$$H_m = x_A x_B (a + b x_B + c x_B^2),$$

where x_A and x_B are the mole fractions of elements A and B, respectively. Maximize the heat of mixing of the solid solution such that $x_A + x_B = 1$.

3.20

$$\max x_1 + x_2$$

such that

$$x_1^2 + x_2^2 = 1.$$

3.21

$$\max \exp(-x_1^2 + 2x_2^2)$$

such that

$$2x_1 + 3x_2 = 4.$$

3.22

$$\max \sin x_1 \cos x_2$$

such that

$$x_1 - x_2 = 0.$$

3.23 For what range of values for c_1 is the solution 4,3 optimal?

$$\max\ c_1x_1 + x_2$$

such that

$$x_1^2 + x_2^2 \leqslant 25,$$

$$x_1 - x_2 \leqslant 1.$$

3.24 For Problem 3.23, illustrate your answer graphically.

Chapter 4

Optimization with Inequality Constraints

In Chapter 3 we introduced the use of Lagrange multipliers to extend the necessary conditions of the calculus to problems involving equality constraints. We also presented a method for using a slack variable to convert problems involving inequality constraints to equalities so that the Lagrange multipliers could be employed. Although these methods have been known for a very long time, it is only recently that a complete theory for treating inequality constraints has been developed. In 1951 the theorem of Kuhn and Tucker provided rigorously a set of necessary conditions for the inequality-constrained problem itself. These necessary conditions, in addition to their generality and completeness, added one additional condition to those which were derived from the slack-variable approach using Lagrange multipliers.

The Kuhn-Tucker Theorem

Consider the optimization problem

$$\min_{\mathbf{x}} f(\mathbf{x})$$

such that

$$g_i(\mathbf{x}) \geqslant 0, \qquad i=1,2,\ldots,r,$$

where

$$\mathbf{x}=\left[x_1,x_2,\ldots,x_n\right]^T$$

and there is no loss in generality in not stating equality constraints directly, since they can always be converted to inequalities. In actuality, however, equality constraints are usually handled by the methods of Chapter 3 and their inclusion here would not add to the discussion. As in the past, we assume that $f(\mathbf{x})$ and $g_i(\mathbf{x})$, $i=1,2,\ldots,r$, are differentiable functions.

The proof of the Kuhn-Tucker necessary conditions proceeds by first defining a Lagrangian-like function

$$L(\mathbf{x},\lambda)=f(\mathbf{x})+\sum_{i=1}^{r}\lambda_i g_i(\mathbf{x})$$

and assuming the existence of the Lagrange multipliers λ_i. Then at the point $\mathbf{x}^*$ satisfying

$$\min_{\mathbf{x}} f(\mathbf{x})=f(\mathbf{x}^*)$$

such that

$$g_i(\mathbf{x}^*)\leqslant 0, \qquad i=1,2,\dots,r,$$

the following conditions must hold:

$$\frac{\partial f}{\partial x_j}(\mathbf{x}^*)+\sum_{i=1}^{r}\lambda_i^*\frac{\partial g_i}{\partial x_j}(\mathbf{x}^*)=0, \qquad j=1,2,\dots,n \tag{4.1}$$

$$g_i(\mathbf{x}^*)\leqslant 0, \qquad i=1,2,\dots,r \tag{4.2}$$

$$(\lambda_i^*)\,g_i(\mathbf{x}^*)=0, \qquad i=1,2,\dots,r, \tag{4.3}$$

$$\lambda_i^*\geqslant 0, \qquad i=1,2,\dots,r, \tag{4.4}$$

or in a more compact notation,

$$\nabla_{\mathbf{x}}L(\mathbf{x}^*,\boldsymbol{\lambda}^*)=\mathbf{0}, \tag{4.5}$$

$$\nabla_{\boldsymbol{\lambda}}L(\mathbf{x}^*,\boldsymbol{\lambda}^*)\leqslant\mathbf{0}, \tag{4.6}$$

$$(\boldsymbol{\lambda}^*)^T\mathbf{g}(\mathbf{x}^*)=\mathbf{0}, \tag{4.7}$$

$$\boldsymbol{\lambda}^*\geqslant\mathbf{0}. \tag{4.8}$$

These results are called the Kuhn-Tucker stationary conditions. Later we shall see another form of the Kuhn-Tucker theorem involving saddle-point conditions.

Although these conditions must hold at the maximum, they are not necessarily very useful in determining the optimum point. In particular, the above set of nonlinear equalities and inequalities presents a very challenging computational problem. For high-dimensional problems, the determination of the optimum by the direct solution of these equations is rarely feasible.

It is instructive to state these conditions explicitly for several other optimization problems.

EXAMPLES

4.1 Determine the Kuhn-Tucker stationarity conditions for the problem

$$\max_{\mathbf{x}} f(\mathbf{x})$$

such that

$$g_i(\mathbf{x}) \leqslant 0, \qquad i=1,2,\ldots,r.$$

The necessary conditions for this problem are obtained by substituting $-f(\mathbf{x})$ for $f(\mathbf{x})$ in the above. Thus

$$-\frac{\partial f}{\partial x_j}(\mathbf{x}^*)+\sum_{i=1}^{r}\lambda_i^*\frac{\partial g_i}{\partial x_j}(\mathbf{x}^*)=0, \qquad j=1,2,\ldots,n,$$

or

$$\frac{\partial f}{\partial x_j}(\mathbf{x}^*)+\sum_{i=1}^{r}(-\lambda_i^*)\frac{\partial g_i}{\partial x_j}(\mathbf{x}^*)=0, \qquad j=1,2,\ldots,n.$$

By simply changing the sign of λ_i^*, the conditions for this problem become:

$$\frac{\partial f}{\partial x_j}(\mathbf{x}^*)+\sum_{i=1}^{r}\lambda_i^*\frac{\partial g_i}{\partial x_j}(\mathbf{x}^*)=0, \qquad j=1,2,\ldots,n, \tag{4.9}$$

$$g_i(\mathbf{x}^*) \leqslant 0, \qquad i=1,2,\ldots,r, \tag{4.10}$$

$$(\lambda_i^*)\, g_i(\mathbf{x}^*)=0, \qquad i=1,2,\ldots,r, \tag{4.11}$$

$$\lambda_i^* \leqslant 0, \qquad i=1,2,\ldots,r. \tag{4.12}$$

4.2 Determine the Kuhn-Tucker conditions for the problem

$$\min_{\mathbf{x}} f(\mathbf{x})$$

such that

$$g_i(\mathbf{x}^*) \geqslant 0, \qquad i=1,2,\ldots,r.$$

In this case we can simply change the sign of $g_i(\mathbf{x})$ in the above conditions. Thus

$$\frac{\partial f}{\partial x_j}(\mathbf{x}^*) + \sum_{i=1}^{n} (-\lambda_i^*) \frac{\partial g_i}{\partial x_j}(\mathbf{x}^*) = 0, \qquad j = 1, 2, \ldots, n$$

$$g_i(\mathbf{x}^*) \geqslant 0, \qquad i = 1, 2, \ldots, r,$$

so the conditions become

$$\frac{\partial f}{\partial x_j}(\mathbf{x}^*) + \sum_{i=1}^{r} \lambda_i^* \frac{\partial g_i}{\partial x_j}(\mathbf{x}^*) = 0, \qquad j = 1, 2, \ldots, n, \tag{4.13}$$

$$g_i(\mathbf{x}^*) \geqslant 0, \qquad i = 1, 2, \ldots, r, \tag{4.14}$$

$$(\lambda_i^*)\, g_i(\mathbf{x}^*) = 0, \qquad i = 1, 2, \ldots, r, \tag{4.15}$$

$$\lambda_i^* \leqslant 0, \qquad i = 1, 2, \ldots, r. \tag{4.16}$$

4.3 Determine the Kuhn-Tucker conditions for the problem

$$\max_{\mathbf{x}} f(\mathbf{x})$$

such that

$$g_i(\mathbf{x}) \geqslant 0, \qquad i = 1, 2, \ldots, r.$$

Changing the signs of both $f(\mathbf{x})$ and $g_i(\mathbf{x})$ in our canonical problem yields

$$-\frac{\partial f}{\partial x_j}(\mathbf{x}^*) - \sum_{i=1}^{r} \lambda_i^* \frac{g_i}{x_j}(\mathbf{x}^*) = 0.$$

So the conditions become

$$\frac{\partial f}{\partial x_j}(\mathbf{x}^*) + \sum_{i=1}^{r} \lambda_i^* \frac{\partial g_i}{\partial x_j}(\mathbf{x}^*) = 0, \qquad j = 1, 2, \ldots, n, \tag{4.17}$$

$$g_i(\mathbf{x}^*) \geqslant 0, \qquad i = 1, 2, \ldots, r, \tag{4.18}$$

$$\lambda_i^* g_i(\mathbf{x}^*) = 0, \qquad i = 1, 2, \ldots, r, \tag{4.19}$$

$$\lambda_i^* \geqslant 0, \qquad i = 1, 2, \ldots, r. \tag{4.20}$$

4.4 Apply the Kuhn-Tucker stationarity conditions to Example 3.7, page 64.

$$\min_x (x-a)^2 + b$$

such that

$$x \geqslant c.$$

Define

$$L(x,\lambda) = (x-a)^2 + b + \lambda(x-c).$$

Then

$$\frac{\partial L}{\partial x} = 2(x-a) + \lambda = 0, \tag{4.21}$$

$$\frac{\partial L}{\partial \lambda} = x - c \geqslant 0, \tag{4.22}$$

$$\lambda(x-c) = 0, \tag{4.23}$$

$$\lambda \leqslant 0. \tag{4.24}$$

Condition (4.23) is called the *complementary slackness condition* and requires that $\lambda^* = 0$ or $x^* = c$ (or both). Since the former case corresponds to a solution $x^* = a$, either

$$\lambda^* = 0 \quad \text{and} \quad x^* = a$$

or

$$\lambda^* = 2(a - x^*) \quad \text{and} \quad x^* = c.$$

Thus if $a \geqslant c$, the solution is

$$x^* = a \quad \text{and} \quad \lambda^* = 0.$$

If $a < c$, the solution is

$$x^* = c \quad \text{and} \quad \lambda^* = 2(a-c) < 0.$$

This solution is identical to the one obtained in Example 3.8, page 66. However, we now have the additional condition

$$\lambda^* \leqslant 0,$$

which we see is satisfied.

The Kuhn-Tucker stationarity conditions are necessary and sufficient if $f(\mathbf{x})$ and $g_i(\mathbf{x})$ are convex functions. This is not surprising, since these conditions ensure that a local minimum is also global. If $f(\mathbf{x})$ is strictly convex, the minimum is also unique.

The Constraint Qualification

It is clear from the above that λ is very similar to the Lagrange multiplier discussed in Chapter 3, except that λ can have either sign when associated with an equality constraint and it must be sign definite (take on only one sign) when associated with an inequality constraint.

Previously we assumed the existence of λ. We now state an additional necessary condition called the *constraint qualification* which ensures that λ exists (is finite).

> *Constraint qualification*: There must exist a vector $\mathbf{h} \in E^n$ such that at each *binding* constraint,
>
> $$(\nabla g_i(\mathbf{x}^*), \mathbf{h}) < 0. \tag{4.25}$$
>
> That is, a *single* $\mathbf{h}$ must exist for all binding constraints.

In reality, the constraint qualification is difficult to verify *a priori*, since it requires a knowledge of the optimum $\mathbf{x}^*$ to evaluate. However, there are several conditions under which the constraint qualification is always satisfied. These are:

1. All inequality constraints are convex, and there exists at least one $\mathbf{x}$ which is strictly inside the feasible region (i.e., not on the boundary).

2. The rank of the matrix of all binding constraints at $\mathbf{x}^*$ is maximal. (Recall this is the existence condition for Lagrange multipliers in equality-constrained problems.)

3. The constraints are linear.

For optimization problems involving equality constraints of the form

$$g_i(\mathbf{x}) \geqslant 0, \qquad i = 1, 2, \ldots, r,$$

whether minimization problems or maximization problems, the constraint qualification given above simply reverses the inequality (4.25).

In the following examples we show geometrical properties of the constraint qualification and that the Kuhn-Tucker stationarity conditions do not hold at the optimum when this condition is not satisfied.

EXAMPLES

4.5 Show that the constraint qualification is not satisfied and that the multipliers do not exist at the optimum for the following problem:

$$\min_{x_1,x_2} (x_1-2)^2+x_2^2$$

such that

$$g_1(x_1,x_2)=x_1 \geqslant 0,$$
$$g_2(x_1,x_2)=x_2 \geqslant 0,$$
$$g_3(x_1,x_2)=(1-x_1)^3-x_2 \geqslant 0.$$

From Fig. 4.1, we can see that the optimum solution $x_1^*=1$, $x_2^*=0$. The binding constraints at that point are $g_2(x_1,x_2)=0$ and $g_3(x_1,x_2)=0$, and their gradient vectors are

$$\nabla g_2(1,0)=\begin{bmatrix} \dfrac{\partial g_2}{\partial x_1} \\ \dfrac{\partial g_2}{\partial x_2} \end{bmatrix}_{1,0} = \begin{bmatrix} 0 \\ 1 \end{bmatrix},$$

$$\nabla g_3(1,0)=\begin{bmatrix} \dfrac{\partial g_3}{\partial x_1} \\ \dfrac{\partial g_3}{\partial x_2} \end{bmatrix}_{1,0} = \begin{bmatrix} -3(1-x_1)^2 \\ -1 \end{bmatrix}_{1,0} = \begin{bmatrix} 0 \\ -1 \end{bmatrix}.$$

The constraint qualification requires the existence of a single vector **h** which satisfies

$$(\nabla g_2(1,0),\mathbf{h})<0$$

and

$$(\nabla g_3(1,0),\mathbf{h})>0.$$

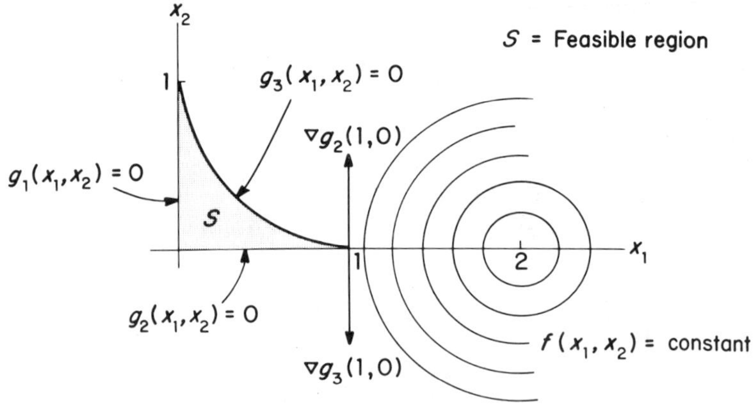

FIGURE 4.1

Writing the inner product in another form, we can get some geometrical insight into the constraint qualification. Equivalently,

$$\|\nabla g_2(1,0)\| \|\mathbf{h}\| \cos\theta_1 > 0,$$

$$\|\nabla g_3(1,0)\| \|\mathbf{h}\| \cos\theta_2 > 0,$$

where θ_1 is the angle between $\nabla g_2(1,0)$ and $\mathbf{h}$, and θ_2 is the angle between $\nabla g_3(1,0)$ and $\mathbf{h}$. These inequalities are obviously satisfied for any $\mathbf{h}$ for which $\cos\theta_1 > 0$ and $\cos\theta_2 > 0$, i.e., for

$$-90° < \theta_1 < 90° \quad \text{and} \quad -90° < \theta_2 < 90°.$$

However, there is no single $\mathbf{h}$ which can satisfy both of these requirements, and the constraint qualification is not satisfied (see Fig 4.2).

Since $g_2(x_1, x_2) \geqslant 0$ and $g_3(x_1, x_2) \geqslant 0$ are equality constraints at $\mathbf{x}^* = [1,0]^T$, the existence of Lagrange multipliers requires that the Jacobian be of maximal rank (i.e., of rank 2) at this point. But

$$J(1,0) = \begin{bmatrix} \dfrac{\partial g_2}{\partial x_1} & \dfrac{\partial g_2}{\partial x_2} \\ \dfrac{\partial g_3}{\partial x_1} & \dfrac{\partial g_3}{\partial x_2} \end{bmatrix}_{1,0} = \begin{bmatrix} 0 & 1 \\ 0 & -1 \end{bmatrix},$$

which has rank 1, so the Lagrange multipliers do not exist, as expected.

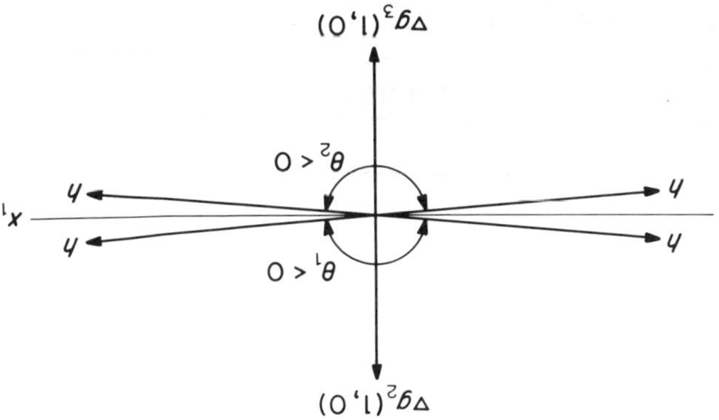

FIGURE 4.2

4.6 For Example 4.5, determine the effect of not satisfying the constraint qualification on the remaining Kuhn-Tucker necessary conditions.

Forming the Lagrangian,

$$L(\mathbf{x},\boldsymbol{\lambda})=(x_1-2)^2+x_2^2+\lambda_1 x_1+\lambda_2 x_2+\lambda_3\left[(1-x_1)^3-x_2\right].$$

The necessary conditions are:

$$\frac{\partial L}{\partial x_1}=2(x_1-2)+\lambda_1-3(1-x_1)^2\lambda_3=0, \tag{4.26}$$

$$\frac{\partial L}{\partial x_2}=2x_2+\lambda_2-\lambda_3=0, \tag{4.27}$$

$$x_1\geqslant 0, \qquad x_2\geqslant 0, \qquad (1-x_1)^3-x_2\geqslant 0, \tag{4.28}$$

$$\lambda_1 x_1=0, \qquad \lambda_2 x_2=0, \qquad \lambda_3\left[(1-x_1)^3-x_2\right]=0, \tag{4.29}$$

$$\lambda_i\leqslant 0, \qquad i=1,2,3. \tag{4.30}$$

At $\mathbf{x}^*=(1,0)$, Eq. (4.25) yields

$$\lambda_1=2,$$

but from (4.29),

$$\lambda_1=0,$$

which leads to a contradiction, and so the Kuhn-Tucker conditions are not satisfied at the minimum.

Saddle-Point Conditions

Two disadvantages of the Kuhn-Tucker stationarity conditions are that (1) they are only necessary conditions for nonconvex functions $f(\mathbf{x})$ and $g_i(\mathbf{x})$, and (2) they apply only when the functions $f(\mathbf{x})$ and $g_i(\mathbf{x})$ are differentiable.

As was shown in Chapter 3, the Lagrangian has a saddle point at $(\mathbf{x}^*, \boldsymbol{\lambda}^*)$, and the Kuhn-Tucker stationarity conditions can also be stated in terms of the saddle-value properties of $L(\mathbf{x}, \boldsymbol{\lambda})$.

Theorem 4.1 *A point* $(\mathbf{x}^*, \boldsymbol{\lambda}^*)$ *with* $\boldsymbol{\lambda}^* \geqslant \mathbf{0}$ *is a saddle point of the Lagrangian* $L(\mathbf{x}, \boldsymbol{\lambda})$ *associated with the primal problem* (*see below*) *if and only if the following conditions hold*:

1. $\mathbf{x}^*$ *minimizes* $L(\mathbf{x}, \boldsymbol{\lambda}^*)$ *over all* $\mathbf{x}$.

2. $g_i(\mathbf{x}^*) \leqslant 0$, $i = 1, 2, \ldots, r$.

3. $\lambda_i^* g_i(\mathbf{x}^*) = 0$, $i = 1, 2, \ldots, r$.

The *primal problem* is defined as

$$\min_{\mathbf{x}} f(\mathbf{x})$$

such that

$$g_i(\mathbf{x}) \leqslant 0, \qquad i = 1, 2, \ldots, r,$$

where $\mathbf{x} = [x_1, x_2, \ldots, x_n]$, and $f(\mathbf{x})$ and $g_i(\mathbf{x})$ are real-valued functions. There are no differentiability or convexity requirements on these functions. Notice also that condition 1 above calls for a *global*, not local, minimization of the Lagrangian. This theorem is proved in Problem 4.11.

In order to establish the connection between the determination of the saddle point for the Lagrangian and the primal nonlinear programming problem, we need the following theorem.

Theorem 4.2 *If the point* $(\mathbf{x}^*, \boldsymbol{\lambda}^*)$ *is a saddle point for the Lagrangian associated with the primal problem, then* $\mathbf{x}^*$ *solves the primal problem.*

This theorem is proved in Problem 4.12.

If the $g_i(\mathbf{x})$ are convex, and if there exists at least one $\mathbf{x}$ which satisfies all the inequalities strictly, [i.e., such that $g_i(\mathbf{x}) < 0$, $i = 1, 2, \ldots, r$], then the saddle-point condition (Theorem 4.2) is both necessary and sufficient. This requirement on $\mathbf{x}$ is a form of *constraint qualification*.

The major advantage of these two theorems is that they provide sufficient conditions for solving optimization problems which are neither differentiable nor convex. The computational problem which arises is the global minimization of $L(\mathbf{x},\boldsymbol{\lambda}^*)$ over all $\mathbf{x}$. This task can be accomplished by a number of direct search algorithms. The difficulty with this minimization is, of course, that it requires knowledge of $\boldsymbol{\lambda}^*$, which is generally not known *a priori*. This saddle-point relation will form the basis for Lagrangian decomposition and a computational algorithm to be discussed in Chapter 13.

Duality

The concept of duality is highly refined in linear programming and leads to several very efficient computional algorithms in that case. A brief review of linear programming and its duality theorem is given in the Appendix.

A similar duality theory exists for nonlinear programming problems. However, the dual nonlinear programs are usually much more difficult to solve than the corresponding primal. The dual does, however, often lead to computationally efficient algorithms when combined with conditions arising in the primal problem. These matters are further discussed in Chapter 13.

We define the *dual function* in nonlinear programming as

$$h(\boldsymbol{\lambda})=\min_{\mathbf{x}} L(\mathbf{x},\boldsymbol{\lambda}),$$

where the minimization of $L(\mathbf{x},\boldsymbol{\lambda})$ is performed over all $\mathbf{x}$ with $\boldsymbol{\lambda}$ as a fixed parameter. We note that the minimizing values of $\mathbf{x}$ may not satisfy the constraints $g_i(\mathbf{x})\leqslant 0$, $i=1,2,\ldots,r$, and in fact, the minimum may not exist for all values of the parameter $\boldsymbol{\lambda}$. We now define the set of $\boldsymbol{\lambda}$ such that $L(\mathbf{x},\boldsymbol{\lambda})$ has a finite minimum with respect to $\mathbf{x}$ as

$$D=\{\boldsymbol{\lambda}|h(\boldsymbol{\lambda}) \text{ exists, and } \boldsymbol{\lambda}\geqslant \mathbf{0}\}.$$

From Theorem 4.1 we recall that $\mathbf{x}^*$ minimizes $L(\mathbf{x},\boldsymbol{\lambda}^*)$. We now wish to explore how $h(\boldsymbol{\lambda})$ can be used to determine $\boldsymbol{\lambda}^*$.

Theorem 4.3 *The dual function $h(\boldsymbol{\lambda})\leqslant f(\mathbf{x})$ for all $\mathbf{x}$ satisfying $g_i(\mathbf{x})\leqslant 0$, $i=1,2,\ldots,r$, and all $\boldsymbol{\lambda}\in D$; i.e., the dual function provides a lower bound on the function $f(\mathbf{x})$.*

Proof: By definition,

$$h(\boldsymbol{\lambda})=\min_{\mathbf{x}} L(\mathbf{x},\boldsymbol{\lambda})\leqslant L(\mathbf{x},\boldsymbol{\lambda}),$$

or

$$h(\boldsymbol{\lambda}) \leqslant f(\mathbf{x}) + \sum_{i=1}^{r} \lambda_i g_i(\mathbf{x}).$$

By hypothesis $\mathbf{x}$ satisfies $g_i(\mathbf{x}) \leqslant 0$, $i = 1, 2, \ldots, r$, and $\boldsymbol{\lambda} \geqslant \mathbf{0}$. Thus

$$\sum_{i=1}^{r} \lambda_i g_i(\mathbf{x}) \leqslant 0$$

and

$$h(\boldsymbol{\lambda}) \leqslant f(\mathbf{x}) \quad \text{for} \quad \boldsymbol{\lambda} \in D \quad \text{and} \quad \mathbf{x} \in \{\mathbf{x} \mid g_i(\mathbf{x}) \leqslant 0,\ i = 1, 2, \ldots, r\}. \square$$

Since the dual function provides a lower bound for $f(\mathbf{x})$, the greatest lower bound must occur at the maximum value of $h(\boldsymbol{\lambda})$ where $\boldsymbol{\lambda} \in D$. We now define the dual problem as

$$\max_{\boldsymbol{\lambda} \in D} h(\boldsymbol{\lambda}).$$

Theorem 4.4 (Duality theorem) *The point* $(\mathbf{x}^*, \boldsymbol{\lambda}^*)$ *is a saddle point of the Lagrangian defined below if and only if*

1. $\mathbf{x}^*$ *solves the primal problem,*

2. $\boldsymbol{\lambda}^*$ *solves the dual problem,*

3. $f(\mathbf{x}^*) = h(\boldsymbol{\lambda}^*)$.

The following terms are defined:

Primal problem:

$$\min_{\mathbf{x}} f(\mathbf{x})$$

such that

$$g_i(\mathbf{x}) \leqslant 0, \qquad i = 1, 2, \ldots, r,$$

$$\mathbf{x} = [x_1, x_2, \ldots, x_n]^T,$$

and $f(\mathbf{x})$, $g_i(\mathbf{x})$ are real-valued functions.

Lagrangian:

$$L(\mathbf{x}, \boldsymbol{\lambda}) = f(\mathbf{x}) + \sum_{i=1}^{r} \lambda_i g_i(\mathbf{x}).$$

Dual function:

$$h(\boldsymbol{\lambda}) = \min_{\mathbf{x}} L(\mathbf{x}, \boldsymbol{\lambda})$$

$$D = \{\boldsymbol{\lambda} | h(\boldsymbol{\lambda}) \text{ exists, and } \boldsymbol{\lambda} \geq 0\}.$$

Dual problem:

$$\max_{\boldsymbol{\lambda} \in D} h(\boldsymbol{\lambda}).$$

This important theorem is proved in Problem 4.15.

The following theorem gives us some insight into the nature of the dual function which will be helpful in designing computational algorithms.

Theorem 4.5 *The dual function $h(\boldsymbol{\lambda})$ is concave over any convex subset of its domain D.*

This theorem is proved in Problem 4.16.

In this chapter, several important theoretical results have been obtained which will be useful in solving optimization problems. Depending on the nature of the functions $f(\mathbf{x})$ and $g_i(\mathbf{x})$, $i = 1, 2, \ldots, r$, these results are either necessary or sufficient, or both. For convenience, these conditions are summarized in Table 4.1 for the primal problem.

Table 4.1
Summary of Conditions on the Primal Problem

$f(\mathbf{x}), g(\mathbf{x})$ DIFFERENTIABLE	$f(\mathbf{x}), g(\mathbf{x})$ CONVEX	STATIONARITY CONDITIONS	SADDLE-POINT CONDITIONS	DUALITY CONDITIONS
Yes	Yes	Necessary and sufficient	Necessary and sufficient	Necessary and sufficient
Yes	No	Necessary only	Sufficient only	Sufficient only
No	Yes	Not applicable	Necessary and sufficient (with constraint qualification)	Necessary and sufficient
No	No	Not applicable	Sufficient only	Sufficient only

Solved Problems

Kuhn-Tucker Theorem

4.1 Without using the Kuhn-Tucker conditions, determine necessary conditions for a minimum of the problem

$$\min_{\mathbf{x}} f(\mathbf{x})$$

such that

$$x_i \geqslant 0, \qquad i=1,2,\ldots,n.$$

Recalling our derivation of classical necessary conditions in Chapter 2, we expand $f(\mathbf{x})$ in a Taylor series about $\mathbf{x}^*$,

$$f(\mathbf{x}^*+h\,\Delta)=f(\mathbf{x}^*)+h\frac{\partial f}{\partial \mathbf{x}}(\mathbf{x}^*)\Delta\mathbf{x}+\tfrac{1}{2}h^2(\Delta\mathbf{x})^T\mathbf{H}(x^*+\theta h\,\Delta\mathbf{x})\Delta\mathbf{x},$$

where $\mathbf{H}$ is the Hessian matrix and $0<\theta<1$.

Now for a minimum at $\mathbf{x}^*$ we require

$$f(\mathbf{x}^*)\leqslant f(\mathbf{x}^*+h\,\Delta\mathbf{x}),$$

which yields the *fundamental inequality*

$$h\frac{\partial f}{\partial \mathbf{x}}(\mathbf{x}^*)\Delta\mathbf{x}+\tfrac{1}{2}(\Delta\mathbf{x})^T\mathbf{H}(\mathbf{x}^*+\theta h\,\Delta\mathbf{x})\Delta\mathbf{x}\geqslant 0,$$

which is a necessary condition for a local minimum at $\mathbf{x}^*$. If $\mathbf{x}^*$ is an interior solution $\mathbf{x}^*>\mathbf{0}$, then the fundamental inequality leads to the first-order necessary conditions

$$\frac{\partial f}{\partial x_j}(\mathbf{x}^*)=0, \qquad j=1,2,\ldots,n,$$

that were derived in Chapter 2. Suppose, however, that some variable has its optimum value on the boundary $x_j^*=0$. Then assuming all other partials equal zero, the fundamental inequality yields

$$\frac{\partial f}{\partial x_j}(\mathbf{x}^*)\Delta x_j \geqslant 0.$$

But Δx_j must be positive (since $x_j^*=0$), which leads to the necessary condition

$$\frac{\partial f}{\partial x_j}(\mathbf{x}^*)\geqslant 0$$

if x_j^* is on the constraint boundary $x_j^*=0$. Since all derivatives vanish at an interior solution, we can write

$$\frac{\partial f}{\partial x_j}(\mathbf{x}^*)x_j^* = 0, \qquad j = 1, 2, \ldots, n.$$

Summarizing the necessary conditions in a compact form, we have

1. $\nabla f(\mathbf{x}^*) \geqslant 0$,
2. $(\mathbf{x}^*)^T \nabla f(\mathbf{x}^*) = 0$,
3. $\mathbf{x}^* \geqslant \mathbf{0}$.

We see from Fig. 4.3(a) that condition 1 above is reasonable, since if $\partial f(\mathbf{x}^*)/\partial x_j < 0$ when $x_j^* = 0$, we could move to an adjacent point $x_j > 0$ having a smaller value of $f(\mathbf{x})$, in which case $x_j^* = 0$ would not be optimal. The other two possible cases are summarized in Fig. 4.3(b, c) for completeness.

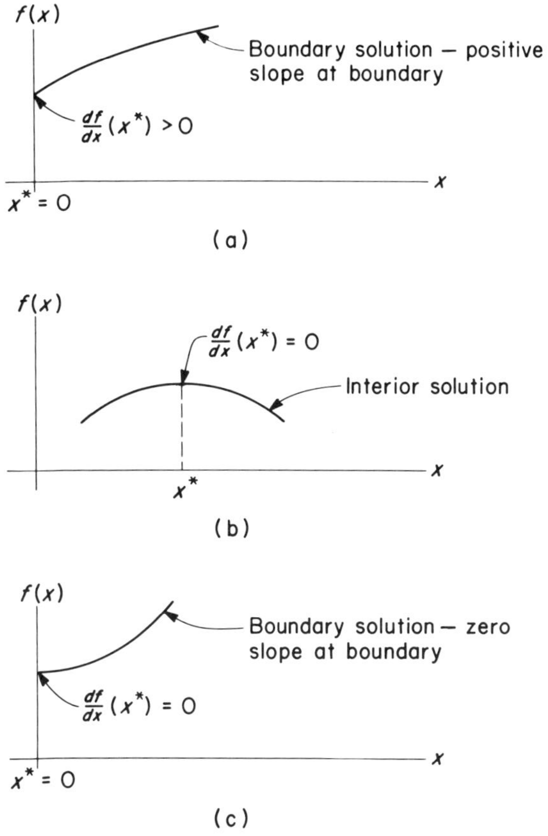

FIGURE 4.3

4.2 Prove the Kuhn-Tucker stationarity conditions (4.5)–(4.8).

For the problem

$$\min_{\mathbf{x}} f(\mathbf{x})$$

such that

$$g_i(\mathbf{x}) \leqslant 0, \qquad i=1,2,\ldots,r,$$

we first convert the inequality constraints to equality constraints by adding slack variables $s_i \geqslant 0$. Thus

$$s_i + g_i(\mathbf{x}) = 0, \qquad i=1,2,\ldots,r,$$

$$s_i \geqslant 0.$$

Now form the Lagrangian

$$L(\mathbf{x},\mathbf{s},\boldsymbol{\lambda}) = f(\mathbf{x}) + \sum_{i=1}^{r} \lambda_i\left[s_i + g_i(\mathbf{x})\right],$$

apply the necessary conditions of Chapter 3 to $\mathbf{x}$ and $\boldsymbol{\lambda}$, and apply the necessary conditions developed in Problem 4.1 to the variables s_i because of the nonnegativity constraint. These conditions yield:

From Chapter 3,

$$\frac{\partial L}{\partial x_j}(\mathbf{x}^*,\mathbf{s}^*,\boldsymbol{\lambda}^*) = \frac{\partial f}{\partial x_j}(\mathbf{x}^*) + \sum_{i=1}^{r} \lambda_i^* \frac{\partial g_i}{\partial x_j}(\mathbf{x}^*) = 0, \qquad j=1,2,\ldots,n,$$

$$\frac{\partial L}{\partial \lambda_i}(\mathbf{x}^*,\mathbf{s}^*,\boldsymbol{\lambda}^*) = s_i^* + g_i(\mathbf{x}^*) = 0, \qquad i=1,2,\ldots,r.$$

From Problem 4.1,

$$\frac{\partial L}{\partial s_i}(\mathbf{x}^*,\mathbf{s}^*,\boldsymbol{\lambda}^*) = \lambda_i^* \geqslant 0, \qquad i=1,2,\ldots,r,$$

$$s_i^* \frac{\partial L}{\partial s_i}(\mathbf{x}^*,\mathbf{s}^*,\boldsymbol{\lambda}^*) = s_i^*\lambda_i^* = 0, \qquad i=1,2,\ldots,r,$$

$$s_i^* \geqslant 0, \qquad i=1,2,\ldots,r,$$

Eliminating s_i^* in the above equations,

$$\frac{\partial f}{\partial x_j}(\mathbf{x}^*)+\sum_{i=1}^{r}\lambda_i^*\frac{\partial g_i}{\partial x_j}(\mathbf{x}^*)=0, \qquad j=1,2,\ldots,n,$$

$$g_i(\mathbf{x}^*)\leqslant 0, \qquad i=1,2,\ldots,r,$$

$$\lambda_i^* g_i(\mathbf{x}^*)=0, \qquad i=1,2,\ldots,r,$$

$$\lambda_i^*\geqslant 0, \qquad i=1,2,\ldots,r,$$

which are the Kuhn-Tucker stationarity conditions (4.5)–(4.8).

4.3 Develop a set of Kuhn-Tucker conditions for the problem

$$\min_{\mathbf{x}} f(\mathbf{x})$$

such that

$$g_i(\mathbf{x})\leqslant 0, \qquad i=1,2,\ldots,r$$

$$x_j\geqslant 0, \qquad j=1,2,\ldots,n$$

Define the Lagrangian

$$L(\mathbf{x},\boldsymbol{\lambda},\boldsymbol{\mu})=f(\mathbf{x})+\sum_{i=1}^{r}\lambda_i g_i(\mathbf{x})+\sum_{j=1}^{n}\mu_j x_j,$$

and attempt to eliminate $\boldsymbol{\mu}^*$ from the necessary conditions

$$\frac{\partial L}{\partial x_j}(\mathbf{x}^*,\boldsymbol{\lambda}^*,\boldsymbol{\mu}^*)=\frac{\partial f}{\partial x_j}(\mathbf{x}^*)+\sum_{i=1}^{r}\lambda_i^*\frac{\partial g_i}{\partial x_j}(\mathbf{x}^*)+\mu_j^*=0, \qquad j=1,2,\ldots,n,$$

$$\mu_j^* x_j^*=0, \qquad \mu_j\leqslant 0.$$

The remaining necessary conditions are not functions of $\boldsymbol{\mu}^*$. We now define

$$L(\mathbf{x},\boldsymbol{\lambda})=f(\mathbf{x})+\sum_{i=1}^{r}\lambda_i g_i(\mathbf{x}),$$

and the above conditions can be written

$$\frac{\partial L}{\partial x_j}(\mathbf{x}^*,\boldsymbol{\lambda}^*)=-\mu_j^*\geqslant 0, \qquad j=1,2,\ldots,n,$$

$$\frac{\partial L}{\partial x_j}(\mathbf{x}^*,\boldsymbol{\lambda}^*)x_j^*=0.$$

The variables μ_j^* have now been eliminated from the necessary conditions, which are summarized below:

$$\nabla_{\mathbf{x}} L(\mathbf{x}^*,\boldsymbol{\lambda}^*) \geq \mathbf{0}, \qquad \nabla_{\boldsymbol{\lambda}} L(\mathbf{x}^*,\boldsymbol{\lambda}^*) \leq \mathbf{0},$$

$$(\mathbf{x}^*)^T \nabla_{\mathbf{x}} L(\mathbf{x}^*,\boldsymbol{\lambda}^*) = 0, \qquad (\boldsymbol{\lambda}^*)^T \nabla_{\boldsymbol{\lambda}} L(\mathbf{x}^*,\boldsymbol{\lambda}^*) = 0,$$

$$\mathbf{x}^* \geq \mathbf{0}, \qquad \boldsymbol{\lambda}^* \geq \mathbf{0}.$$

4.4 Write out all the necessary stationarity conditions for the problem

$$\min\{ f(\mathbf{x}) | \mathbf{x} \in S_1 \cap S_2 \cap S_3 \},$$

where

$S_1 = (E^n)^+,$

$S_2 = \{\mathbf{x} | w_i(\mathbf{x}) = 0,\ i = 1,2,\ldots,n_w\},$

$S_3 = \{x | g_i(x) \leq 0,\ i = 1,2,\ldots,n_g\},$

$\mathbf{x}$ is an n-dimensional vector of real variables, and

$f(\mathbf{x})$, $w_i(\mathbf{x})$, and $g_i(\mathbf{x})$ are real-valued functions defined

on S_1 and differentiable everywhere.

Define

$$L(\mathbf{x},\boldsymbol{\lambda},\boldsymbol{\mu}) = f(\mathbf{x}) + \boldsymbol{\lambda}^T \mathbf{w}(x) + \boldsymbol{\mu}^T \mathbf{g}(\mathbf{x})$$

The necessary conditions are

$$\nabla_{\mathbf{x}} L(\mathbf{x}^*,\boldsymbol{\lambda}^*,\boldsymbol{\mu}^*) \geq \mathbf{0}, \qquad \mathbf{g}(\mathbf{x}^*) \leq \mathbf{0},$$

$$\mathbf{x}^* \geq \mathbf{0}, \qquad \boldsymbol{\mu}^* \geq \mathbf{0},$$

$$(\mathbf{x}^*)^T \nabla_{\mathbf{x}} L(\mathbf{x}^*,\boldsymbol{\lambda}^*,\boldsymbol{\mu}^*) = 0, \qquad (\boldsymbol{\mu}^*)^T \mathbf{g}(\mathbf{x}^*) = 0,$$

$$\mathbf{w}(\mathbf{x}^*) = \mathbf{0}.$$

4.5 Using the Kuhn-Tucker stationarity conditions, solve the problem

$$\min_{\mathbf{x}} \left\{ f(\mathbf{x}) = (x_1 - 1)^2 + (x_2 - 2) \right\}$$

such that

$$x_2 - x_1 = 1,$$
$$x_1 + x_2 \leqslant 2,$$
$$x_1 \geqslant 0, \qquad x_2 \geqslant 0,$$

and check the answer graphically.

The necessary conditions were stated in Problem 4.4.
Define

$$L(\mathbf{x},\lambda,\mu) = (x_1 - 1)^2 + (x_2 - 2)^2 + \lambda(x_2 - x_1 - 1) + \mu(x_1 + x_2 - 2).$$

Now

$$\frac{\partial L}{\partial x_1} = 2(x_1 - 1) - \lambda + \mu \geqslant 0, \tag{4.31}$$

$$\frac{\partial L}{\partial x_2} = 2(x_2 - 2) + \lambda + \mu \geqslant 0, \tag{4.32}$$

$$x_1 \geqslant 0, \quad x_2 \geqslant 0, \tag{4.33}$$

$$x_1[2(x_1 - 1) - \lambda + \mu] = 0, \tag{4.34}$$

$$x_2[2(x_2 - 2) + \lambda + \mu] = 0, \tag{4.35}$$

$$x_1 + x_2 - 2 \leqslant 0, \tag{4.36}$$

$$\mu \geqslant 0, \tag{4.37}$$

$$\mu[x_1 + x_2 - 2] = 0, \tag{4.38}$$

$$x_2 - x_1 - 1 = 0. \tag{4.39}$$

To solve these equations, first assume $x_1 \neq 0$, $x_2 \neq 0$, and $\mu = 0$. Then solve (4.33), (4.34), and (4.38), i.e.,

$$2(x_1 - 1) - \lambda = 0,$$
$$2(x_2 - 2) + \lambda = 0,$$
$$x_2 - x_1 - 1 = 0,$$

yielding

$$x_1^* = 1, \quad x_2^* = 2, \quad \text{and} \quad \lambda^* = 0.$$

But this solution does not satisfy the inequality (4.35), i.e.,

$$1+2-2\leqslant 0, \qquad \text{or} \quad 1\leqslant 0.$$

Next let $\mu\neq 0$ and solve (4.33), (4.34), (4.37), and (4.38) simultaneously:

$$x_1+x_2-2=0,$$

$$x_2-x_1-1=0,$$

$$2(x_1-1)-\lambda+\mu=0,$$

$$2(x_2-2)+\lambda+\mu=0,$$

yielding

$$x_1^*=\tfrac{1}{2}, \qquad x_2^*=\tfrac{3}{2}, \qquad \mu^*=1, \quad \lambda^*=0,$$

and

$$f(x_1^*,x_2^*)=\tfrac{1}{2}.$$

Checking (4.32) and (4.36), we see that

$$x_1^*\geqslant 0, \quad x_2^*\geqslant 0, \qquad \text{and} \quad \mu\geqslant 0,$$

so the solution is obtained. The solution and the feasible region are shown in Fig. 4.4.

4.6 Consider a curve-fitting problem where we wish to determine a set of parameters $\mathbf{x}=[x_1,x_2,\ldots,x_m]$ such that the sum of the squared errors between our model and the observed data is minimized, i.e.,

$$\min_{\mathbf{x}} \sum_{i=1}^{N} (\phi_i-p_i)^2, \tag{4.40}$$

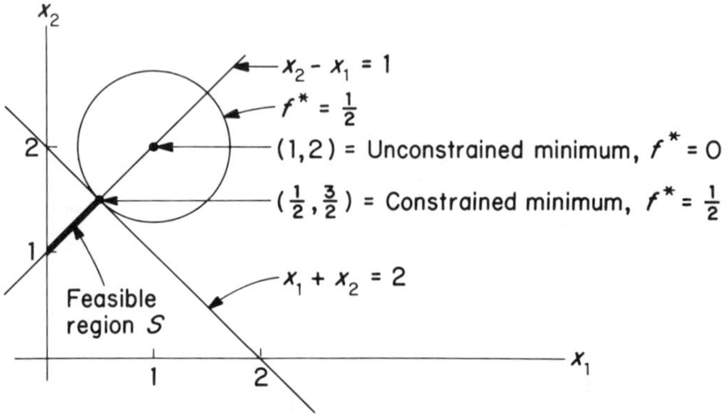

FIGURE 4.4

where

$$\phi_i = f(\mathbf{x};\mathbf{y}_i)$$

$$p_i = i\text{th observation of } \phi_i$$

$$\mathbf{y}_i = \text{an } n\text{-dimensional vector which is specified for each observation.}$$

From physical reasoning, it is known that $x_j \geqslant 0, j = 1,2,\ldots,m$.

(a) Formulate the Kuhn-Tucker stationarity conditions.

(b) Are these conditions necessary, sufficient, or both for this problem, and why?

(c) Show how to solve the Kuhn-Tucker conditions for the optimal values of $\mathbf{x}$.

Substituting for ϕ_i in (4.39) yields the nonlinear programming problem

$$\min_{\mathbf{x}} \left\{ F(\mathbf{x}) = \sum_{i=1}^{N} \left[f(\mathbf{x};\mathbf{y}_i) - p_i \right]^2 \right\},$$

such that

$$x_j \geqslant 0, \qquad j = 1,2,\ldots,m.$$

(a) From Problem 4.4, the necessary conditions are

$$\nabla_{\mathbf{x}} F(\mathbf{x}^*) \geqslant \mathbf{0},$$

$$(\mathbf{x}^*)^T \nabla_{\mathbf{x}} F(\mathbf{x}^*) = 0,$$

$$\mathbf{x}^* \geqslant \mathbf{0}.$$

Thus either

$$\sum_{i=1}^{N} 2\left[f(\mathbf{x}^*;\mathbf{y}_i) - p_i \right] \frac{\partial f}{\partial x_j}(\mathbf{x}^*) = 0, \qquad x_j^* \geqslant 0, \qquad j = 1,2,\ldots,m,$$

or

$$\sum_{i=1}^{N} 2\left[f(\mathbf{x}^*;\mathbf{y}_i) - p_i \right] \frac{\partial f}{\partial x_j}(\mathbf{x}^*) > 0, \qquad x_j^* = 0, \qquad j = 1,2,\ldots,m.$$

(b) Although the objective function is convex in $f(\mathbf{x};\mathbf{y}_i)$, it is not necessarily convex in $\mathbf{x}$ until $f(\mathbf{x};\mathbf{y}_i)$ is defined. Hence the Kuhn-Tucker conditions are *necessary* only.

(c) An algorithm for solving these equations is defined below where we denote

$$\sum_{i=1}^{N} 2[f(\mathbf{x};\mathbf{y}_i)-p_i]\frac{\partial f}{\partial x_j}(\mathbf{x})=0$$

as equation I_j.

1. Assume $x_j > 0$, $j = 1,2,\dots,m$,

2. Solve I_j for all j such that $x_j^* \neq 0$ using the Newton-Raphson or another suitable algorithm.

3. If $x_j^* \geqslant 0$, all j, we are finished. If some $x_k < 0$, set $x_k^* = 0$ and eliminate I_k from the set of simultaneous equations. Return to step 2.

Although this algorithm represents a feasible way to solve these equations, there is no guarantee of convergence for arbitrary functions $f(\mathbf{x};\mathbf{y}_i)$.

4.7 Solve the problem

$$\max x$$

such that

$$(x^2-4)^2-25 \leqslant 0.$$

Define

$$L = x + \lambda\left[(x^2-4)^2-25\right].$$

Then the stationarity conditions are

$$\frac{\partial L}{\partial x} = 1 + 2\lambda(x^2-4)2x = 0,$$

$$(x^2-4)^2-25 \leqslant 0,$$

$$\lambda \leqslant 0,$$

$$\lambda\left[(x^2-4)^2-25\right] = 0.$$

Assume $\lambda < 0$ and solve

$$1 + 4\lambda x(x^2-4) = 0, \tag{4.41}$$

$$(x^2-4)^2-25 = 0. \tag{4.42}$$

Equation (4.41) yields

$$x^2-4 = \pm 5,$$

$$x = \pm 3.$$

Substituting in Eq. (4.40) yields

$$\lambda^* = -\frac{1}{20x} \leqslant 0,$$

so

$$x^* = 3 \quad \text{and} \quad \lambda^* = -\tfrac{1}{60}.$$

Constraint Qualification

4.8 Show that if the Kuhn-Tucker stationarity conditions are satisfied, including the constraint qualification, then

$$(\nabla_{\mathbf{x}} f(\mathbf{x}^*), \mathbf{h}) \geqslant 0.$$

For the problem

$$\min_{\mathbf{x}} f(x)$$

such that

$$g_i(\mathbf{x}) \leqslant 0, \qquad i = 1, 2, \ldots, r,$$

we define

$$L(\mathbf{x}, \boldsymbol{\lambda}) = f(\mathbf{x}) + \sum_{i=1}^{r} \lambda_i g_i(\mathbf{x})$$

and

$$\nabla_{\mathbf{x}} L(\mathbf{x}^*, \boldsymbol{\lambda}^*) = \nabla_{\mathbf{x}} f(\mathbf{x}^*) + \sum_{i=1}^{r} \lambda_i^* \nabla_{\mathbf{x}} g_i(\mathbf{x}^*) = 0.$$

Consider only those constraints which are binding at $\mathbf{x}^*$ (since $\lambda_i^* = 0$ otherwise) and form the inner product

$$(\nabla_{\mathbf{x}} L(\mathbf{x}^*, \boldsymbol{\lambda}^*), \mathbf{h}),$$

or

$$(\nabla_{\mathbf{x}} f(\mathbf{x}^*), \mathbf{h}) + \sum_{i=1}^{r} \lambda_i^* (\nabla_{\mathbf{x}} g_i(\mathbf{x}^*), \mathbf{h}) = 0.$$

Then

$$(\nabla_{\mathbf{x}} f(\mathbf{x}^*), \mathbf{h}) = -\sum_{i=1}^{r} \lambda_i^* (\nabla_{\mathbf{x}} g_i(\mathbf{x}^*), \mathbf{h}).$$

But

$$\lambda_i^* \geqslant 0,$$

and by the constraint qualifications,

$$(\nabla_{\mathbf{x}} g_i(\mathbf{x}^*), h) < 0,$$

so

$$(\nabla_{\mathbf{x}} f(\mathbf{x}^*), \mathbf{h}) \geqslant 0.$$

4.9 Show that the constraint qualification is not satisfied at the optimum for the problem

$$\max x_1$$

such that

$$g_1(x_1, x_2) = (x_2 - 2) + (x_1 - 1)^3 \leqslant 0,$$

$$g_2(x_1, x_2) = (x_1 - 1)^3 - (x_2 - 2) \leqslant 0,$$

$$x_1 \geqslant 0, \qquad x_2 \geqslant 0.$$

From Fig. 4.5, the solution is (1,2) and

$$\nabla_{\mathbf{x}} g_1(1,2) = \begin{bmatrix} 3(x_1 - 1)^2 \\ 1 \end{bmatrix}_{1,2} = \begin{bmatrix} 0 \\ 1 \end{bmatrix},$$

$$\nabla_{\mathbf{x}} g_2(1,2) = \begin{bmatrix} 3(x_1 - 1)^2 \\ -1 \end{bmatrix}_{1,2} = \begin{bmatrix} 0 \\ -1 \end{bmatrix}.$$

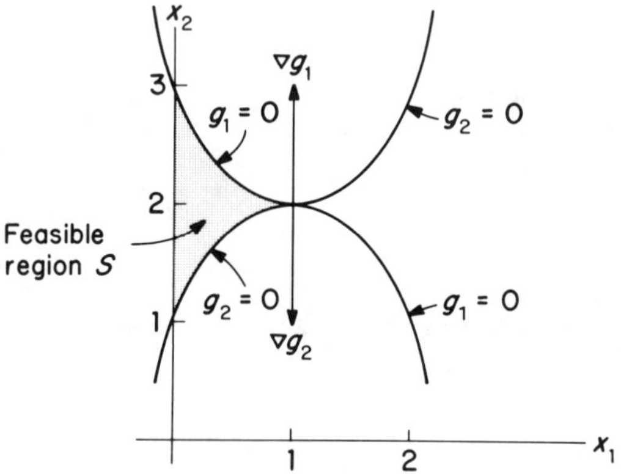

FIGURE 4.5

The constraint qualification requires

$$[0,1]\begin{bmatrix} h_1 \\ h_2 \end{bmatrix} < 0 \quad \text{and} \quad [0,-1]\begin{bmatrix} h_1 \\ h_2 \end{bmatrix} < 0,$$

or

$$h_2 < 0 \quad \text{and} \quad -h_2 < 0,$$

which is clearly impossible, so the constraint qualification fails, and therefore the stationarity conditions fail. The reader should verify this.

4.10 Examine the constraint qualification and the inequality of Problem 4.8 for

$$\min x_1$$

such that

$$g_1(x_1,x_2) = x_2 - (x_1 - 2) \leqslant 0,$$

$$g_2(x_1,x_2) = -x_2 - (x_1 - 2) \leqslant 0.$$

From Fig. 4.6 we see that the optimal solution is $(2,0)$ and

$$\nabla_{\mathbf{x}} g_1(2,0) = \begin{bmatrix} -1 \\ 1 \end{bmatrix},$$

$$\nabla_{\mathbf{x}} g_2(2,0) = \begin{bmatrix} -1 \\ -1 \end{bmatrix},$$

$$\nabla_{\mathbf{x}} f(2,0) = \begin{bmatrix} 1 \\ 0 \end{bmatrix}.$$

In order to satisfy the constraint qualification, we must have

$$\|\nabla_{\mathbf{x}} g_1\| \, \|\mathbf{h}\| \cos\theta_1 < 0$$

and

$$\|\nabla_{\mathbf{x}} g_2\| \, \|\mathbf{h}\| \cos\theta_2 < 0,$$

i.e.,

$$90° < \theta_1 < -90°,$$

$$90° < \theta_2 < -90°,$$

which are both satisfied by any $\mathbf{h}$ strictly inside the feasible region S. Thus the constraint qualification is satisfied. Now

$$\|\nabla_{\mathbf{x}} f\| \, \|\mathbf{h}\| \cos\xi \leqslant 0$$

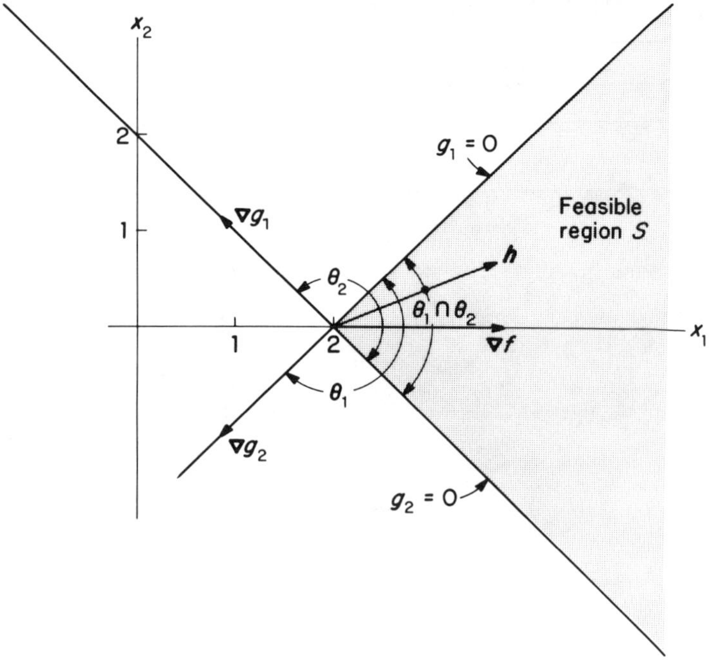

FIGURE 4.6

requires that

$$-90° \leqslant \xi \leqslant 90°,$$

and clearly every $\mathbf{h}$ strictly inside S also satisfies this inequality.

Saddle-Point Conditions

4.11 Prove Theorem 4.1, p. 90.

By definition of a saddle point,

$$L(\mathbf{x}^*, \boldsymbol{\lambda}) \leqslant L(\mathbf{x}^*, \boldsymbol{\lambda}^*) \leqslant L(\mathbf{x}, \boldsymbol{\lambda}^*).$$

We first prove the necessity that if $(\mathbf{x}^*, \boldsymbol{\lambda}^*)$ is a saddle point of $L(\mathbf{x}, \boldsymbol{\lambda})$ with $\boldsymbol{\lambda}^* \geqslant 0$, then conditions 1 to 3 hold.

The right-hand inequality above yields condition 1 directly:

$$\mathbf{x}^* \text{ minimizes } L(\mathbf{x}, \boldsymbol{\lambda}^*) \quad \text{over all } \mathbf{x}.$$

Expanding the left-hand inequality gives

$$f(\mathbf{x}^*)+\sum_{i=1}^{r}\lambda_i g_i(\mathbf{x}^*)\leqslant f(\mathbf{x}^*)+\sum_{i=1}^{r}\lambda_i^* g_i(\mathbf{x}^*),$$

or

$$\sum_{i=1}^{r}(\lambda_i-\lambda_i^*)\,g_i(\mathbf{x}^*)\leqslant 0$$

for all $\boldsymbol{\lambda}\geqslant\mathbf{0}$. Now if any $g_i(\mathbf{x}^*)$ is positive, we can select an arbitrarily large λ_i such that the inequality is violated. Hence condition 2 must hold, i.e.,

$$g_i(\mathbf{x}^*)\leqslant 0,\qquad i=1,2,\dots,r.$$

The above inequality must hold for $\boldsymbol{\lambda}=\mathbf{0}$, yielding

$$-\sum_{i=1}^{r}\lambda_i^* g_i(\mathbf{x}^*)\leqslant 0,$$

or equivalently

$$\sum_{i=1}^{r}\lambda_i^* g_i(\mathbf{x}^*)\geqslant 0.$$

However, $\lambda_i^*\geqslant 0$ and $g_i(\mathbf{x}^*)\leqslant 0$ require that

$$\sum_{i=1}^{r}\lambda_i^* g_i(\mathbf{x}^*)\leqslant 0,$$

and the only way of satisfying both of these inequalities is to require

$$\lambda_i^* g_i(\mathbf{x}^*)=0,\qquad i=1,2,\dots,r,$$

which is condition 3.

To show sufficiency, we must prove that these conditions imply that $L(\mathbf{x},\boldsymbol{\lambda})$ has a saddle point at $(\mathbf{x}^*,\boldsymbol{\lambda}^*)$.

Condition 1 implies immediately that

$$L(\mathbf{x}^*,\boldsymbol{\lambda}^*)\leqslant L(\mathbf{x},\boldsymbol{\lambda}^*)$$

for all $\mathbf{x}$, which is half of the saddle-point definition. The other half is obtained by expanding

$$L(\mathbf{x}^*,\boldsymbol{\lambda})=f(\mathbf{x}^*)+\sum_{i=1}^{r}\lambda_i g_i(\mathbf{x}^*).$$

Then from condition 2 and $\lambda_i\geqslant 0$,

$$\lambda_i g_i(\mathbf{x}^*)\leqslant 0,$$

and so

$$L(\mathbf{x}^*,\boldsymbol{\lambda}) \leqslant f(\mathbf{x}^*).$$

Now from condition 3,

$$f(\mathbf{x}^*) = L(\mathbf{x}^*,\boldsymbol{\lambda}^*),$$

and the sufficiency is proven.

4.12 Prove Theorem 4.2, page 90.

If $(\mathbf{x}^*,\boldsymbol{\lambda}^*)$ is a saddle point of the Lagrangian, then Theorem 4.1 holds and the inequality constraints of the primal problem are satisfied at $\mathbf{x}^*$. It only remains to show that

$$f(\mathbf{x}^*) \leqslant f(\mathbf{x}) \qquad \text{for all } \mathbf{x}$$

such that

$$g_i(\mathbf{x}) \leqslant 0, \qquad i=1,2,\ldots,r.$$

From the saddle-point definition,

$$f(\mathbf{x}^*) + \sum_{i=1}^{r} \lambda_i^* g_i(\mathbf{x}^*) \leqslant f(\mathbf{x}) + \sum_{i=1}^{r} \lambda_i^* g_i(\mathbf{x})$$

for all $\mathbf{x}$. By condition 3 of Theorem 4.1 the left-hand side reduces to $f(\mathbf{x}^*)$. Now for any $\mathbf{x}$ which satisfies $g_i(\mathbf{x}) \leqslant 0$, since $\lambda_i^* \geqslant 0$, $i=1,2,\ldots,r$,

$$\sum_{i=1}^{r} \lambda_i^* g_i(\mathbf{x}) \leqslant 0.$$

Thus the inequality can be written

$$f(\mathbf{x}^*) \leqslant f(\mathbf{x})$$

for all $\mathbf{x}$ such that $g_i(\mathbf{x}) \leqslant 0$, $i=1,2,\ldots,r$.

4.13 Are the Kuhn-Tucker conditions satisfied at $x=6$ for the following problem? Does the Lagrangian have a saddle point there?

$$\max_x (x-4)^2$$

such that

$$1 \leqslant x \leqslant 6.$$

Write

$$g_1(x) = x-6 \leqslant 0,$$

$$g_2(x) = 1-x \leqslant 0.$$

Then

$$L(x,\lambda_1,\lambda_2)=(x-4)^2+\lambda_1(x-6)+\lambda_2(1-x),$$

and the stationarity conditions are

$$\frac{\partial L}{\partial x}=2(x-4)+\lambda_1-\lambda_2=0,$$

$$x\leqslant 6, \qquad x\geqslant 1,$$

$$\lambda_1\leqslant 0, \qquad \lambda_2\leqslant 0,$$

$$\lambda_1(x-6)=0,$$

$$\lambda_2(1-x)=0.$$

At $x=6$, we have $\lambda_1=-4$ and $\lambda_2=0$, and the stationarity conditions are satisfied.

We now examine the nature of the Lagrangian at $(6,-4,0)$. Holding (λ_1,λ_2) fixed, the Hessian with respect to x is

$$\frac{\partial^2 L}{\partial x^2}(6,-4,0)=2>0,$$

which implies that $L(6,-4,0)$ has a minimum with respect to x (instead of a maximum as required).

When the primal problem is to be maximized, the dual function is defined as

$$h(\boldsymbol{\lambda})=\max_x L(x,\boldsymbol{\lambda}).$$

Since $L(x,\lambda_1,\lambda_2)$ does not possess a maximum at $(6,-4,0)$, the dual function does not exist at this stationarity point, and therefore no saddle point exists either.

Thus the stationarity conditions are satisfied at $x=6$, but the Lagrangian does not have a saddle point there. The reason for this phenomenon is that we are asked to minimize a convex function. The stationarity conditions, being necessary conditions for a local maximum only, are satisfied, while the saddle-point conditions, being sufficient, are not.

4.14 For Problem 4.5, show that

$$L(x_1^*,x_2^*,\lambda,\mu)\leqslant L(x_1^*,x_2^*,\lambda^*,\mu^*)\leqslant L(x_1,x_2,\lambda^*,\mu^*).$$

From Problem 4.5,

$$L(x_1,x_2,\lambda,\mu)=(x_1-1)^2+(x_2-2)^2+\lambda(x_2-x_1-1)+\mu(x_1+x_2-2)$$

and

$$x_1^*=\tfrac{1}{2}, \qquad x_2^*=\tfrac{3}{2}, \qquad \lambda^*=0, \qquad \mu^*=1.$$

Then

$$L(x_1^*,x_2^*,\lambda^*,\mu^*)=\left(-\tfrac{1}{2}\right)^2+\left(-\tfrac{1}{2}\right)^2=\tfrac{1}{2}$$

and

$$L(x_1,x_2,\lambda^*,\mu^*)=(x_1-1)^2+(x_2-2)^2+x_1+x_2-2,$$

$$\frac{\partial L}{\partial x_1}=2(x_1-1)+1=0,$$

$$\frac{\partial L}{\partial x_2}=2(x_2-2)+1=0.$$

Therefore, $x_1^*=\frac{1}{2}$ and $x_2^*=\frac{3}{2}$.

The Hessian matrix of the function $L(x_1,x_2,\lambda^*,\mu^*)$ is

$$\mathbf{H}=\begin{bmatrix}2 & 0\\ 0 & 2\end{bmatrix},$$

which is positive definite, so $(\frac{1}{2},\frac{3}{2})$ is the absolute minimum of $L(x_1,x_2,\lambda^*,\mu^*)$ and

$$L(x_1^*,x_2^*,\lambda^*,\mu^*)\leqslant L(x_1,x_2,\lambda^*,\mu^*).$$

Now determine the dual function

$$h(\lambda,\mu)=\min_{\mathbf{x}} L(x_1,x_2,\lambda,\mu).$$

we have

$$\frac{\partial L}{\partial x_1}=2(x_1-1)-\lambda+\mu=0,$$

so

$$x_1=1+\frac{1}{2}(\lambda-\mu),$$

and

$$\frac{\partial L}{\partial x_2}=2(x_2-1)+\lambda+\mu=0,$$

so

$$x_2=2-\tfrac{1}{2}(\lambda+\mu).$$

Substituting into $L(x_1,x_2,\lambda,\mu)$, we have

$$h(\lambda,\mu)=-\tfrac{1}{2}(\lambda^2+\mu^2)+\mu,$$

$$\frac{\partial h}{\partial \lambda}=-\lambda=0,\qquad \frac{\partial h}{\partial \mu}=-\mu+1=0,$$

yielding $\lambda^*=0$ and $\mu^*=1$. The Hessian matrix of $h(\lambda,\mu)$ is

$$\mathbf{H}=\begin{bmatrix} -1 & 0 \\ 0 & -1 \end{bmatrix},$$

and since $-\mathbf{H}$ is positive definite, $h(\lambda,\mu)$ has a maximum at $\lambda^*=0$, $\mu^*=1$. Thus

$$h(\lambda,\mu)=L(x_1^*,x_2^*,\lambda,\mu)\leqslant L(x_1^*,x_2^*,\lambda^*,\mu^*)=h(\lambda^*,\mu^*).$$

DUALITY

4.15 Prove the duality theorem (Theorem 4.4, p. 92).

To prove necessity, assume that $L(\mathbf{x},\boldsymbol{\lambda})$ has a saddle point at $(\mathbf{x}^*,\boldsymbol{\lambda}^*)$. Then by Theorem 4.1, $\mathbf{x}^*$ solves the primal problem, which is condition 1.

By definition, $h(\lambda)=\min_x L(\mathbf{x},\boldsymbol{\lambda})$, so

$$h(\boldsymbol{\lambda}^*)=f(\mathbf{x}^*)+\sum_{i=1}^{r}\lambda_i^* g_i(\mathbf{x}^*).$$

But

$$\lambda_i^* g_i(\mathbf{x}^*)=0, \qquad i=1,2,\ldots,r$$

by Theorem 4.1 so

$$h(\boldsymbol{\lambda}^*)=f(\mathbf{x}^*)$$

and condition 3 holds. Now by Theorem 4.3,

$$h(\boldsymbol{\lambda})\leqslant f(\mathbf{x}^*)=h(\boldsymbol{\lambda}^*) \qquad \text{for all} \quad \boldsymbol{\lambda}\in D.$$

Thus $\boldsymbol{\lambda}^*$ maximizes $h(\boldsymbol{\lambda})$ over D, which is condition 2 and completes the proof of necessity.

To prove sufficiency, we must prove that conditions 1–3 imply that $L(\mathbf{x},\boldsymbol{\lambda})$ has a saddle point at $(\mathbf{x}^*,\boldsymbol{\lambda}^*)$, or equivalently that conditions 1–3 of Theorem 4.1 hold.

Since $\mathbf{x}^*$ solves the primal problem, $g_i(\mathbf{x}^*)\leqslant 0$ and $\lambda_i^* g_i(\mathbf{x}^*)=0$, $i=1,2,\ldots,r$, which are conditions 2 and 3 of Theorem 4.1. Since $\boldsymbol{\lambda}^*$ solves the dual problem, $\lambda_i^*\geqslant 0$, $i=1,2,\ldots,r$, which is a hypothesis in Theorem 4.1. It now remains to demonstrate that condition 1 holds, i.e., that $\mathbf{x}^*$ minimizes $L(\mathbf{x},\boldsymbol{\lambda}^*)$.

To show this, we first assume the contrary, i.e., that there exists some $\hat{\mathbf{x}}\neq\mathbf{x}^*$ such that

$$L(\hat{\mathbf{x}},\boldsymbol{\lambda}^*)<L(\mathbf{x}^*,\boldsymbol{\lambda}^*).$$

By definition,

$$h(\boldsymbol{\lambda}^*)=L(\hat{\mathbf{x}},\boldsymbol{\lambda}^*)=f(\hat{\mathbf{x}})+\sum_{i=1}^{r}\lambda_i^* g_i(\hat{\mathbf{x}})<f(\mathbf{x}^*)+\sum_{i=1}^{r}\lambda_i^* g_i(\mathbf{x}^*).$$

But $h(\boldsymbol{\lambda}^*)=f(\mathbf{x}^*)$ by hypothesis, so

$$\sum_{i=1}^{r} \lambda_i^* g_i(\mathbf{x}^*) > 0.$$

However, since $\mathbf{x}^*$ solves the primal problem, $\lambda_i^* g_i(\mathbf{x}^*)=0$, which leads to a contradiction. Hence $\hat{\mathbf{x}}=\mathbf{x}^*$, and $\mathbf{x}^*$ minimizes $L(\mathbf{x},\boldsymbol{\lambda}^*)$, which is condition 1 of Theorem 4.1, and sufficiency is proved.

4.16 Prove Theorem 4.5, p. 93.

Let D^* be a convex subset of D, and consider any two points in D^*, say $\boldsymbol{\lambda}^1$ and $\boldsymbol{\lambda}^2$. Then any point $\boldsymbol{\lambda}$ on the line connecting these points can be written

$$\boldsymbol{\lambda}=\alpha\boldsymbol{\lambda}^1+(1-\alpha)\boldsymbol{\lambda}^2, \qquad 0\leqslant\lambda\leqslant 1,$$

and $\boldsymbol{\lambda}\in D^*$, since D^* is convex.

By definition,

$$h(\boldsymbol{\lambda})=\min_{\mathbf{x}} L(\mathbf{x},\boldsymbol{\lambda})$$

but $L(\mathbf{x},\boldsymbol{\lambda})$ is linear in $\boldsymbol{\lambda}$, so

$$h(\boldsymbol{\lambda})=\min_{\mathbf{x}}\left[\alpha L(\mathbf{x},\boldsymbol{\lambda}^1)+(1-\alpha)L(\mathbf{x},\boldsymbol{\lambda}^2)\right], \qquad 0\leqslant\alpha\leqslant 1.$$

But the minimum of a sum is greater than or equal to the sum of the minima, so

$$\min_{\mathbf{x}}\left[\alpha L(\mathbf{x},\boldsymbol{\lambda}^1)+(1-\alpha)L(\mathbf{x},\boldsymbol{\lambda}^2)\right]\geqslant\alpha\min_{\mathbf{x}} L(\mathbf{x},\boldsymbol{\lambda}^1)+(1-\alpha)\min_{\mathbf{x}} L(\mathbf{x},\boldsymbol{\lambda}^2).$$

Hence

$$h(\boldsymbol{\lambda})\geqslant\alpha h(\boldsymbol{\lambda}^1)+(1-\alpha)h(\boldsymbol{\lambda}^2),$$

which implies that $h(\boldsymbol{\lambda})$ is concave over D^*.

4.17 Solve the following problem P and formulate the dual problem D. Also verify that $\min P=\max D$.

$$P\colon\ \min(-x^2-x^3) \qquad \text{such that} \quad x^2\leqslant 1.$$

Since the constraint implies that $x^*\in[-1,1]$, a graphical solution $x^*=1$ is easily obtained from Fig. 4.7. Now, form the Lagrangian

$$L(x,\lambda)=-x^2-x^3+\lambda(x^2-1).$$

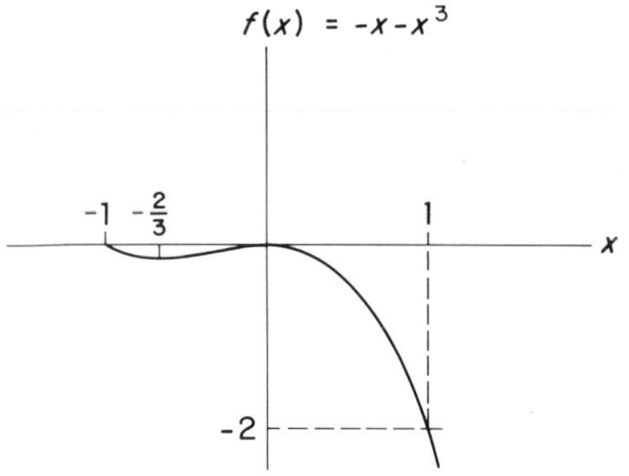

FIGURE 4.7

The stationarity conditions are

$$\frac{\partial L}{\partial x} = -2x - 3x^2 + \lambda(2x) = 0, \tag{4.43}$$

$$x^2 \leqslant 1,$$

$$\lambda \geqslant 0,$$

$$\lambda(x^2 - 1) = 0.$$

If $\lambda = 0$, $x(-2-3x) = 0$; so $x = -\frac{2}{3}$ or $x = 0$.
If $x = 1$, $\lambda = \frac{5}{2}$.
If $x = -1$, $\lambda = -\frac{1}{2}$, which does not satisfy $\lambda \geqslant 0$.
Thus there are three possible solutions,

$$x = -\tfrac{2}{3}, \qquad x = 0, \qquad x = 1$$

where

$$L\left(-\tfrac{2}{3}, 0\right) = -\tfrac{4}{27}, \qquad L(0,0) = 0, \qquad L\left(1, \tfrac{5}{2}\right) = -2,$$

so $x^* = 1$ is the solution which minimizes $L(x, \lambda^*)$, and this agrees with the graphical solution.

From (4.42) for $x \neq 0$,

$$x = \tfrac{2}{3}(\lambda - 1).$$

Substituting in $L(x, \lambda)$ yields

$$h(\lambda) = \tfrac{4}{27}\lambda^3 - \tfrac{4}{9}\lambda^2 - \tfrac{5}{9}\lambda - \tfrac{4}{27}.$$

The dual problem is

$$\max h(\lambda)$$

such that

$$\lambda \geqslant 0.$$

Thus

$$\frac{\partial h}{\partial \lambda} = \tfrac{4}{9}\lambda^2 - \tfrac{8}{9}\lambda - \tfrac{5}{9} = 0,$$

from which

$$\lambda = -\tfrac{1}{2} \quad \text{or} \quad \lambda = \tfrac{5}{2}.$$

Only the positive value satisfies the constraint, so $\lambda^* = \frac{5}{2}$. Now

$$h\left(\tfrac{5}{2}\right) = \tfrac{4}{27}\left(\tfrac{125}{8}\right) - \tfrac{4}{9}\left(\tfrac{25}{4}\right) - \tfrac{5}{9}\left(\tfrac{5}{2}\right) - \tfrac{4}{27} = -2.$$

and

$$f(1) = -1^2 - 1^3 = -2.$$

Supplementary Problems

4.18 Formulate the dual problem and show that $f(\mathbf{x}^*) = h(\boldsymbol{\lambda}^*)$ when the primal is

$$\min_{x_1, x_2} (x_1 - 1)^2 + (x_2 - 2)^2$$

such that

$$x_2 - x_1 = 1,$$

$$x_1 + x_2 \leqslant 2,$$

$$x_1 \geqslant 0, \qquad x_2 \geqslant 0.$$

4.19 Find the minimum distance from the origin to the convex set

$$x_1 + x_2 \geqslant 4,$$

$$2x_1 + x_2 \geqslant 5.$$

Solve graphically and show that the stationarity conditions hold at the minimum.

4.20 Apply the Kuhn-Tucker theory to the linear programming problem

$$\min_{\mathbf{x}} \mathbf{c}^T\mathbf{x}$$

such that

$$\mathbf{Ax} \geqslant \mathbf{b},$$

$$\mathbf{x} \geqslant 0.$$

4.21 Write the Lagrangian and obtain the dual problem for Problem 4.20.

4.22 Investigate whether the Kuhn-Tucker stationarity conditions hold at the optimizing point for the problem

$$\max x$$

such that

$$-x^2 \geqslant 0.$$

4.23 For Problem 4.22, show that the Lagrangian does not have a saddle point at the optimum. Why not?

4.24 Are the Kuhn-Tucker conditions satisfied for the following problem?

$$\min_{x_1, x_2} (x_1 - 3)^2 + x_2$$

such that

$$-x_1 + (x_2 - 1)^2 \geqslant 0, \qquad x_1 \geqslant 0, \qquad x_2 \geqslant 0.$$

4.25 Is the constraint qualification satisfied in Problem 4.24?

4.26 Solve the following problem using the Kuhn-Tucker theory

$$\max_{x_1, x_2} 25(x_1 - 2)^2 + (x_2 - 2)^2$$

such that

$$x_1 + x_2 \geqslant 2,$$

$$x_1 - x_2 \geqslant -3,$$

$$x_1 + x_2 \leqslant 6,$$

$$x_1 - 3x_2 \leqslant 2,$$

$$x_1 \geqslant 0, \qquad x_2 \geqslant 0.$$

4.27 Solve Problem 4.26 graphically.

4.28 Use the stationarity conditions to determine a range of values for a_1 over which the solution $x_1 = 4$, $x_2 = 3$ is optimal

$$\max_{x_1, x_2} a_1 x_1 + x_2$$

such that

$$x_1^2 + x_2^2 \leqslant 25,$$

$$x_1 - x_2 \leqslant 1,$$

$$x_1 \geqslant 0, \qquad x_2 \geqslant 0.$$

4.29 Show the results of Problem 4.28 graphically.

4.30 Prove that the minimum of a sum is greater than or equal to the sum of the minima (Hint: see Problem 4.16).

Chapter 5

One-Dimensional Search Techniques

In this chapter we consider the simplest type of function minimization problem. Letting $f(x)$ denote a real-valued function of a *single*, real variable x, we assume that $f(x)$ has a minimum at x^* in some interval $[a^0,b^0]$. The one-dimensional problem is to develop algorithms for finding x^* such that at the nth iteration we have reduced the initial interval to a smaller interval $[a^n,b^n]$ containing x^*. Thus $[a^n,b^n]\subset[a^0,b^0]$ and $x^*\in[a^n,b^n]$. The interval $[a^n,b^n]$ containing x^* is called the *interval of uncertainty* after the nth iteration and will be denoted by the symbol L_n.

For problems of this type, the functional form of $f(x)$ may not be known. Hence the algorithms to be discussed here require only a point-by-point evaluation of $f(x)$ in the interval $[a,b]$. In order to compare the efficiency of various methods, we are especially interested in the number of function evaluations or experimental trials required to obtain a specified reduction in the interval of uncertainty. In particular, we shall seek the method which obtains the most efficient use of these experiments.

Note that these experiments consist of function evaluations only; in particular, derivatives of $f(x)$ need not be computed. However, the applicability of these algorithms is limited to functions which are unimodal in $[a^0,b^0]$.

Unimodal Functions

Assuming that $f(x)$ has a minimum in the interval of uncertainty, $f(x)$ is *unimodal* on $[a^0,b^0]$ if, for every $x\in[a^0,b^0]$, $f(x)$ decreases as x tends to x^*. Simply stated, a function having only one local minimum is unimodal. Two examples of functions which are unimodal on the interval $[a^0,b^0]$ are shown in Fig. 5.1. Note that we have not required the continuity or differentiability of the function $f(x)$.

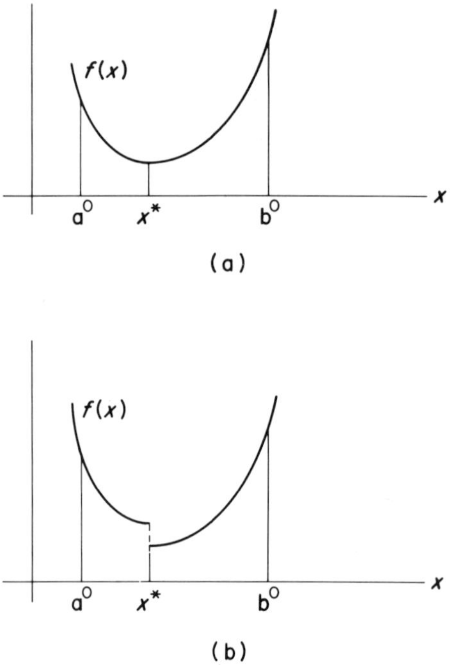

FIGURE 5.1

Lemma 5.1 *Suppose $f(x)$ is unimodal in $[a^0,b^0]$ and attains a minimum at $x^* \in [a^0,b^0]$. Then in order to obtain a subinterval of $[a^0,b^0]$ containing x^* it is necessary to evaluate $f(x)$ at least twice in $[a^0,b^0]$.*

Proof: Let us choose x_1 such that $a^0 < x_1 < b^0$ and compute $f(x_1)$ (see Fig. 5.2). Then either $x^* \in [a^0,x_1]$ or $x^* \in [x_1,b^0]$. Now we choose $x_2 > x_1$ and compute $f(x_2)$. If $f(x_2) > f(x_1)$, then $x^* \in [a^0,x_2]$; otherwise $x^* \in [x_1,b^0]$. ■

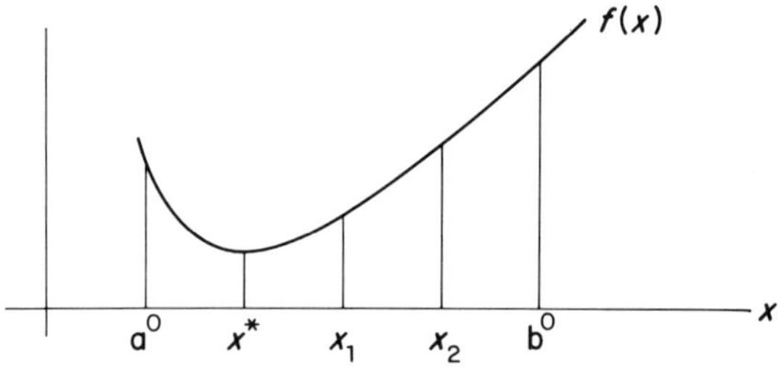

FIGURE 5.2

Dichotomous Search

The *dichotomous search* algorithm is the simplest algorithm for reducing the length of the interval of uncertainity containing x^*. Let c^n be the midpoint of the interval $[a^n, b^n]$, i.e., $c^n = (a^n + b^n)/2$; we then choose a small number $\varepsilon > 0$ and set $x_1^n = c^n - \varepsilon/2$ and $x_2^n = c^n + \varepsilon/2$. If $f(x_2^n) > f(x_1^n)$, then according to Lemma 5.1, $a^{n+1} = a^n$ and $b^{n+1} = x_2^n = c^n + \varepsilon/2$, as illustrated in Fig. 5.3. Otherwise $a^{n+1} = x_1^n = c^n - \varepsilon/2$ and $b^{n+1} = b^n$. In theory we can choose ε to be the smallest interval for which $f(x)$ and $f(x+\varepsilon)$ are distinguishable. Evidently, in each iteration, the length of the interval of uncertainity is reduced by a factor of approximately 0.5. A more detailed analysis of this reduction is given in Problem 5.1.

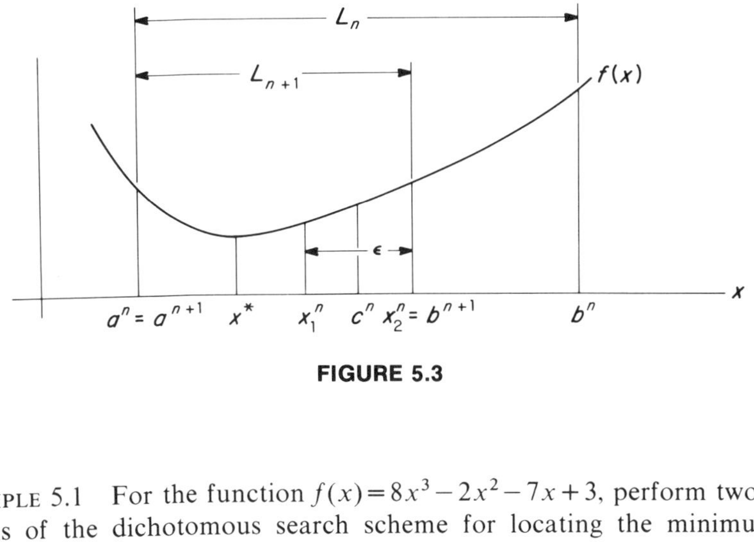

FIGURE 5.3

EXAMPLE 5.1 For the function $f(x) = 8x^3 - 2x^2 - 7x + 3$, perform two iterations of the dichotomous search scheme for locating the minimum in $[0, 1]$.

We can analytically verify that in the interval $[-1, 1]$, $f(x)$ has a maximum of 5 at $x = -0.464$ and a minimum of -0.2 at $x = 0.63$. The graph of $f(x)$ is shown in Fig. 5.4.

We arbitrarily choose $\varepsilon = 0.1$ and set $a^0 = 0$, $b^0 = 1$. Then $c^0 = 0.5$, $x_1^0 = 0.45$, and $x_2^0 = 0.55$. By direct computation we obtain $f(0.45) = 0.52$ and $f(0.55) = -0.124$. Since $f(x_1^0) > f(x_2^0)$, we set $a^1 = x_1^0 = 0.45$ and $b^1 = b^0 = 1$. Then $c^1 = 0.725$, $x_1^1 = 0.675$, and $x_2^1 = 0.775$. Once again, by direct computation $f(0.675) = -0.17$ and $f(0.775) = 0.076$. Since $f(x_2^1) > f(x_1^1)$, we set $a^2 = a^1 = 0.45$ and $b^2 = x_2^1 = 0.775$.

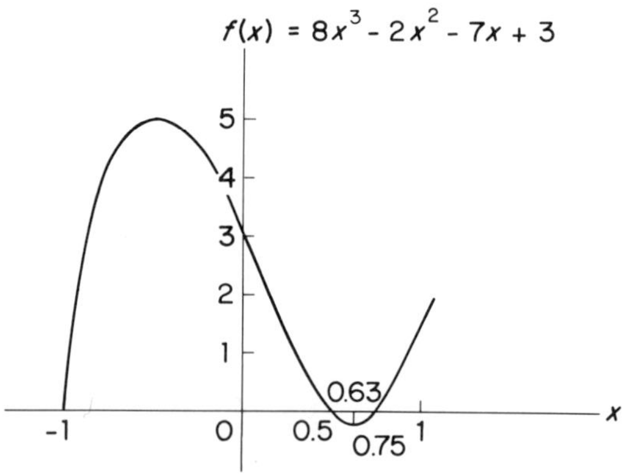

FIGURE 5.4

Equal-Interval Search

In the dichotomous search, we chose x_1^n and x_2^n to be symmetric with respect to the midpoint of $[a^n, b^n]$. However, this is not the only possible choice; in fact, the choice of x_1^n and x_2^n in $[a^n, b^n]$ is perfectly arbitrary. For example, we can choose x_1^n and x_2^n so as to divide $[a^n, b^n]$ into three equal subintervals. We can also use more than two points in each iteration. Thus we can divide $[a^n, b^n]$ into any number of subintervals of equal length. Of course, this will necessitate more function evaluations in each iteration; however, we can economize on function evaluations by a proper choice of the points in the interval. This economy results from the fact that some of the function values computed in one iteration can also be used in the next iteration. In the dichotomous search, this economy was not present, since the values of $f(x_1^n)$ and $f(x_2^n)$ were not used in the $(n+1)$st iteration.

For the simplest equal-interval search scheme, we choose two points $x_2^n > x_1^n$ to divide $[a^n, b^n]$ into three subintervals of equal length. This method is called the *two-point equal-interval* search. The procedure is illustrated in Fig. 5.5, where $x_1^n = a^n + (b^n - a^n)/3$ and $x_2^n = b^n - (b^n - a^n)/3$. According to Lemma 5.1, $b^{n+1} = x_2^n$ and $a^{n+1} = a^n$. Note that $x_1^n \in [a^{n+1}, b^{n+1}]$, but still $f(x_1^n)$ is not useful in computing the $(n+1)$st iteration. Also $b^{n+1} - a^{n+1} = 2(b^n - a^n)/3$.

For a more economical use of the function values consider the *three-point equal-interval* search scheme shown in Fig. 5.6. In Fig. 5.6 we choose the points x_1^n, x_2^n, and x_3^n to divide $[a^n, b^n]$ into four subintervals of equal length, and then compute $f(x_1^n)$, $f(x_2^n)$, and $f(x_3^n)$. We retain that subinterval whose midpoint corresponds to the minimum of these three function

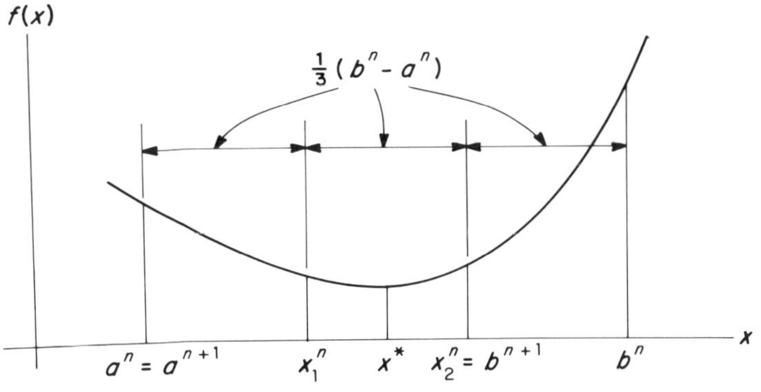

FIGURE 5.5

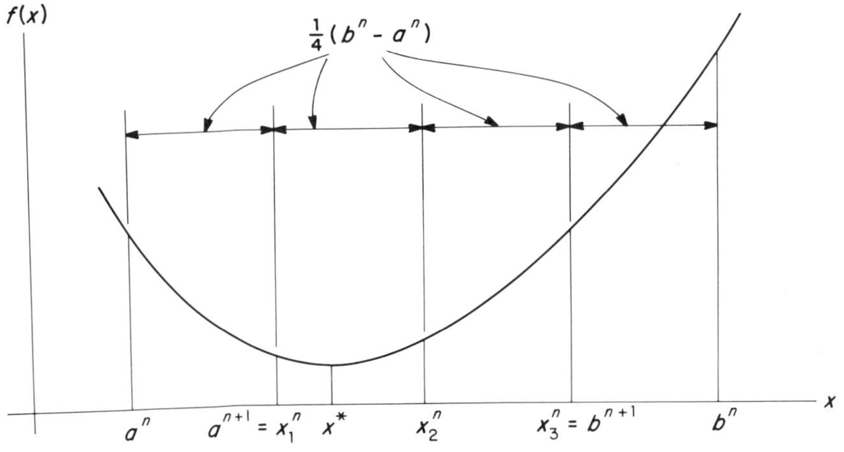

FIGURE 5.6

values. Thus in Fig. 5.6, $f(x_2^n) = \min\,[f(x_1^n), f(x_2^n), f(x_3^n)]$. Hence $a^{n+1} = x_1^n$ and $b^{n+1} = x_3^n$. Note that $x_2^n \in [a^{n+1}, b^{n+1}]$ and is in fact the midpoint of $[a^{n+1}, b^{n+1}]$. Therefore, $f(x_2^n)$ can be used again in the $(n+1)$st iteration, which results in a more economical use of function values than in the two-point method. More details of this algorithm are discussed in Problem 5.3.

We can continue to divide $[a^n, b^n]$ into more and more subintervals of equal length. However, it can be shown that the three-point equal-interval search scheme is the best as far as the efficient use of function evaluations is concerned.

EXAMPLES

5.2 Consider $f(x)$ as given in Example 5.1 in the interval [0, 1]. Apply the three-point equal-interval search scheme and compare the results with those of Example 5.1.

In this case $a^0=0$ and $b^0=1$. Hence we choose $x_1^0=0.25$, $x_2^0=0.5$, and $x_3^0=0.75$. By direct computation $f(0.25)=1.24$, and $f(0.5)=f(0.75)=0$. Thus we can set either $a^1=0.5$ and $b^1=1$ or $a^1=0.25$ and $b^1=0.75$, and both these intervals contain x^*. In this case $b^1-a^1=0.50$, compared with $b^1-a^1=1-0.45=0.55$ in Example 5.1. We can further reduce the length of the interval containing x^* by noting that $f(x)$ is unimodal and $f(0.5)=f(0.75)<f(0.25)$. Thus we can set $a^1=0.5$ and $b^1=0.75$, which results in $b^1-a^1=0.25$ rather than 0.5. However, in this case we must evaluate $f(x)$ three times in the next iteration.

5.3 For the function $f(x)=e^x-5x$, perform two iterations of the three-point equal-interval search scheme for locating x^* in [1,2]. Since $d^2f/dx^2=e^x>0$, $f(x)$ is strictly convex on all finite intervals and hence is unimodal. This function actually attains a minimum at $x^*=\ln 5=1.609$.

We are given $a^0=1$ and $b^0=2$. We choose $x_1^0=1.25$, $x_2^0=1.5$, and $x_3^0=1.75$. By direct computation $f(1.25)=-2.75$, $f(1.5)=-3.0$, and $f(1.75)=-2.96$. Evidently $f(1.5)=\min[f(1.25), f(1.5), f(1.75)]$. Hence we set $a^1=1.25$ and $b^1=1.75$. Then $(b^1-a^1)/4=0.125$, and we choose $x_1^1=1.375$, $x_2^1=1.5$, and $x_3^1=1.625$. We have already computed $f(x_2^1)=f(1.5)=-3.0$. As for the other function values, we obtain by direct computation $f(1.375)=-2.95$ and $f(1.625)=-3.03$. Now $f(1.625)=\min[f(1.375), f(1.5), f(1.625)]$. Therefore, we set $a^2=1.5$ and $b^2=1.75$. We can either continue these iterations or take the midpoint of $[a^2,b^2]$, which is 1.625, as an estimate of x^*.

Fibonacci Search

So far we have discussed three different search techniques which use two new function evaluations in each iteration. According to Lemma 5.1, this is the minimum number of function evaluations needed for reducing the length of the interval of uncertainty in each iteration. In the dichotomous and the two-point equal-interval methods, no trials were retained for use in subsequent iterations, while in the three-point equal-interval method one previous trial was used with two new trials. We now seek a method which utilizes one previous trial and only one new trial to comprise the two required experiments. One such method is called the *Fibonacci search*.

Suppose we wish to place n experiments and the minimum separation is ε. Then letting L_n be the desired final interval of uncertainty after n trials and using the dichotomous search, we place the nth experiment symmetrically in the interval L_{n-1} a distance ε from the $(n-1)$st experiment, as shown in Fig. 5.7. In this figure, the cross (×) represents the old trial, while the dot (●) represents the new trial. This symmetrical placement results in an economy of experiments since only one new function evaluation is required. From Fig. 5.7 the relation between L_n and L_{n-1} is given by

$$L_{n-1} = 2L_n - \varepsilon. \tag{5.1}$$

Consider now the optimal placement when two experiments remain. From the definition of dichotomous search, the × in Fig. 5.7 must have been the minimum in the interval L_{n-2}. This situation is shown in Fig. 5.8, which results directly in the relation

$$L_{n-2} = L_{n-1} + L_n.$$

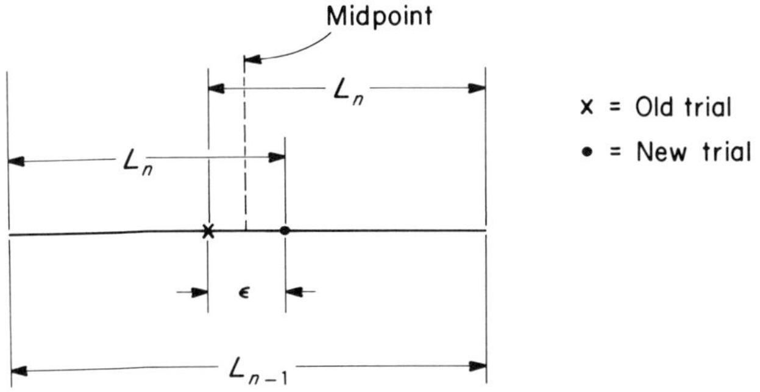

FIGURE 5.7

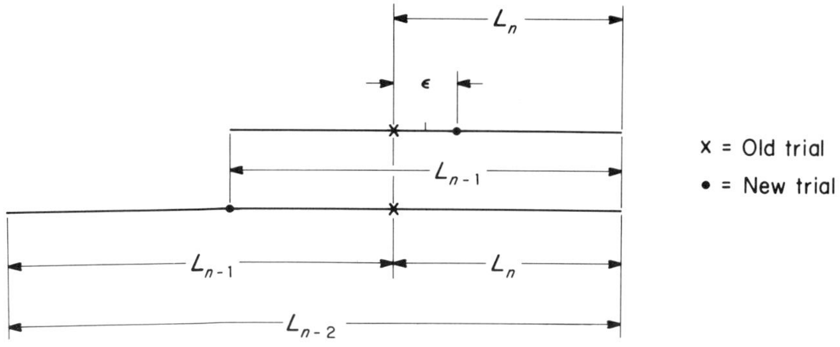

FIGURE 5.8

Similarly we can see that

$$L_{n-3}=L_{n-2}+L_{n-1}$$

or, in general,

$$L_{n-k}=L_{n-(k-1)}+L_{n-(k-2)}, \qquad k=2,3,\dots,n. \tag{5.2}$$

Using (5.1) and (5.2) we see that

$$\begin{aligned} L_{n-2}&=L_{n-1}+L_n=3L_n-\varepsilon,\\ L_{n-3}&=L_{n-2}+L_{n-1}=5L_n-2\varepsilon,\\ L_{n-4}&=L_{n-3}+L_{n-2}=8L_n-3\varepsilon,\\ L_{n-5}&=L_{n-4}+L_{n-3}=13L_n-5\varepsilon,\\ &\cdots\cdots\cdots\cdots\cdots\\ L_{n-k}&=F_{k+1}L_n-F_{k-1}\varepsilon. \end{aligned} \tag{5.3}$$

The coefficients F_k in this sequence have the property that

$$F_k=F_{k-1}+F_{k-2}, \qquad k=2,3,\dots, \tag{5.4}$$

where $F_0=F_1=1$. This sequence of numbers F_k is called the *Fibonacci sequence* after the Italian scientist of that name.

EXAMPLE 5.4 The Italian scientist, Leonard of Pisa, nicknamed Fibonacci, in 1202 came across a certain sequence of integers while considering the reproduction of rabbits. He assumed that it takes one month for rabbits to mature to fertility and another month to produce a litter of two. Letting F_n be the number of pairs of rabbits alive after n months and starting with one pair of rabbits, we see that $F_0=F_1=1$. After two months, the first pair have produced one pair of offspring for a total of two pairs, so $F_2=2$. Continuing in this way $F_3=3$, $F_4=5$, $F_5=8$, $F_6=13$, and $F_k=F_{k-1}+F_{k-2}$, which is the Fibonacci sequence of Eq. (5.4).

Letting $k=n-1$ in (5.3) yields

$$L_n=\frac{L_1+F_{n-2}\varepsilon}{F_n}, \tag{5.5}$$

which relates the original interval L_1 to the final interval L_n after n experiments. Since no interval reduction occurs until two trials are placed, $L_0=L_1$. Similarly for $k=n-2$, Eq. (5.3) yields

$$L_2=F_{n-1}L_n-F_{n-3}\varepsilon,$$

or substituting (5.5),

$$L_2=\frac{F_{n-1}L_1}{F_n}+\frac{(F_{n-1}F_{n-2}-F_{n-3}F_n)\varepsilon}{F_n}.$$

Substituting the identity (5.4) and using the property of the Fibonacci sequence that

$$F_{n-2}^2-F_{n-1}F_{n-3}=(-1)^n$$

yields

$$L_2 = \frac{F_{n-1}L_1}{F_n} + \frac{(-1)^n \varepsilon}{F_n}. \tag{5.6}$$

The *Fibonacci search algorithm* is now summarized as follows:

1. Define the initial interval of uncertainty as
$$L_1 = b^0 - a^0.$$
2. Using the minimal measurement separation ε and the specified number of trials, n, calculate L_2 from (5.6).
3. Place the first trial a^1 exactly L_2 units from one of the ends, say a^0.
4. Place the second trial b^1 exactly L_2 units from the other end, say b^0.
5. Evaluate $f(a^1)$ and $f(b^1)$, and eliminate the interval (a^0, b^1) if $f(a^1) < f(b^1)$, or (a^1, b^0) if $f(a^1) < f(b^1)$.
6. Place the next trial symmetrically on the interval and continue as in step 5.
7. Continue for n trials, at which time the last trial will occur exactly ε units from the existing internal point. The theoretical final interval of uncertainty is then given by (5.5).

EXAMPLE 5.5 Apply the Fibonacci search to the function shown in Fig. 5.9(a) for $n = 4$, $\varepsilon = 0$.

Here $L_1 = 1$, $F_3 = 3$, $F_4 = 5$, and $L_2 = 0.8$. The solution is shown graphically in Fig. 5.9(b). The final interval of uncertainty is given by (5.5) as

$$L_4 = \frac{L_1}{F_4} = 0.2.$$

Compared with other one-dimensional optimization techniques, the Fibonacci search has the following advantages:

1. It is the optimal search plan for a specified number of trials (i.e. the best interval reduction is obtained for each trial).
2. No computation of the function gradient is required.
3. The algorithm is simple and ideal for computer applications.
4. Only four sample points need be stored at each iteration.
5. The final interval of uncertainty is easily calculated in advance.

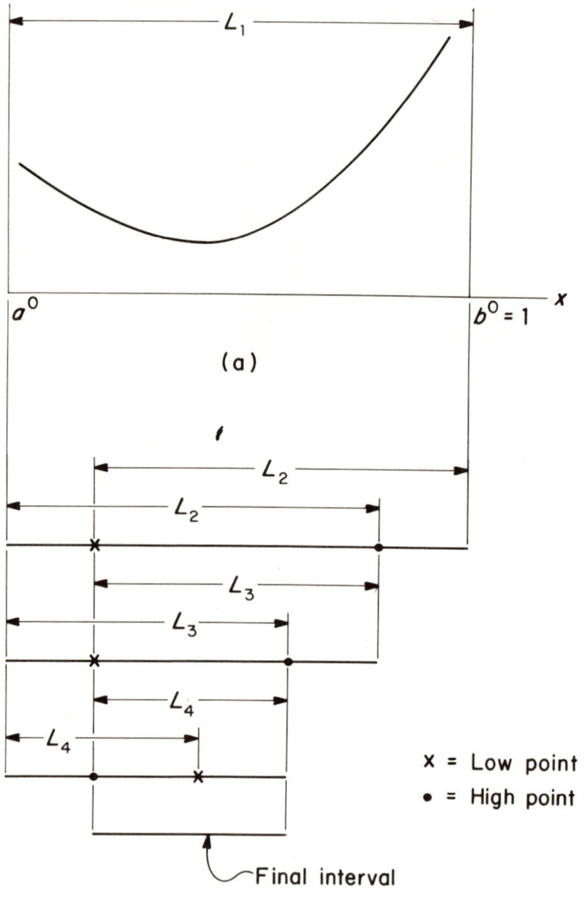

(b)

FIGURE 5.9

Despite these advantages, the method also has disadvantages:

1. The number of trials in the search, n, must be specified in advance.

2. The minimum separation ε must be specified.

Each of these disadvantages can be overcome as seen in the next section if one is willing to sacrifice some computational efficiency.

Golden-Section Search

Suppose we require that the ratio between the ending and the beginning interval of uncertainty remain constant *in each iteration*. The algorithm having this property is called the *golden-section algorithm*. If we let λ denote the unknown constant ratio between the lengths of the ending and the beginning intervals in each iteration, then for the nth iteration this requirement can be written

$$\frac{x_2^n - a^n}{b^n - a^n} = \frac{b^n - x_1^n}{b^n - a^n} = \lambda. \tag{5.7}$$

This requirement corresponds to placing each of the two points x_1^n and x_2^n symmetrically on the beginning interval $[a_1^n, b^n]$. The distances $x_2^n - a^n$ and $b^n - x_1^n$ are equal and are given by $\lambda(b^n - a^n)$ as shown in Fig. 5.10. From (5.7), by simple algebraic manipulations, we obtain

$$\frac{x_1^n - a^n}{x_2^n - a^n} = \frac{1}{\lambda} - 1. \tag{5.8}$$

From Fig. 5.10, $f(x_2^n) > f(x_1^n)$, so we set $b^{n+1} = x_2^n$, $a^{n+1} = a^n$, and we choose $x_2^{n+1} = x_1^n$. Now in order to satisfy (5.7) in the $(n+1)$st iteration we must have

$$\lambda = \frac{x_2^{n+1} - a^{n+1}}{b^{n+1} - a^{n+1}} = \frac{b^{n+1} - x_1^{n+1}}{b^{n+1} - a^{n+1}}, \tag{5.9}$$

which reduces by substitution to

$$\lambda = \frac{x_1^n - a^n}{x_2^n - a^n}. \tag{5.10}$$

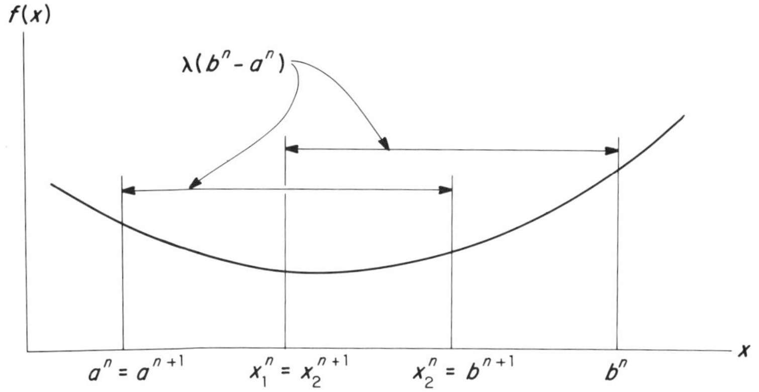

FIGURE 5.10

Combining (5.8) and (5.9), we can now determine λ by solving the quadratic $\lambda^2+\lambda-1=0$. Taking the positive root $\lambda=(\sqrt{5}-1)/2=0.618$, we can choose $x_1^{n+1}=a^{n+1}+(1-\lambda)(b^{n+1}-a^{n+1})$.

Similarly, if $f(x_1^n)>f(x_2^n)$, then we set $b^{n+1}=b^n$, $a^{n+1}=x_1^n$, and choose $x_1^{n+1}=x_2^n$. Once again it follows by similar arguments that $\lambda=0.618$ approximately, and the proper choice of x_2^{n+1} is given by $x_2^{n+1}=b^{n+1}-(1-\lambda)(b^{n+1}-a^{n+1})$. Note that either x_1^n or x_2^n and its corresponding function value is used in the $(n+1)$st iteration.

EXAMPLES

5.6 Apply the golden-section search scheme to locate the minimum of $f(x)=e^x-5x$ in the interval [1,2], and compare the result with that of Example 5.3.

In Example 5.3, we used the three-point equal-interval search scheme. After five function evaluations we attained $\Delta L=0.25$. We will now allow five function evaluations for the golden-section search scheme and obtain the resulting ΔL. Since we are given $a^0=1$ and $b^0=2$, the intermediate points for the golden-section search scheme are $x_1^0=1.382$ and $x_2^0=1.618$. By direct computation $f(1.382)=-2.94$ and $f(1.618)=3.1$. Since $f(x_1^0)>f(x_2^0)$, we set $a^1=x_1^0=1.382$ and $b^1=b^0=2$.

In the next iteration we set $x_1^1=x_2^0=1.618$ and compute $x_2^1=2-(1-0.618)(2-1.382)=1.765$. Then by direct computation, $f(1.765)=-3.0$. Since $f(x_2^1)>f(x_1^1)$, we set $a^2=a^1=1.382$ and $b^2=x_2^1=1.765$.

In the third iteration we set $x_2^2=x_1^1=1.618$ and compute $x_1^2=1.382+0.382(1.765-1.382)=1.527$. By direct computation $f(1.527)=-2.0$ and evidently $f(x_1^2)>f(x_2^2)$, so we set $a^3=x_1^2=1.527$ and $b^3=b^2=1.765$.

In the fourth iteration we set $x_1^3=x_2^2=1.618$ and compute $x_2^3=1.765-0.382(1.765-1.527)=1.674$. Then by direct computation $f(1.674)=-3.06$, and evidently $f(x_2^3)>f(x_1^3)$, so we have $a^4=a^3=1.527$ and $b^4=x_2^3=1.674$. Now we have used up five function evaluations, and we compute $\Delta L=(b^4-a^4)/(b^0-a^0)=0.147$. Comparing this value with 0.25 achieved in Example 5.3 clearly shows the superiority of the golden-section search scheme over the three-point equal-interval search scheme.

5.7 By the symmetric placement of successive trials in the unit interval, the golden-section search has in fact divided the unit interval into two unequal parts such that the ratio of the whole to the larger is equal to the ratio of the larger to the smaller. Denoting the larger part by λ, we see that

$$\frac{1}{\lambda}=\frac{\lambda}{1-\lambda},$$

which reduces to $\lambda^2+\lambda-1=0$.

This way of dividing a line segment was discovered by the geometer Euclid, and came to be known as the *golden section*. This ratio figures prominently in the architecture of ancient Greece.

5.8 Denoting by L_n the final interval of uncertainty after placing n trials and by L_1 the initial interval, we see that for the golden-section search,

n	0	2	3	4	...
L_n	L_1	λL_1	$\lambda^2 L_1$	$\lambda^3 L_1$	...

Denoting by D_n the ratio of the final interval of uncertainty after placing n trials to the initial interval, we have

$$\frac{L_n}{L_{n-1}}=\lambda$$

and

$$D_n=\frac{L_n}{L_1}=\lambda^{n-1}. \tag{5.11}$$

5.9 For $n=11$ compare the interval reductions for the Fibonacci and the golden-section search.

Letting $\varepsilon \approx 0$ for the Fibonacci search, we get from (5.5)

$$\frac{L_{11}}{L_1}=\frac{1}{F_{11}}=\frac{1}{144}=0.006944.$$

For the golden-section search, (5.11) yields

$$\frac{L_{11}}{L_1}=\lambda^{10}\approx(0.618)^{10}=0.008130.$$

Thus after 11 trials the Fibonacci search achieves a smaller interval of uncertainty than the golden-section search by about 12%.

Quadratic Interpolation

The search programs discussed so far can be easily programmed on a computer. They also have easily computable convergence rates, as seen in Problems 5.1 and 5.2 for the dichotomous and three-point equal-interval searches. For the golden-section algorithm, we have shown that the ratio of the lengths of the ending interval to the beginning interval after n iterations is λ^{n-1}, where $\lambda=0.618$ approximately. The main disadvantage of each of these methods is that to achieve reasonable accuracy in the location of x^* they require a large number of function evaluations.

Another approach to locating x^* involves the approximation of $f(x)$ by a polynomial of finite degree. We can then determine the location of the minimum of the polynomial analytically and use the value as an estimate of x^*. The quadratic polynomial is the simplest one to use for this purpose, although the cubic polynomial has been used with success. Before discussing the quadratic interpolation technique, it is worth pointing out that this technique is often used in more complex minimization problems. For example, suppose $\mathbf{x}$ is an n-vector, and $f(\mathbf{x})$ is a function of n variables. Then many algorithms for locating a minimum of $f(\mathbf{x})$ use the following one-dimensional search scheme, sometimes called *minimizing down the line*:

> Let $\mathbf{x}^0$ be a starting point, and let $\mathbf{v}$ be a downhill direction at $\mathbf{x}^0$. Then we define $\mathbf{x}^1=\mathbf{x}^0+\lambda_1\mathbf{v}$, where λ_1 minimizes $f(\mathbf{x}^0+\lambda\mathbf{v})$ with respect to the scalar λ. This last minimization is of course equivalent to minimizing the scalar function $g(\lambda)=f(\mathbf{x}^0+\lambda\mathbf{v})$ with respect to λ. Since $\mathbf{v}$ is downhill at $\mathbf{x}^0$, we have $\lambda \geqslant 0$. An upper bound m is now introduced to avoid an excessively large step size. Thus the one-dimensional search is equivalent to locating a minimum of $g(\lambda)$ for $0 \leqslant \lambda \leqslant m$. These methods are further discussed in Chapter 6.

We will now show how the solution to this one-dimensional search problem can be simplified by using quadratic interpolation. Assuming that the length of the vector $\mathbf{v}$ is unity, we now compute $g(1)$. If $g(1)>g(0)$, compute $g(\lambda)$ for $\lambda=\frac{1}{2},\frac{1}{4},\frac{1}{8},\ldots,$ until $g(\lambda)<g(0)$. We then define $a=0$, $b=\lambda$, and $c=2\lambda$ and compute $\hat{\lambda}$ as

$$\hat{\lambda}=\frac{1}{2}\,\frac{g(a)(c^2-b^2)+g(b)(a^2-c^2)+g(c)(b^2-a^2)}{g(a)(c-b)+g(b)(a-c)+g(c)(b-a)}; \qquad (5.12)$$

this expression is derived in Problem 5.2. If $g(1)\leqslant g(0)$, we then compute $g(\lambda)$ for $\lambda=2,4,8,\ldots,a,b,c$, where c denotes the point where $g(c)>g(b)$for the first time. Again we compute $\hat{\lambda}$ by (5.12), and if $g(\hat{\lambda})<g(b)$ we then accept $\hat{\lambda}$ as an estimate of λ_1, where λ_1 minimizes $f(\mathbf{x}^0+\lambda\mathbf{v})$ with respect to λ. If $g(\hat{\lambda})\geqslant g(b)$, then we accept b as an estimate of λ_1. This quadratic interpolation procedure is illustrated graphically in Figs. 5.11 and 5.12 for the cases where $g(1)>g(0)$ and $g(1)<g(0)$ respectively.

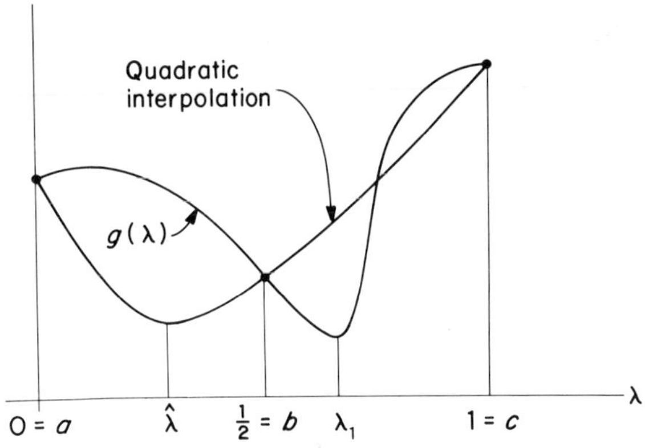

FIGURE 5.11

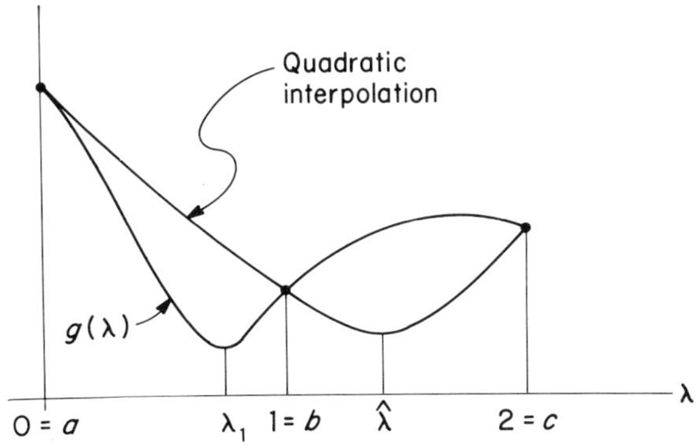

FIGURE 5.12

EXAMPLES

5.10 Find $\hat{\lambda}$ from (5.12) and compare the result with that of Example 5.1 for $g(\lambda)=8\lambda^3-2\lambda^2-7\lambda+3$.

We note that $g(0)=3$ and $g(1)=2$, i.e., $g(1)<g(0)$. We then compute $g(\lambda)$ for $\lambda=2$ and obtain $g(2)=45$. Since $g(2)>g(1)$, we set $a=0$, $b=1$, and $c=2$. Then by (5.12) we have

$$\hat{\lambda}=\frac{1}{2}\,\frac{3(4-1)+2(0-4)+45(1-0)}{3(2-1)+2(0-2)+45(1-0)}=0.52.$$

From Example 5.1 we know that $g(\lambda)$ attains a minimum in $[0,1]$ at

$\lambda^* = 0.63$, and thus $\hat{\lambda} = 0.52$ is a poor estimate of λ^* although it does require only three function evaluations. Most of the other interval search techniques will yield a better estimate of λ^* with four function evaluations. However, this accuracy may not be needed in the initial stages of an optimization algorithm which uses a one-dimensional search in every iteration.

5.11 Find the minimum of $f(x) = e^x - 5x$ in $[1,2]$ by the quadratic interpolation technique.

Using the new variable $\lambda = x - 1$, we obtain $g(\lambda) = e^{\lambda+1} - 5(\lambda + 1)$. By direct computation $g(0) = -2.282$ and $g(1) = -2.611$. Since $g(1) < g(0)$, we compute $g(2) = 5.086$. Now $g(2) > g(1)$, so we set $a = 0$, $b = 1$, and $c = 2$ in (5.12), obtaining

$$\hat{\lambda} = \frac{1}{2}\,\frac{-2.282(4-1) + (-2.611)(-4) + 5.086}{-2.282(2-1) + (-2.611)(-2) + 5.086} = 0.531.$$

The exact solution λ^* is given by $\lambda^* = x^* - 1 = 0.609$.

Exhaustive Search

The interval search techniques discussed so far are sequential in nature, which means that the placement of the trials in the $(n+1)$st iteration is not specified *a priori* but is dependent upon the interval of uncertainty obtained in the nth iteration. However, in certain other interval search schemes, the placement of all the trials can be specified in advance. One such scheme is the *exhaustive search*, in which we divide the original interval of uncertainty $[a^0, b^0]$ into a large number of subintervals of equal length, as shown in Fig. 5.13. We then evaluate $f(x)$ at the endpoints of each subinterval and compute the minimum over this finite set of function values. This minimum is then used as an estimate of x^*. For example, in Fig. 5.13, this procedure leads to x_5 as an estimate of x^*. Clearly from this example, the lengths of the subintervals must be made sufficiently small so that a sharp and narrow minimum is not missed. On the other hand, small subintervals cause an increase in the total number of function evaluations and result in a highly inefficient algorithm. The exhaustive search scheme is most effective on relatively smooth functions which are fairly flat in the vicinity of the minimum.

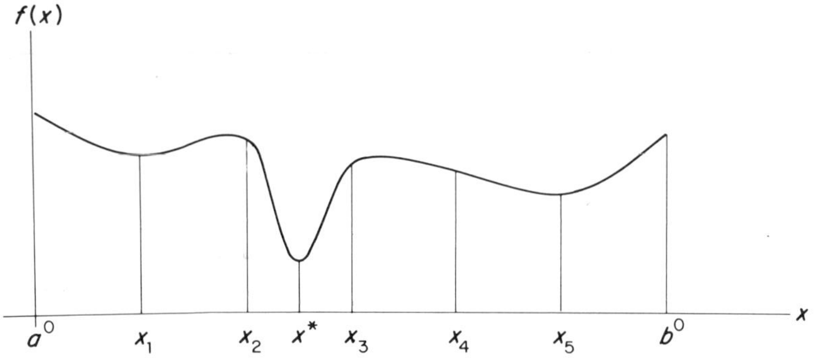

FIGURE 5.13

Solved Problems

5.1 A measure of effectiveness for the dichotomous search scheme can be obtained by letting D_n denote the ratio of the lengths of the final interval of uncertainty to the initial interval containing x^* after n iterations have been performed. Determine an equation of D_n.

Without loss of generality, we assume that the length of the initial interval of uncertainty is unity. This normalization can be easily accomplished by introducing a new variable $y=(x-a^0)/(b^0-a^0)$. Denote the interval of uncertainty after the nth iteration by $L_n=b^n-a^n$. Then for the dichotomous search scheme we have $2L_{n+1}=L_n+\varepsilon$ (see Fig. 5.3). Noting that $L_0=1$ and using the above equation for L_n, we obtain

$$\begin{aligned} L_n &= \frac{1}{2^n}+\frac{\varepsilon}{2^n}+\cdots+\frac{\varepsilon}{2} \\ &= \frac{1}{2^n}+\frac{\varepsilon}{2}\left(1+\frac{1}{2}+\cdots+\frac{1}{2^{n-1}}\right) \\ &= \frac{1}{2^n}+\varepsilon\left(1-\frac{1}{2^n}\right). \end{aligned}$$

After n iterations, we have

$$D_n=L_n=\frac{1}{2^n}+\varepsilon\left(1-\frac{1}{2^n}\right). \tag{5.13}$$

Since each iteration requires two function evaluations, the total number of function evaluations N is given by $N=2n$. Note that as $n\to\infty$, $L_n\to\varepsilon$. In other words, the length of the final interval of uncertainty containing x^* is at least ε for the dichotomous search.

5.2 Compare Eq. (5.13) in Problem 5.1 with the results of Example 5.1.

The numerical result in Example 5.1 resulted in $L_2 = b^2 - a^2 = 0.775 - 0.45 = 0.325$. Using Eq. (5.13), we have also $L_2 = \frac{1}{4} + 0.1(1 - \frac{1}{4}) = 0.325$.

In Problem 5.1 we saw that the final interval of uncertainty is at least ε for the dichotomous search algorithm. From Example 5.1, $\varepsilon = 0.1$ and $L_2 = 0.325$. However, if we approximate the optimum after two iterations by c_2, the midpoint of $[a^2, b^2]$, we see that $c_2 = 0.6125$. Thus the error after only two iterations is $x^* - c_2 = 0.63 - 0.6125 = 0.0175$, which is already much less than ε. This example indicates that L_n represents a worst case for the interval of uncertainty after n iterations, and the actual error may be much less. The example also points out the importance of choosing ε carefully, since in this case it restricts the theoretical accuracy obtainable.

5.3 Perform three iterations of the three-point equal-interval search to minimize $f(x) = x^2 - 4x + 2$ over the interval $0 \leqslant x \leqslant 10$.

Evaluating $f(x)$ at three equidistant points on the interval $[0, 10]$ yields

$$f(2.5) = -6.75, \qquad f(5.0) = -3.00, \qquad f(7.5) = 13.25.$$

We retain the interval $[0, 5]$ and evaluate $f(x)$ at three equidistant points on the interval $[0, 5]$, yielding

$$f(1.25) = -3.94, \qquad f(2.5) = -6.75, \qquad f(3.75) = -6.44.$$

We retain the interval $[2.5, 5]$ and evaluate $f(x)$ at three equidistant points on the interval $[2.5, 5]$, obtaining

$$f(3.125) = -6.98, \qquad f(3.75) = -6.44 \qquad f(4.335) = -5.11.$$

We retain the interval $[2.5, 3.75]$. Thus after three iterations consisting of 9 trials, we have reduced the interval of uncertainty to 12.5% of its initial value.

5.4 For the three-point equal-interval search algorithm, find a lower bound for N, the total number of function evaluations required to obtain a specified ratio ΔL between the final and the initial intervals of uncertainty.

Let $L_n = b^n - a^n$; then from Fig. 5.6 we see that $L_{n+1} = L_n/2$ for the three-point equal-interval search scheme. Thus taking $L_0 = 1$, $L_n/L_0 = (\frac{1}{2})^n$, and in order to satisfy $L_n/L_0 \leqslant \Delta L$ we must have

$$\left(\tfrac{1}{2}\right)^n \leqslant \Delta L, \qquad \text{or} \quad 2^n \geqslant \frac{1}{\Delta L}.$$

Thus

$$n \ln 2 \geqslant \ln(1/\Delta L) = -\ln \Delta L$$

and

$$n \geqslant \frac{-\ln \Delta L}{\ln 2} = -1.441 \ln \Delta L.$$

In the first iteration we compute three function values, and in all subsequent iterations we compute only two function values per iteration. Hence the total number of function evaluations N required in n iterations is $3+2(n-1)$. Thus we have

$$N \geqslant 1-2.882\ln \Delta L. \tag{5.14}$$

5.5 Compare (5.14) with the results obtained in Example 5.4.

From Example 5.4, $\Delta L=(a^2-b^2)/(a^0-b^0)=0.25$. Therefore from (5.14),

$$N \geqslant 1-2.882\ln 0.25=5 \quad \text{(approximately)}$$

and we used 5 function evaluations to obtain the ratio $\Delta L=0.25$. This result agrees with our experience in Example 5.4.

5.6 Find the total number of function evaluations required by the Fibonacci search scheme to obtain $L_n/L_1 \leqslant 10^{-4}$.

From (5.5) and taking $\varepsilon \approx 0$, we have

$$\frac{L_n}{L_1}=\frac{1}{F_n} \leqslant 10^{-4}.$$

Since $F_{19}=6{,}765$ and $F_{20}=10{,}946$, we see that 20 function evaluations are required.

5.7 Derive Eq. (5.12) for $\hat{\lambda}$ in the quadratic interpolation scheme.

First assume that $g(\lambda)$ is approximated by the quadratic form $q(\lambda)=u_0+u_1\lambda+u_2\lambda^2$. For a minimum, $dq/d\lambda=0$ and we have $\hat{\lambda}=-\frac{1}{2}(u_1/u_2)$. If in addition $u_2=\frac{1}{2}(d^2q/d\lambda^2)>0$, then $q(\lambda)$ attains a minimum at $\hat{\lambda}$. We now determine the unknown coeffieients u_0, u_1, and u_2 by approximating $g(\lambda)$ by $q(\lambda)$. Choosing three values of λ denoted by a, b, and c, and equating $g(\lambda)$ to $q(\lambda)$ for these values of λ, we have

$$g(a)=u_0+u_1a+u_2a^2, \tag{5.15a}$$

$$g(b)=u_0+u_1b+u_2b^2, \tag{5.15b}$$

$$g(c)=u_0+u_1c+u_2c^2. \tag{5.15c}$$

We can solve for u_1 and u_2 using Cramer's rule.

$$\Delta=\begin{vmatrix} 1 & a & a^2 \\ 1 & b & b^2 \\ 1 & c & c^2 \end{vmatrix}=(a-b)(b-c)(c-a),$$

$$u_1 = \frac{1}{\Delta}\begin{vmatrix} 1 & g(a) & a^2 \\ 1 & g(b) & b^2 \\ 1 & g(c) & c^2 \end{vmatrix} = \frac{g(a)(b^2-c^2)+g(b)(c^2-a^2)+g(c)(a^2-b^2)}{(a-b)(b-c)(c-a)},$$

$$u_2 = \frac{1}{\Delta}\begin{vmatrix} 1 & a & g(a) \\ 1 & b & g(b) \\ 1 & c & g(c) \end{vmatrix} = \frac{g(a)(c-b)+g(b)(a-c)+g(c)(b-a)}{(a-b)(b-c)(c-a)},$$

and

$$\hat{\lambda} = -\frac{1}{2}\frac{u_1}{u_2} = \frac{1}{2}\frac{g(a)(c^2-b^2)+g(b)(a^2-c^2)+g(c)(b^2-a^2)}{g(a)(c-b)+g(b)(a-c)+g(c)(b-a)}.$$

5.8 Determine a closed-form solution for F_k in Eq. (5.4).

As a trial solution assume that

$$F_k = Ap_1^k + Bp_2^k, \tag{5.17}$$

where A, B, p_1, and p_2 are unknown quantities to be determined. Substituting (5.17) into (5.4) gives

$$Ap_1^k + Bp_2^k = A\left(p_1^{k-1}+p_1^{k-2}\right) + B\left(p_2^{k-1}+p_2^{k-2}\right),$$

or

$$A\left(p_1^k - p_1^{k-1} - p_1^{k-2}\right) + B\left(p_2^k - p_2^{k-1} - p_2^{k-2}\right) = 0.$$

Thus for any k, p_1 and p_2 must satisfy

$$p^2 - p - 1 = 0,$$

and hence

$$p_1 = \frac{1+\sqrt{5}}{2} = 1.618, \qquad p_2 = \frac{1-\sqrt{5}}{2} = -0.618.$$

We obtain A and B using the initial conditions $F_0 = F_1 = 1$. Thus $A+B=1$ and $Ap_1 + Bp_2 = 1$, so

$$A = \frac{1-p_2}{p_1-p_2}, \qquad B = \frac{p_1-1}{p_1-p_2}.$$

Finally,

$$F_k = \left(\frac{1-p_2}{p_1-p_2}\right)p_1^k + \left(\frac{p_1-1}{p_1-p_2}\right)p_2^k.$$

But

$$1-p_2=1-\frac{1-\sqrt{5}}{2}=\frac{1+\sqrt{5}}{2}=p_1$$

and

$$p_1-1=\frac{1+\sqrt{5}}{2}-1=\frac{-1+\sqrt{5}}{2}=-p_2.$$

So

$$F_k=\frac{1}{p_1-p_2}\left[p_1^{k+1}-p_2^{k+1}\right],$$

or

$$F_k=\frac{1}{\sqrt{5}}\left[\left(\frac{1+\sqrt{5}}{2}\right)^{k+1}-\left(\frac{1-\sqrt{5}}{2}\right)^{k+1}\right].$$

5.9 Show that the golden-section search and the Fibonacci search have the same asymptotic convergence rate.

We define the asymptotic convergence rate as

$$\lim_{n\to\infty}\frac{L_n}{L_{n-1}}.$$

For the golden-section search (from Example 5.8),

$$\frac{L_n}{L_{n-1}}=\lambda=0.618$$

for all n. Taking $\varepsilon=0$ for the Fibonacci search,

$$\frac{L_n}{L_1}=\frac{1}{F_n}\quad\text{and}\quad\frac{L_{n-1}}{L_1}=\frac{1}{F_{n-1}}.$$

Thus

$$\lim_{n\to\infty}\frac{L_n}{L_{n-1}}=\lim_{n\to\infty}\frac{F_{n-1}}{F_n}=\lim_{n\to\infty}\frac{Ap_1^{n-1}+Bp_2^{n-1}}{Ap_1^n+Bp_2^n},\tag{5.18}$$

where $p_1=1.618$ and $p_2=-0.618$. Since $|p_2|<1$, Eq. (5.18) reduces to

$$\lim_{n\to\infty}\frac{p_1^{n-1}}{p_1^n}=\frac{1}{p_1}=\frac{1}{1.618}=0.618,$$

which is identical to the asymptotic convergence rate for the golden-section search.

5.10 Minimize $f(x)=x^2-6x+2$ over the range $0\leqslant x\leqslant 10$ so that the final interval is less than 3% of the initial interval, using the Fibonacci search.

To obtain

$$\frac{L_n}{L_1} = \frac{1}{F_n} < 3\%$$

we require 8 trials, since $1/F_8 = 1/34 < 0.03$. Taking $\varepsilon \approx 0$,

$$L_2 = \frac{F_{n-1}L_1}{F_n} = \tfrac{21}{34}(10) = 6.18,$$

$$f(6.18) = 3.112,$$

$$f(10 - 6.18) = f(3.82) = 6.328.$$

Since $f(3.82) < f(6.18)$, we discard the interval [6.18, 10] and place the next trial symmetrically at $6.18 - 3.82 = 2.36$. Then since $f(2.36) = -6.578 < f(3.82)$, we discard the interval [3.82, 6.18] and place the next trial symmetrically at $3.82 - 2.36 = 1.46$. Then since $f(1.46) = -4.659 > f(2.36)$, we discard the interval [0, 1.47] and place the next trial at $3.82 - 2.36 + 1.47 = 2.93$. Then since $f(2.93) = -6.996 < f(1.47)$, we discard [1.47, 2.36] and place the next trial at $3.82 - 2.93 + 2.36 = 3.25$. Then since $f(3.25) = -6.947 > f(2.94)$, we discard [3.25, 3.82] and place the next trial at $3.25 - 2.94 + 2.36 = 2.67$. Then since $f(2.67) = -6.87 > f(2.94)$, we discard [2.36, 2.67] and place the last trial at $3.25 - 2.94 + 2.67 = 2.98$. Finally $f(2.98) = -6.9 < f(2.67)$, and the final interval of uncertainty is [2.98, 3.25], which is less than 3% of the initial interval. The minimum actually occurs at $x^* = 3.0$, which is contained in the final interval as expected.

Supplementary Problems

5.11 Define unimodality when $f(x)$ has a maximum in $[a^0, b^0]$.

5.12 Show that $F_{n-2}^2 - F_{n-1}F_{n-3} = (-1)^n$ for the Fibonacci sequence.

5.13 Derive Eqs. (5.8) and (5.10).

5.14 Sketch $g(\lambda)$ and its quadratic interpolation for Example 5.10.

5.15 Sketch $g(\lambda)$ and its quadratic interpolation for Example 5.11.

5.16 Locate the Fibonacci trials for Problem 5.10 as in Fig. 5.9(b).

5.17 Apply the Fibonacci search technique with $n = 6$ to locate the minimum of $f(x) = e^x - 5x$ over the interval [1, 2]. Compare the results with those of Examples 5.3, 5.6, and 5.11.

5.18 Find the minimum number of function evaluations needed to satisfy the requirement that $\Delta L \leqslant 0.125$ using (a) the three-point equal search and (b) the Fibonacci search.

5.19 Use quadratic interpolation to estimate the location of the minimum for the function $f(x) = x^4 - 2x^2 + 1$ over the interval [0, 2].

Chapter 6

Unconstrained Gradient Techniques

In Chapters 2 to 4, we dealt with *indirect methods* of solving optimization problems. These methods are called indirect because they provide only necessary conditions for an optimum. These conditions, in the form of sets of equations and inequalities, must still be solved in order to determine numerical solutions.

In Chapter 5, we introduced *direct methods* for optimizing unimodal functions having only one degree of freedom. These methods are called direct because they are usually iterative numerical procedures or *algorithms* which operate on the objective function directly to obtain numerical answers. In this case, there is no formal intermediate step of specifying necessary conditions for an optimum, although the algorithms themselves may be derived using these conditions.

In this chapter we begin to discuss direct methods for solving multidimensional problems. The *gradient algorithm*, discussed first, is the simplest of many such methods. This algorithm forms the basis for many direct methods used in optimizing both constrained and unconstrained problems. Although these methods often pose problems of convergence, they are by far the most widely applied because of their simplicity. In this chapter we discuss the gradient algorithm and its application to unconstrained problems.

The Gradient Direction

In Chapter 1, we defined the gradient of a differentiable function of n variables, $f(\mathbf{x})$, as the n-dimensional vector

$$\nabla f(\mathbf{x}) = \left[\frac{\partial f}{\partial x_1}(\mathbf{x}), \frac{\partial f}{\partial x_2}(\mathbf{x}), \ldots, \frac{\partial f}{\partial x_n}(\mathbf{x}) \right]^T.$$

This vector defines a direction called the *gradient direction*, which is the basis of all gradient algorithms. Before deriving the gradient algorithm formally, it is useful to examine intuitively why the gradient direction is a good candidate for solving optimization problems. We obtain some insight by considering an alternative name for the gradient direction; namely, the direction of *steepest descent* for minimization problems or the direction of *steepest ascent* for maximization problems.

Suppose, for example, that we wish to climb to the top of a (unimodal) mountain peak. It is intuitively obvious that if we take each step in the direction of steepest ascent, we will reach the top along the shortest possible path. Since this path conserves shoe leather (although perhaps not energy), it seems somehow efficient and therefore worthy of further consideration.

EXAMPLES

6.1 Consider a nearsighted fly who is perched on the side of a bowl and wishes to reach the bottom. The surface of the bowl is described by the function $f(\mathbf{x})$. If the fly were not nearsighted, it would immediately see the bottom of the bowl and fly there. However, its nearsightedness requires that it walk along the surface of the bowl. Since the fly can take only fly-sized steps of fixed length, it wants to choose the direction of each step carefully to minimize the time and distance to reach the bottom. By choosing each step in the negative gradient direction defined by $-\nabla f(\mathbf{x})$, this minimization can be accomplished. Thus at each step, the fly chooses from among all possible downward directions the one which goes down the most. In other words, it chooses the path of steepest descent. We notice that the nearsighted fly and the gradient vector both utilize only *local* information at each point. Hence they will both achieve only *local optima*.

6.2 Consider a small body resting on the rim of a bowl of any shape. If the body were to slide down the inside of the bowl with enough friction to make inertia negligible, then it would follow the negative gradient direction at each point, and it would trace out the path of steepest descent in reaching the bottom.

We now derive the path of steepest descent for the problem

$$\min_{\mathbf{x}} f(\mathbf{x}),$$

where $\mathbf{x}$ is an n-dimensional vector and $f(\mathbf{x})$ is a differentiable function.

Choosing some arbitrary starting point $\mathbf{x}_0$, we wish to determine a neighboring point $\mathbf{x}_0 + d\mathbf{x}$ such that

$$f(\mathbf{x}_0 + d\mathbf{x}) < f(\mathbf{x}_0).$$

In fact, we want to choose $d\mathbf{x}$ so that the difference

$$f(\mathbf{x}_0) - f(\mathbf{x}_0 + d\mathbf{x})$$

is maximized. The vector $d\mathbf{x}$ will describe an incremental distance ds along the path of steepest descent. The incremental path is defined by

$$(ds)^2 = \sum_{i=1}^{n} (dx_i)^2,$$

or

$$1 - \sum_{i=1}^{n} \left(\frac{dx_i}{ds} \right)^2 = 0,$$

where the dx_i/ds are the direction cosines.

We now choose dx_i/ds so that the rate of change df/ds is maximized. Since

$$\frac{df}{ds} = \sum_{i=1}^{n} \frac{\partial f}{\partial x_i} \frac{dx_i}{ds},$$

the path of steepest descent is obtained by choosing dx_i/ds to find

$$\max_{dx_i/ds} \sum_{i=1}^{n} \frac{\partial f}{\partial x_i} \frac{dx_i}{ds}$$

such that

$$1 - \sum_{i=1}^{n} \left(\frac{dx_i}{ds} \right)^2 = 0.$$

We can solve this problem by using Lagrange multipliers, where

$$L\left(\frac{dx_i}{ds}, \lambda \right) = \sum_{i=1}^{n} \frac{\partial f}{\partial x_i} \frac{dx_i}{ds} + \lambda \left[1 - \sum_{i=1}^{n} \left(\frac{dx_i}{ds} \right)^2 \right].$$

The necessary conditions are

$$\nabla_{dx_i/ds} L = \frac{\partial f}{\partial x_i} - 2\lambda \frac{dx_i}{ds} = 0, \qquad i = 1, 2, \ldots, n, \tag{6.1}$$

$$\nabla_\lambda L = 1 - \sum_{i=1}^{n} \left(\frac{dx_i}{ds} \right)^2 = 0. \tag{6.2}$$

Solving (6.1) and (6.2) yields

$$\frac{dx_i}{ds} = \frac{1}{2\lambda} \frac{\partial f}{\partial x_i},$$

where

$$\lambda = \pm \frac{1}{2} \left[\sum_{i=1}^{n} \left(\frac{\partial f}{\partial x_i} \right)^2 \right]^{1/2},$$

or

$$\frac{dx_i}{ds}(\mathbf{x}) = \pm \frac{\dfrac{\partial f}{\partial x_i}(\mathbf{x})}{\left[\displaystyle\sum_{i=1}^{n} \left(\frac{\partial f}{\partial x_i}(\mathbf{x}) \right)^2 \right]^{1/2}}, \qquad i = 1, 2, \ldots, n. \tag{6.3}$$

Since the denominator is simply a normalizing factor, we see that the path of steepest descent is described by the gradient vector. For maximization problems the positive gradient direction is chosen, and for minimization problems the negative gradient direction is chosen.

EXAMPLES

6.3 Show that the objective function decreases monotonically along the gradient path.

$$\frac{df}{ds} = \sum_{i=1}^{n} \frac{\partial f}{\partial x_i} \frac{dx_i}{ds} = \pm \frac{\displaystyle\sum_{i=1}^{n} \frac{\partial f}{\partial x_i} \frac{\partial f}{\partial x_i}}{\left[\displaystyle\sum_{i=1}^{n} \left(\frac{\partial f}{\partial x_i} \right)^2 \right]^{1/2}} = \pm \left[\sum_{i=1}^{n} \left(\frac{\partial f}{\partial x_i} \right)^2 \right]^{1/2}.$$

Since the terms inside the brackets are nonnegative, the square root exists and

$$\frac{df}{ds} \geqslant 0 \qquad \text{when maximizing,}$$

$$\frac{df}{ds} \leqslant 0 \qquad \text{when minimizing.}$$

6.4 Show that along the gradient path, the optimum of $f(\mathbf{x})$ is not attained in a finite distance.

From Example 6.3,

$$\frac{df}{ds} = \pm\left[\sum_{i=1}^{n}\left(\frac{\partial f}{\partial x_i}\right)^2\right]^{1/2}.$$

However, at the optimum $\mathbf{x}^*$,

$$\frac{\partial f}{\partial x_i}(\mathbf{x}^*) = 0, \qquad i = 1, 2, \ldots, n.$$

Thus as the optimum is approached, $\partial f/\partial x_i$ approaches zero and the rate of convergence df/ds decreases. In the limit as

$$\frac{\partial f}{\partial x_i}(\mathbf{x}) \to 0, \qquad i = 1, 2, \ldots, n,$$

the convergence rate

$$\frac{df}{ds}(\mathbf{x}) \to 0$$

and the optimum is not attained precisely in a finite distance. When traveling at a constant speed v, $ds = v\,dt$, and the optimum is also not attained precisely in a finite time.

6.5 Derive the gradient path for minimizing the elliptic paraboloid

$$f(\mathbf{x}) = x_1^2 + 2x_2^2$$

The gradient is

$$\nabla f(\mathbf{x}) = \begin{bmatrix} 2x_1 \\ 4x_2 \end{bmatrix}$$

From Eq. (6.3),

$$\frac{dx_1}{ds} = -K\frac{\partial f}{\partial x_1} = -2Kx_1,$$

$$\frac{dx_2}{ds} = -K\frac{\partial f}{\partial x_2} = -4Kx_2,$$

where K is the normalizing factor

$$K = \left[\sum_{i=1}^{n}\left(\frac{\partial f}{\partial x_i}(\mathbf{x})\right)^2\right]^{-1/2}.$$

Thus

$$\frac{dx_2}{dx_1} = \frac{dx_2}{ds}\frac{ds}{dx_1} = \frac{2x_2}{x_1},$$

or

$$\frac{dx_2}{x_2} = 2\frac{dx_1}{x_1}.$$

Integrating yields

$$\ln x_2 = 2\ln x_1 + c$$

or

$$\ln\frac{x_2}{x_1^2} = c$$

or

$$x_2 = e^c x_1^2.$$

Thus the gradient path is a parabola as shown in Fig. 6.1.

6.6 Show that the gradient path in Example 6.5 is orthogonal to the level curves.

Two curves are orthogonal if the inner product of their normal vectors equals zero. Thus for the level curves the normal vector is

$$\mathbf{n}_1 = \begin{bmatrix} 2x_1 \\ 4x_2 \end{bmatrix},$$

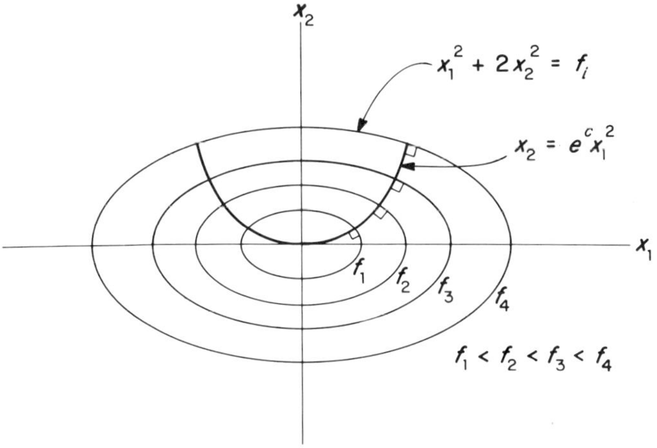

FIGURE 6.1

and for the gradient the normal vector is

$$\mathbf{n}_2 = \begin{bmatrix} 2e^c x_1 \\ -1 \end{bmatrix}.$$

The inner product of $\mathbf{n}_1$ and $\mathbf{n}_2$ is

$$(\mathbf{n}_1, \mathbf{n}_2) = [2x_1, 4x_2]\begin{bmatrix} 2e^c x_1 \\ -1 \end{bmatrix} = 4(e^c x_1^2 - x_2),$$

which equals zero, as shown in Example 6.5.

The Gradient Algorithm

The continuous equation for the path of steepest descent is given by Eq. (6.3). In practice, a problem could be optimized by implementing this equation on an analog computer and calculating $\partial f / \partial x_i$, $i = 1, 2, \ldots, n$, continuously as a function of $\mathbf{x}$. However, as a practical matter, one usually solves such problems on a digital computer, and hence an iterative form of Eq. (6.3) is required. This iterative form is called the *gradient algorithm*.

In general, an *algorithm* is a rule which takes a point $\mathbf{x}_k \in S \subset E^n$ (or points $\mathbf{x}_k, \mathbf{x}_{k-1}, \ldots$) and defines a new point $\mathbf{x}_{k+1} \in S \subset E^n$.

The gradient algorithm is obtained from Eq. (6.3) by writing

$$dx_i = \pm ds \frac{\partial f^n}{\partial x_i}(\mathbf{x}), \qquad i = 1, 2, \ldots, n,$$

where $\partial f^n / \partial x_i$ indicates the ith component of the *normalized* gradient. Regarding ds as an arbitrary step size k to be chosen, the gradient algorithm can be written

$$\mathbf{x}^{p+1} = \mathbf{x}^p + k\nabla f(\mathbf{x}^p), \tag{6.4}$$

where

$$k > 0 \qquad \text{for maximization,}$$
$$k < 0 \qquad \text{for minimization,}$$

and $\nabla f(\mathbf{x}^p)$ is understood to be the normalized gradient evaluated at $\mathbf{x}^p$.

If k is small, the discrete algorithm will closely approximate the gradient path, but convergence will be slow, since the gradient must be calculated many times. On the other hand, if k is large, convergence will initially be very fast, but the algorithm will oscillate about the optimum.

EXAMPLE 6.7 For Example 6.5, show how the gradient algorithm behaves for $k = 4$ and a starting point $\mathbf{x}_0 = (4, 4)$.

The solution is shown in Fig. 6.2.

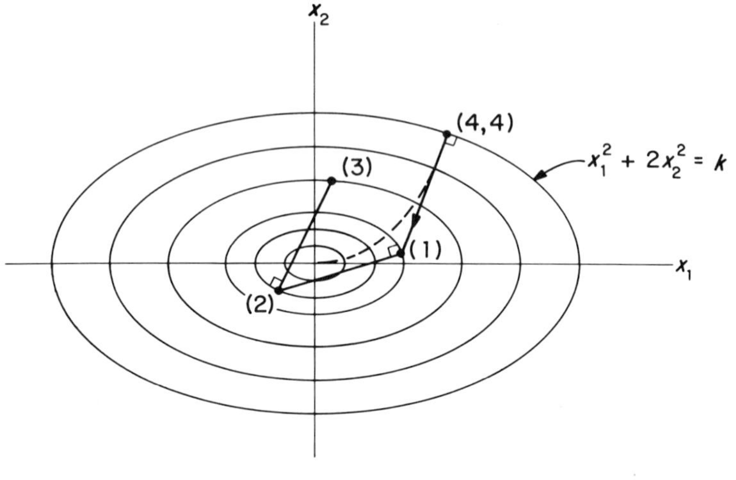

FIGURE 6.2

Thus it is clear that some mechanism must be found for reducing the step size as the optimum is approached. In general, we want large steps when we are far away from the optimum, and a decrease in the step size when oscillations are about to occur. Many schemes can be employed, such as reducing the step size by a fixed amount after every mth step or when the objective function fails to decrease. Although these *ad hoc* methods are often employed successfully, there is one method for choosing the step size which tends to minimize the number of steps required. This method is discussed in the next section.

The Optimal Gradient

In the *optimal-gradient* method we follow the gradient direction defined at $\mathbf{x}^p$ until a local minimum is reached. Since the objective function would increase if we continued past this point, a new gradient determination $\nabla f(\mathbf{x}^{p+1})$ is made here. This method tends to reduce the number of steps taken; however, the amount of computation at each step is increased due to the minimization problem which must be solved to determine each new step size. Since $f(\mathbf{x}^{p+1})$ is usually not known analytically as a function of the step size k, the one-dimensional minimization with respect to k is performed by one of the methods described in Chapter 5.

The one-dimensional problem to be solved at each step is to determine the step size k which minimizes $f(\mathbf{x}^{p+1})$ along the line $\mathbf{x}^{p+1} - \mathbf{x}^p = \nabla f(\mathbf{x}^p)$. This problem can be written as

$$\min_k f(\mathbf{x}^p + k\nabla f(\mathbf{x}^p)),$$

which is a function of k only, once $\mathbf{x}^p$ is specified. The necessary condition for this minimum is

$$\frac{\partial f}{\partial k}(\mathbf{x}^p + k\nabla f(\mathbf{x}^p)) = 0.$$

EXAMPLE 6.8 Determine the optimal step size for the problem

$$\min_{\mathbf{x}} \left[f(\mathbf{x}) = \tfrac{1}{2}\mathbf{x}^T\mathbf{Q}\mathbf{x} \right].$$

In this case the optimal step size can be determined analytically as follows:

$$\nabla f(\mathbf{x}^p) = \mathbf{Q}\mathbf{x}^p$$

and

$$\mathbf{x}^{p+1} = \mathbf{x}^p + k\mathbf{Q}\mathbf{x}^p.$$

So we wish to minimize

$$f(\mathbf{x}^{p+1}) = \tfrac{1}{2}(\mathbf{x}^p + k\mathbf{Q}\mathbf{x}^p)^T Q\,(\mathbf{x}^p + k\mathbf{Q}\mathbf{x}^p)$$
$$= \tfrac{1}{2}\left[\mathbf{x}^T\mathbf{Q}\mathbf{x} + k\mathbf{x}^T\mathbf{Q}^2\mathbf{x} + k\mathbf{x}^T\mathbf{Q}^T\mathbf{Q}\mathbf{x} + k^2\mathbf{x}^T\mathbf{Q}^T\mathbf{Q}^2\mathbf{x}\right]_{\mathbf{x}^p}$$

and

$$\frac{\partial f}{\partial k} = \tfrac{1}{2}\left[\mathbf{x}^T\mathbf{Q}^2\mathbf{x} + \mathbf{x}^T\mathbf{Q}^T\mathbf{Q}\mathbf{x} + 2k\mathbf{x}^T\mathbf{Q}^T\mathbf{Q}^2\mathbf{x}\right]_{\mathbf{x}^p} = 0.$$

For a quadratic form, $\mathbf{Q} = \mathbf{Q}^T$ and

$$k^* = -\left[\frac{\mathbf{x}^T\mathbf{Q}^2\mathbf{x}}{\mathbf{x}^T\mathbf{Q}^3\mathbf{x}}\right]_{\mathbf{x}^p}. \qquad (6.5)$$

Convergence Considerations

We say that an algorithm converges if

$$\lim_{p\to\infty} \nabla f(\mathbf{x}^p) = \mathbf{0}.$$

For quadratic forms, the gradient algorithm in Eq. (6.3) converges if $\mathbf{Q}$ is positive definite. However, as was shown in Example 6.4, the method does not attain the optimum in a finite number of steps. When the optimal-step-size method is employed, the gradient algorithm improves $f(\mathbf{x}^p)$ geometrically, i.e.,

$$\lim_{p\to\infty}\left[\frac{f(\mathbf{x}^{p+1}) - f(\mathbf{x}^*)}{f(\mathbf{x}^p) - f(\mathbf{x}^*)}\right] = \lim_{p\to\infty}\frac{f(\mathbf{x}^{p+1})}{f(\mathbf{x}^p)} = a,$$

where $0 \leqslant a < 1$. This convergence rate is linear in $f(\mathbf{x}^p)$ and is considerably slower than the quadratically convergent algorithms discussed in Chapter 9.

Occasionally the rate of convergence of the gradient algorithm can be accelerated by rescaling the objective function to a more symmetrical form.

EXAMPLE 6.9

$$\min_{\mathbf{x}} x_1^2 + 2x_2^2.$$

The elliptical level curves for this problem are shown in Fig. 6.1 along with the gradient path, which would require an infinite number of discrete steps to attain the minimum. If we let

$$y^2 = 2x_2^2,$$

then the problem is transformed to

$$\min_{x_1, y} x_1^2 + y^2$$

and the level curves are circles, so the gradient algorithm attains the minimum exactly in one step beginning from any point.

Since the gradient algorithm uses local gradient information only, it converges to a local minimum. Which one it converges to depends upon the initial point.

EXAMPLE 6.10 Consider the surface shown in Fig. 6.3, which represents two mountain peaks connected by a ridge. Examine the behavior of the gradient algorithm if the initial points are at A, B, and C, respectively.

The gradient path from point A will converge to the local maximum $f_A = 1000$, and the gradient algorithm from point B will converge to the local maximum $f_B = 2000$. The gradient path from point C moves across the ridge to C', and the gradient direction at C' points back toward C. Thus in two steps, almost no progress is made toward either maximum. This behavior along a ridge (or valley) is one of the drawbacks of the gradient algorithm. Note, however, that the optimal gradient would choose the top of the ridge to reevaluate the gradient and then would move along the ridge in an efficient manner.

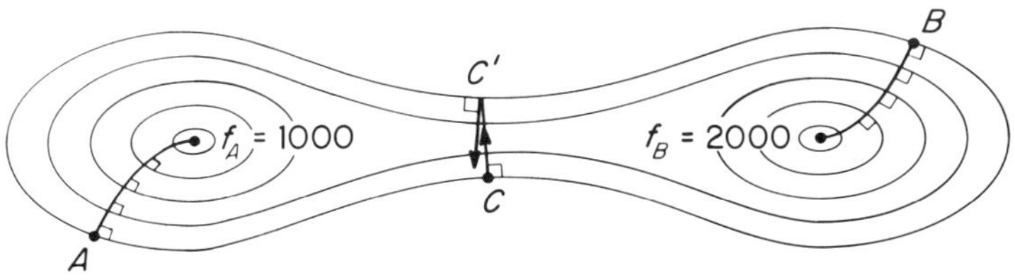

FIGURE 6.3

Several methods have been used to accelerate the convergence of the gradient algorithm along ridges. Typical of these is the algorithm

$$\mathbf{x}^{p+1}=\mathbf{x}^p+k_1\nabla f(\mathbf{x}^p)+k_2\nabla f(\mathbf{x}^{p-1}),$$

which uses gradient directions from the last two steps. If k_1 and k_2 are the optimal step sizes, this algorithm is called the *method of parallel tangents* or PARTAN (see Problem 6.11).

It should be clear that the number of steps required by the gradient algorithm will depend primarily on the location of the initial point. In general, the closer this point is to the optimum, the better. Often a direct enumeration of $f(\mathbf{x})$ over a coarse grid is warranted for choosing the proper initial point. This is especially true for surfaces which are not unimodal.

Of equal importance is the ending point, since by definition an algorithm will continue indefinitely unless it is given a stopping rule. Although one would like to stop as close to the optimum as possible, this precious point is not known, so no direct measure of

$$f(\mathbf{x}^p)-f(\mathbf{x}^*)$$

is possible. The most commonly used rule is to stop when the rate of improvement in $f(\mathbf{x}^p)$ falls below some positive threshold ε. Thus for minimization problems the rule is to stop when

$$f(\mathbf{x}^p)-f(\mathbf{x}^{p+1})\leqslant\varepsilon,$$

and for maximization problems the rule is to stop when

$$f(\mathbf{x}^{p+1})-f(\mathbf{x}^p)\leqslant\varepsilon.$$

Since the necessary condition

$$\nabla f(\mathbf{x}^*)=0$$

holds at $\mathbf{x}^*$, two alternative rules are to stop when

$$\max_i\left|\frac{\partial f}{\partial x_i}(\mathbf{x}^p)\right|<\varepsilon$$

or when

$$\sum_{i=1}^{n}\left(\frac{\partial f}{\partial x_i}(\mathbf{x}^p)\right)^2<\varepsilon.$$

A different stopping procedure is to abandon the gradient algorithm in the vicinity of the optimum and fit a quadratic surface through the last several points. The optimum is then approximated by applying the required necessary conditions to the quadratic surface (see Problem 6.12 and also "Quadratic Interpolation," pp. 130 ff.).

Throughout this chapter it has been assumed that the gradient vector could be determined analytically and then evaluated at some desired point. In case the gradient is not known analytically, it can be determined approximately at each point $\mathbf{x}^p$ by making small perturbations or experiments. Thus

$$\frac{\partial f}{\partial x_2}(\mathbf{x}^p) \approx \frac{f(\mathbf{x}^p + \Delta\mathbf{x}_i) - f(\mathbf{x}^p)}{\Delta x_i},$$

where $\Delta\mathbf{x}_i$ is a vector of zeros except for the ith element, which has the value Δx_i. Even when the gradient is known analytically but is very complex, this experimental procedure is often preferred for reasons of computational efficiency.

Solved Problems

6.1 For the surface

$$f(\mathbf{x}) = 3x_1 + 4x_2^2,$$

sketch the level curves and indicate the continuous gradient path for minimizing $f(\mathbf{x})$, beginning at some point (x_1^0, x_2^0) in the first quadrant.

The solution is shown in Fig. 6.4.

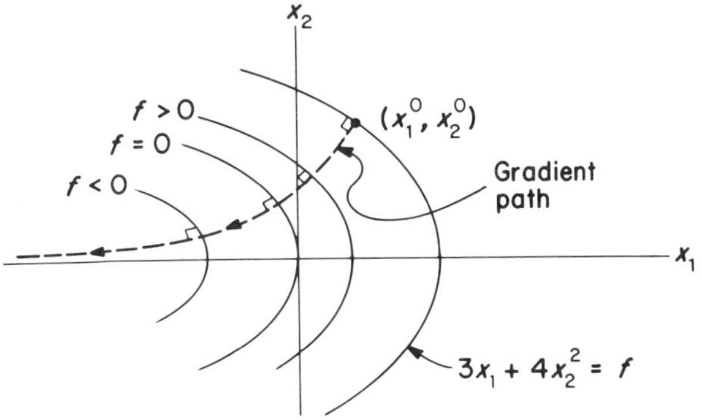

FIGURE 6.4

6.2 For Problem 6.1 determine analytically the continuous gradient path.

The gradient path is given by

$$\frac{dx_i}{ds} = -k\frac{\partial f}{\partial x_i}, \qquad i=1,2,$$

so

$$\frac{dx_1}{ds} = -k3, \qquad \frac{dx_2}{ds} = -k8x_2, \qquad \frac{dx_2}{dx_1} = \frac{8}{3}x_2.$$

Then

$$\int_{x_2^0}^{x_2}\frac{dx_2}{x_2} = \frac{8}{3}\int_{x_1^0}^{x_1}dx_1,$$

or

$$\ln\frac{x_2}{x_2^0} = \frac{8}{3}(x_1 - x_1^0).$$

So the gradient path is

$$x_2 = x_2^0\exp\left[\tfrac{8}{3}(x_1 - x_1^0)\right].$$

6.3 For Problem 6.1 determine the discrete form of the gradient algorithm

$$\mathbf{x}^{p+1} = \mathbf{x}^p + k\nabla f(\mathbf{x}^p), \qquad k<0.$$

Normalizing the gradient yields

$$\nabla f(\mathbf{x}^p) = \frac{1}{\sqrt{9+64x_2^2}}\begin{bmatrix} 3 \\ 8x_2 \end{bmatrix}_{\mathbf{x}^p}.$$

So

$$x_1^{p+1} = x_1^p + \frac{3k}{\sqrt{9+64(x_2^p)^2}}, \qquad k<0,$$

$$x_2^{p+1} = x_2^p + \frac{8kx_2^p}{\sqrt{9+64(x_2^p)^2}}, \qquad k<0.$$

6.4 Determine the step size k_p such that a specified percentage reduction in $f(\mathbf{x})$ is made at the pth iteration.

For the gradient algorithm,

$$\mathbf{x}^{p+1}=\mathbf{x}^p+\Delta\mathbf{x}^p,$$

where

$$\Delta\mathbf{x}^p=k_p\nabla f(\mathbf{x}^p).$$

To a first-order approximation,

$$\Delta f(\mathbf{x})=\frac{\partial f}{\partial x_1}(\mathbf{x})\Delta x_1+\frac{\partial f}{\partial x_2}(\mathbf{x})\Delta x_2+\cdots+\frac{\partial f}{\partial x_n}(\mathbf{x})\Delta x_n,$$

or

$$\begin{aligned}\Delta f(\mathbf{x}^p)&=[\nabla f(\mathbf{x}^p)]^T\Delta\mathbf{x}^p\\&=k_p[\nabla f(\mathbf{x}^p)]^T\nabla f(\mathbf{x}^p).\end{aligned}$$

If we wish $\nabla f(\mathbf{x}^p)$ to be some fixed proportion c of $f(\mathbf{x}^p)$, then

$$\Delta f(\mathbf{x}^p)=cf(\mathbf{x}^p),\qquad 0<c<1,$$

and

$$k_p=\frac{cf(\mathbf{x}^p)}{[\nabla f(\mathbf{x}^p)]^T\ \nabla f(\mathbf{x}^p)}.$$

It is reasonable to choose $c=0.1$, representing a 10% reduction in $f(\mathbf{x}^p)$ at each step.

6.5 Determine an approximation to the optimal step size k_p^* when $f(\mathbf{x})$ is twice differentiable.

Expanding $f(\mathbf{x})$ in a Taylor series about $\mathbf{x}^p$ and ignoring terms higher than second order gives

$$f(\mathbf{x}^p+\Delta\mathbf{x})=f(\mathbf{x}^p)+[\nabla f(\mathbf{x}^p)]^T\Delta\mathbf{x}+\tfrac{1}{2}\Delta\mathbf{x}^T\mathbf{H}(\mathbf{x}^p)\Delta\mathbf{x},$$

where $\mathbf{H}(\mathbf{x}^p)$ is the Hessian matrix. Along the gradient direction,

$$\Delta\mathbf{x}=k_p\nabla f(\mathbf{x}^p)=k_p\nabla f$$

So

$$f(\mathbf{x}^p+k_p\nabla f)=f(\mathbf{x}^p)+k_p(\nabla f)^T\nabla f+\tfrac{1}{2}k_p^2(\nabla f)^T\mathbf{H}(\mathbf{x}^p)\nabla f$$

$$\frac{\partial f}{\partial k_p}=(\nabla f)^T\nabla f+k_p(\nabla f)^T\mathbf{H}(\mathbf{x}^p)\nabla f=0$$

and

$$k_p^* = \left[\frac{-(\nabla f)^T \ \nabla f}{(\nabla f)^T \mathbf{H}(\mathbf{x}^p)\nabla f} \right]_{x_p}.$$

When

$$f(\mathbf{x}) = \tfrac{1}{2}\mathbf{x}^T\mathbf{Q}\mathbf{x},$$

we have

$$\nabla f(\mathbf{x}) = \mathbf{Q}\mathbf{x} \quad \text{and} \quad H(\mathbf{x}) = \mathbf{Q}.$$

Then the above expression for k_p is identical to the form determined in Example 6.8.

6.6 Under what conditions does the optimum step size in Example 6.9 converge in a finite number of steps?

In Example 6.9,

$$f(\mathbf{x}) = \tfrac{1}{2}\mathbf{x}^T\mathbf{Q}\mathbf{x} \quad \text{and} \quad k^* = -\frac{\mathbf{x}^T\mathbf{Q}^2\mathbf{x}}{\mathbf{x}^T\mathbf{Q}^3\mathbf{x}}.$$

So

$$\begin{aligned} f(\mathbf{x}^{p+1}) &= \tfrac{1}{2}(\mathbf{x} + k\mathbf{Q}\mathbf{x})^T\mathbf{Q}(\mathbf{x} + k\mathbf{Q}\mathbf{x}) \\ &= \tfrac{1}{2}[\mathbf{x}^T\mathbf{Q}\mathbf{x} + k\mathbf{x}^T\mathbf{Q}\mathbf{x} + k\mathbf{x}^T\mathbf{Q}^2\mathbf{x} + k\mathbf{x}^T\mathbf{Q}^3\mathbf{x}]_{\mathbf{x}^p}. \end{aligned}$$

Substituting for k yields

$$f(\mathbf{x}^{p+1}) = \frac{1}{2}\left[\mathbf{x}^T\mathbf{Q}\mathbf{x} - \frac{(\mathbf{x}^T\mathbf{Q}^2\mathbf{x})^2}{\mathbf{x}^T\mathbf{Q}^3\mathbf{x}} \right]_{\mathbf{x}^p}$$

If $\mathbf{Q}$ is positive definite, a minimum exists at the origin and $\mathbf{Q}$ can be written as a diagonal matrix having positive eigenvalues λ_i. Thus

$$f(\mathbf{x}^{p+1}) = \frac{1}{2}\left[\sum_{i=1}^{n} \lambda_i x_i^2 - \frac{\left(\sum_{i=1}^{n} \lambda_i^2 x_i^2 \right)^2}{\sum_{i=1}^{n} \lambda_i^3 x_i^2} \right]_{\mathbf{x}^p}$$

In order to reach the origin for some finite p,

$$f(\mathbf{x}^{p+1}) = 0,$$

which requires that

$$\left(\sum_{i=1}^{n} \lambda_i x_i^2\right)\left(\sum_{i=1}^{n} \lambda_i^3 x_i^2\right)=\left(\sum_{i=1}^{n} \lambda_i^2 x_i^2\right)^2.$$

This equality is satisfied if all but one x_i^p is 0, i.e., if the point $\mathbf{x}^p$ lies on one of the coordinates. In general, this will not occur unless the initial point is chosen on the coordinate. The one exception occurs when $\lambda_i=\lambda$ for all i, in which case the method converges in one step for any value of $\mathbf{x}^p$. The level curves in two dimensions are circles for this case.

6.7 Show that successive steps of the optimal-gradient algorithm are orthogonal.

The normal vectors at $\mathbf{x}^p$ and $\mathbf{x}^{p+1}$ are given by $\nabla f(\mathbf{x}^p)$ and $\nabla f(\mathbf{x}^{p+1})$ respectively. These two vectors are orthogonal if

$$(\nabla f(\mathbf{x}^p), \nabla f(\mathbf{x}^{p+1}))=0,$$

or, using the gradient algorithm, if

$$(\nabla f(\mathbf{x}^p), \nabla f(\mathbf{x}^p + k\nabla f(\mathbf{x}^p)))=0.$$

For the optimal-gradient algorithm, it is necessary that

$$\frac{\partial f}{\partial k}(\mathbf{x}^{p+1})=0,$$

or equivalently that

$$\frac{\partial f}{\partial k}(\mathbf{x}^p + k\nabla f(\mathbf{x}^p))=0.$$

This partial differentiation yields

$$(\nabla f(\mathbf{x}^p + k\nabla f(\mathbf{x}^p)), \nabla f(\mathbf{x}^p))=0,$$

which is the required relation for orthogonality.

6.8 For the quadratic form

$$f(x)=\tfrac{1}{2}\mathbf{x}^T\mathbf{Q}\mathbf{x},$$

show that successive steps of the optimal-gradient algorithm are orthogonal.

From Problem 6.7,

$$(\nabla f(\mathbf{x}^p), \nabla f(\mathbf{x}^{p+1}))=0,$$

where

$$\nabla f(\mathbf{x}^p)=\mathbf{Qx},$$

$$\nabla f(\mathbf{x}^{p+1})=\mathbf{Q}(\mathbf{x}+k\mathbf{Qx}),$$

$$k^*=-\frac{\mathbf{x}^T\mathbf{Q}^2\mathbf{x}}{\mathbf{x}^T\mathbf{Q}^3\mathbf{x}}.$$

Substituting into the orthogonality relation yields

$$(\mathbf{Qx},\mathbf{Qx}+k\mathbf{Q}^2\mathbf{x})=\mathbf{x}^T\mathbf{Q}^2\mathbf{x}-\left(\frac{\mathbf{x}^T\mathbf{Q}^2\mathbf{x}}{\mathbf{x}^T\mathbf{Q}^3\mathbf{x}}\right)\mathbf{x}^T\mathbf{Q}^3\mathbf{x}$$

$$=\mathbf{x}^T\mathbf{Q}^2\mathbf{x}-\mathbf{x}^T\mathbf{Q}^2\mathbf{x}=0.$$

6.9 Apply the optimal-gradient algorithm to the quadratic form

$$f(\mathbf{x})=\tfrac{1}{2}(x_1^2+2x_2^2)$$

beginning at the point (4,4), and show that

$$(\nabla f(\mathbf{x}^0),\nabla f(\mathbf{x}^1))=0.$$

In matrix form,

$$f(\mathbf{x})=\tfrac{1}{2}\mathbf{x}^T\mathbf{Qx},$$

where

$$\mathbf{Q}=\begin{bmatrix}1&0\\0&2\end{bmatrix}.$$

Then

$$\mathbf{Q}^2=\begin{bmatrix}1&0\\0&4\end{bmatrix},\quad \mathbf{Q}^3=\begin{bmatrix}1&0\\0&8\end{bmatrix},$$

and

$$\nabla f(\mathbf{x})=\begin{bmatrix}x_1\\2x_2\end{bmatrix},$$

$$k^0=\frac{-[4,4]\begin{bmatrix}1&0\\0&4\end{bmatrix}\begin{bmatrix}4\\4\end{bmatrix}}{[4,4]\begin{bmatrix}1&0\\0&8\end{bmatrix}\begin{bmatrix}4\\4\end{bmatrix}}=-\frac{5}{9}.$$

So

$$\mathbf{x}^1 = \begin{bmatrix} 4 \\ 4 \end{bmatrix} - \frac{5}{9}\begin{bmatrix} 4 \\ 8 \end{bmatrix} = \begin{bmatrix} \frac{16}{9} \\ -\frac{4}{9} \end{bmatrix},$$

$$\nabla f(\mathbf{x}^0) = \begin{bmatrix} 4 \\ 8 \end{bmatrix},$$

$$\nabla f(\mathbf{x}^1) = \begin{bmatrix} \frac{16}{9} \\ -\frac{8}{9} \end{bmatrix},$$

and

$$(\nabla f(\mathbf{x}^0), \nabla f(\mathbf{x}^1)) = [4, 8]\begin{bmatrix} \frac{16}{9} \\ -\frac{8}{9} \end{bmatrix} = 0.$$

From Fig. 6.5 we see that the gradient direction at $\mathbf{x}^1$ is also orthogonal to the level curve at that point.

6.10 Perform four iterations of the optimal-gradient algorithm to find

$$\min_{\mathbf{x}} \; 2x_1^2 + 2x_1x_2 + 5x_2^2$$

beginning at the point 2, −2.

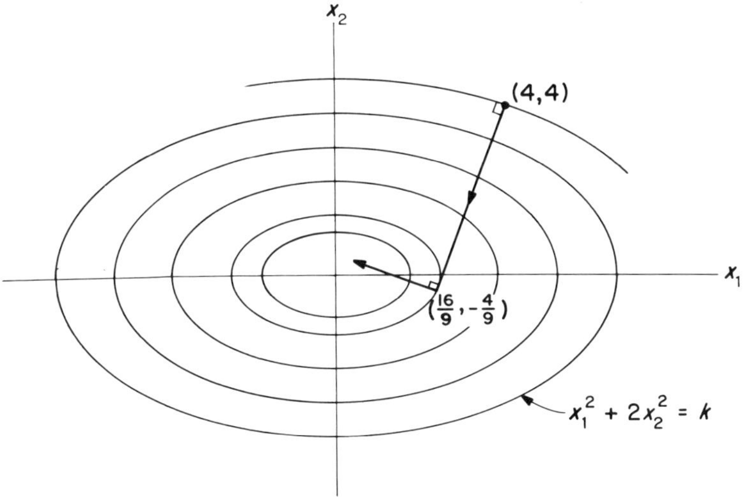

FIGURE 6.5

In matrix form,

$$f(\mathbf{x})=\tfrac{1}{2}\mathbf{x}^T\begin{bmatrix}4 & 2\\ 2 & 10\end{bmatrix}\mathbf{x},\qquad \nabla f(\mathbf{x})=\mathbf{Q}\mathbf{x}=\begin{bmatrix}4x_1+2x_2\\ 2x_1+10x_2\end{bmatrix}.$$

Since

$$Q^2=\begin{bmatrix}20 & 28\\ 28 & 104\end{bmatrix}\quad\text{and}\quad Q^3=\begin{bmatrix}136 & 320\\ 320 & 1096\end{bmatrix},$$

by Eq. (6.5) we have

$$k^*=\frac{-(20x_1^2+56x_1x_2+104x_2^2)}{136x_1^2+640x_1x_2+1096x_2^2},$$

and at the initial point 2, −2,

$$k_0^*=-0.122.$$

So

$$\mathbf{x}^1=\begin{bmatrix}2\\ -2\end{bmatrix}-0.122\begin{bmatrix}4\\ -16\end{bmatrix}=\begin{bmatrix}1.511\\ -0.046\end{bmatrix}.$$

In a similar manner,

$$k_1^*=-0.156,\qquad x_1^2=0.577,\qquad x_2^2=-0.448,$$

$$k_2^*=-0.141,\qquad x_1^3=0.377,\qquad x_2^3=0.022,$$

$$k_3^*=-0.134,\qquad x_1^4=0.183,\qquad x_2^4=-0.101.$$

6.11 For a quadratic function of n variables, the method of parallel tangents (PARTAN) locates the optimal point in $2n-1$ steps. In PARTAN, after choosing the starting point $\mathbf{x}^0$, the next two points $\mathbf{x}^1$ and $\mathbf{x}^2$ are determined by the optimal-gradient algorithm. Then $\mathbf{x}^3$ is located along the direction $\mathbf{x}^0-\mathbf{x}^2$, and this process continues.

For the two-dimensional quadratic form

$$f(\mathbf{x})=x_1^2+2x_2^2,$$

show that the PARTAN method will converge exactly from an arbitrary initial point 5, 5 to the optimum 0, 0 in three steps.

In matrix form,

$$f(\mathbf{x})=\tfrac{1}{2}\mathbf{x}^T\begin{bmatrix}2 & 0\\ 0 & 4\end{bmatrix}\mathbf{x},$$

$$\nabla f(\mathbf{x})=\begin{bmatrix}2x_1\\ 4x_2\end{bmatrix},\qquad \mathbf{Q}^2=\begin{bmatrix}4 & 0\\ 0 & 16\end{bmatrix},\qquad \mathbf{Q}^3=\begin{bmatrix}8 & 0\\ 0 & 64\end{bmatrix}.$$

The optimal step size is

$$k^* = \frac{-(4x_1^2 + 16x_2^2)}{8x_1^2 + 64x_2^2},$$

so

$$k_0^* = -\tfrac{20}{72}$$

and

$$x^1 = \begin{bmatrix} 5 \\ 5 \end{bmatrix} - 0.2778 \begin{bmatrix} 10 \\ 20 \end{bmatrix} = \begin{bmatrix} \frac{20}{9} \\ -\frac{5}{9} \end{bmatrix},$$

$$k_1^* = -\tfrac{5}{12},$$

$$\mathbf{x}^2 = \begin{bmatrix} \frac{20}{9} \\ -\frac{5}{9} \end{bmatrix} - \frac{5}{12} \begin{bmatrix} \frac{40}{9} \\ -\frac{20}{9} \end{bmatrix} = \frac{1}{108} \begin{bmatrix} 40 \\ 40 \end{bmatrix}.$$

Now

$$\mathbf{x}^0 - \mathbf{x}^2 = \begin{bmatrix} 5 \\ 5 \end{bmatrix} - \frac{1}{108} \begin{bmatrix} 40 \\ 40 \end{bmatrix} = \frac{1}{108} \begin{bmatrix} 500 \\ 500 \end{bmatrix},$$

and $\mathbf{x}^3$ will be located along $\mathbf{x}^0 - \mathbf{x}^2$. Hence

$$\mathbf{x}^3 = \mathbf{x}^2 + k_3(\mathbf{x}^0 - \mathbf{x}^2) = \frac{1}{108} \begin{bmatrix} 40 - 500k_3 \\ 40 - 500k_3 \end{bmatrix}.$$

Now

$$f(x^3) = \frac{1}{(108)^2} \left[(40 - 500k_3)^2 + 2(40 - 500k_3)^2 \right],$$

and the necessary condition $\partial f / \partial k_3 = 0$ yields $k_3^* = 40/500$. Therefore

$$\mathbf{x}^3 = \frac{1}{108} \begin{bmatrix} 40 - 40 \\ 40 - 40 \end{bmatrix} = \begin{bmatrix} 0 \\ 0 \end{bmatrix},$$

and the PARTAN algorithm has converged exactly to the minimum in three steps. These three steps are shown in Fig. 6.6.

6.12 The gradient algorithm is rather inefficient in the vicinity of the optimum, due to the small step size that is required there. One procedure for speeding convergence in this region is to fit a quadratic surface through several points in the region and then approximate the optimum by minimizing the quadratic surface.

Consider the application of the gradient algorithm with fixed step size $k = 0.03$ to the problem

$$\min_x \left[f(x) = \exp(2x^2 + x + 1) \right].$$

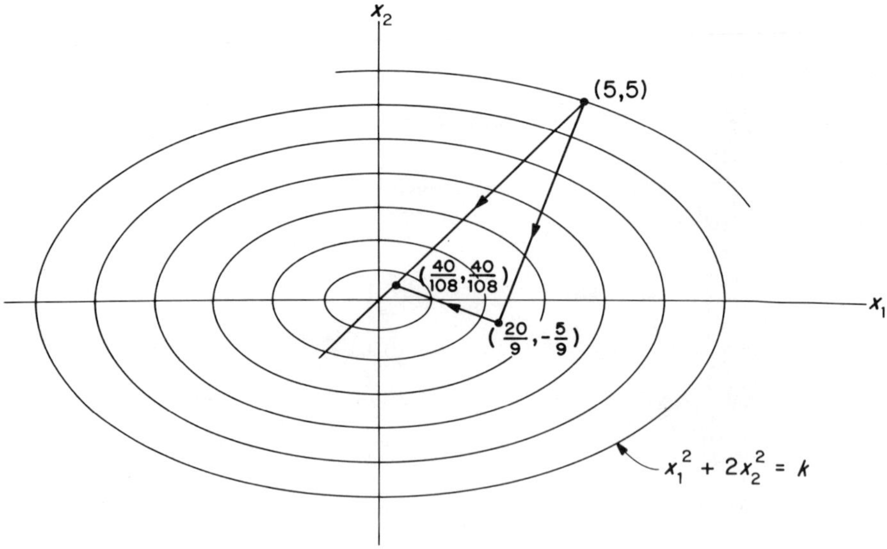

FIGURE 6.6

Suppose the last three iterations were

$$x = 0.03, \qquad f(0.03) = \exp(1.0318),$$

$$x = 0.00, \qquad f(0.00) = \exp(1.0),$$

$$x = -0.03, \qquad f(-0.03) = \exp(0.9718).$$

We now fit the quadratic

$$\hat{f}\ (x) = ax^2 + bx + c$$

through the three points to uniquely determine a, b and c. This procedure yields

$$a = 5.55, \qquad b = 2.83, \qquad c = 2.72.$$

Minimizing $\hat{f}(x)$,

$$\frac{\partial \hat{f}}{\partial x} = 2ax + b = 0,$$

$$\hat{x}^* = \frac{-b}{2a} = -0.255.$$

The actual minimum is found by minimizing $f(x)$. Thus

$$\frac{\partial f}{\partial x} = (4x+1)\exp(2x^2 + x + 1) = 0,$$

and

$$x^* = -0.250.$$

In this case, $f(\hat{x}^*) = 2.400$ and $f(x^*) = 2.400$ (calculated to three decimal places of accuracy), so the error is less than 0.1%. Note that another seven gradient steps of length 0.03 would be required to reach $x = -0.24$, at which point $f(-0.24)$ would still yield comparable accuracy. A reduction in the step size and further iterations would be required to achieve greater accuracy.

Supplementary Problems

6.13 Use the gradient algorithm in Eq. (6.4) to find

$$\max_{\mathbf{x}} 36 - 9(x_1 - 6)^2 - 4(x_2 - 6)^2$$

beginning at the point (7,9), and using several different values of k.

6.14 Perform four iterations of the optimal gradient algorithm to find

$$\min_{\mathbf{x}} x_1^2 + x_2^2 + 2x_3^2$$

beginning at the point $(2, -2, 1)$.

6.15 Solve Problem 6.14 for

$$\min_{\mathbf{x}} x_1^2 + x_2^2 + x_3^2 + x_4^2$$

beginning at the point $(2, -2, 1, -1)$.

6.16 For the quadratic form

$$f(\mathbf{x}) = x_1^2 + 5x_2^2,$$

derive the continuous gradient path from an arbitrary initial point $\mathbf{x}^0$. Sketch the level curves and the gradient path.

6.17 For Problem 6.16 derive the discrete gradient algorithm using a fixed step size L. Determine five gradient steps for $L = 2$ beginning at the point $(5, 10)$.

6.18 Repeat Problem 6.17 for $L = 4$, but reduce L by one-half whenever $f(\mathbf{x}^p)$ starts to increase.

6.19 Show that the optimal-gradient method converges for the quadratic form

$$f(\mathbf{x}) = \tfrac{1}{2}\mathbf{x}^T\mathbf{Q}\mathbf{x}$$

if $\mathbf{Q}$ is positive definite.

6.20 In Chapter 4 we showed how Lagrange multipliers could be used to solve problems of the form

$$\min_{\mathbf{x}} f(\mathbf{x})$$

such that

$$g_i(\mathbf{x}) \leqslant 0, \qquad i = 1, 2, \ldots, r.$$

Show how the gradient algorithm can be used to find

$$\min_{\mathbf{x}} \left\{ L(\mathbf{x}, \lambda^*) = f(\mathbf{x}) + \sum_{i=1}^{r} \lambda_i^* g_i(\mathbf{x}) \right\}.$$

6.21 Determine the minimum of

$$f(\mathbf{x}) = 3x_1^2 + 4x_2^2$$

exactly in three steps using the PARTAN algorithm beginning at the point (6, 6).

Chapter 7

Constrained Gradient Techniques

There are many algorithms which have been proposed to solve constrained optimization problems using gradient techniques. Although these algorithms are too numerous to attempt a comprehensive treatment here, they can generally be divided into two categories: (1) boundary-following methods, and (2) penalty-function methods. The former are treated in this chapter, and the latter are treated in Chapter 8. Several algorithms from each category are discussed.

Within the category of boundary-following methods, there are many algorithms which apply to particular classes of problems. One such class occurs when either the objective function or the constraints are linear. In this case, many algorithms utilize linear approximations to the nonlinear equations and may imbed a linear-programming solution in an algorithm which successively improves on the approximations. These methods can generally be adapted with some extra effort to problems having nonlinearities in both the objective function and the constraints. The *feasible-directions method* and the *gradient-projection method*, discussed in this chapter, are both of this type.

For the special case when the constraints are linear and the objective function is quadratic, a solution can be easily obtained using linear programming techniques. This special case is called quadratic programming and is described in Chapter 10.

Although the gradient algorithm discussed in Chapter 6 is not known for its rapid convergence, it is simple to apply and is applicable to most nonlinear functions which are differentiable. For these reasons, we confine our attention in this chapter to methods which use the gradient algorithm in the solution of constrained optimization problems.

Boundary Following with Inequality Constraints

In Chapter 6, the feasible region for applying the gradient algorithm was assumed to be the entire space E^n. When inequality constraints are added, the feasible region is constrained to be in the interior or on the boundary of some set $S \subset E^n$. By choosing a starting point inside the feasible region, the gradient path can be followed until a constraint boundary is encountered. The task then is to determine an alternative method for improving the value of the objective function without leaving the feasible region. This usually requires a deviation from the gradient path in favor of a move along the boundary—hence the name *boundary-following* methods.

Perhaps the simplest boundary-following technique is to switch directions from the gradient of the objective function to the gradient of the violated constraint whenever a move outside the feasible region occurs. This method is known as *hemstitching*, since the path zigzags along the constraint boundary, alternating directions between the gradient of the objective function and the gradient of the constraint.

EXAMPLE 7.1 Apply the hemstitching algorithm to

$$\min x_1^2 + 2x_2^2$$

such that

$$x_1 + x_2 \geqslant 4.$$

Begin at the feasible point (1,4.5) and take fixed steps of length 1.0.

The normalized gradient paths are

$$\nabla f(\mathbf{x}) = \left(4x_1^2 + 16x_2^2\right)^{-1/2} \begin{bmatrix} 2x_1 \\ 4x_2 \end{bmatrix}$$

for the objective function, and

$$\nabla g(\mathbf{x}) = \frac{1}{\sqrt{2}} \begin{bmatrix} 1 \\ 1 \end{bmatrix}$$

for the constraint.

The steps are calculated from the algorithm

$$\mathbf{x}^{p+1} = \mathbf{x}^p - 1.0 \nabla f(\mathbf{x}^p)$$

if $\mathbf{x}^p$ is inside the feasible region, and

$$\mathbf{x}^{p+1} = \mathbf{x}^p + 1.0 \nabla g(\mathbf{x}^p)$$

if $\mathbf{x}^p$ is outside the feasible region. Using this method, the steps are:

$$
\begin{array}{lll}
x_1^0=1, & x_2^0=4.50 & \text{(feasible)},\\
x_1^1=0.89, & x_2^1=3.50 & \text{(feasible)},\\
x_1^2=0.76, & x_2^2=2.50 & \text{(unfeasible)},\\
x_1^3=1.47, & x_2^3=3.21 & \text{(feasible)},\\
x^41=1.05, & x_2^4=2.30 & \text{(unfeasible)},\\
x^51=1.76, & x_2^5=3.01 & \text{(feasible)},\\
x_1^6=1.37, & x_2^6=2.10 & \text{(unfeasible)},\\
x_1^7=2.08, & x_2^7=2.81 & \text{(feasible)}.
\end{array}
$$

These steps in the hemstitching algorithm are shown in Fig. 7.1.

The minimum lies on the constraint $x_2=4-x_1$ at the point which satisfies

$$\frac{\partial}{\partial x_1}\left[x_1^2+2(4-x_1)^2\right]=0,$$

or $x_1^*=2.667$, $x_2^*=1.333$. At the optimum,

$$-\nabla f(\mathbf{x}^*)=\begin{bmatrix}-0.707\\-0.707\end{bmatrix} \quad \text{and} \quad \nabla g(\mathbf{x}^*)=\begin{bmatrix}0.707\\0.707\end{bmatrix}.$$

Thus the gradients of the objective function and the constraint are equal and opposite at the optimum, as expected.

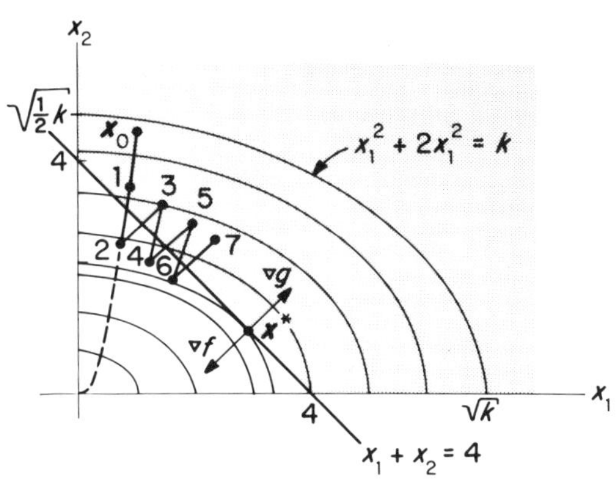

FIGURE 7.1

The hemstitching method is not very efficient, since many steps are required to move a small distance along the boundary. However, the algorithm does provide an automatic stopping rule, since progress along the boundary ceases when the gradient vectors are collinear. In this case, the rule is to stop when

$$-\nabla f(\mathbf{x}^*) = \nabla g(\mathbf{x}^*) + \varepsilon,$$

where ε is chosen to suitably small.

Since the gradient algorithm uses local information only, there is always the possibility that it will determine a local optimum which is not the global optimum. This can occur whenever the feasible region (or the objective function) is nonconvex.

EXAMPLE 7.2 Show the path taken by the hemstitching algorithm on the level curves in Fig. 7.2. Determine two paths beginning at the points A and B respectively.

The path from A obtains a local optimum which is not global at A', and the path from B obtains a local optimum which is also global at B'.

As we discussed in Chapter 2, the only ways to guarantee a global optimum are either to specify that all functions are convex or to choose a sufficient number of different starting points so that all local optima are obtained. In general, the number of different starting points required is not known *a priori* when the number of stationary points is not known. In this

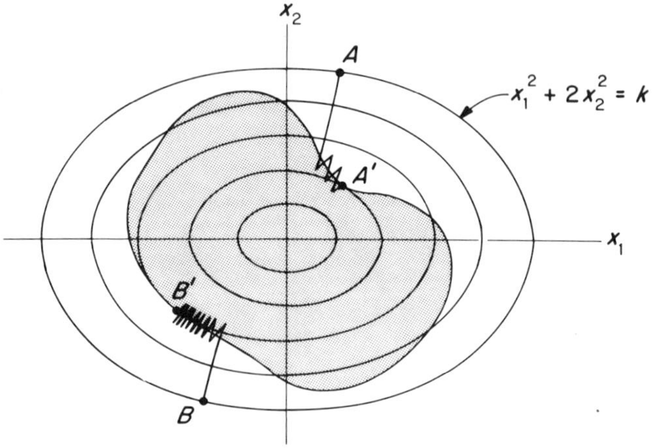

FIGURE 7.2

case, a grid search of $f(\mathbf{x})$ over the feasible region is warranted in order to get some advance information about the nature of the surface.

For inequality-constrained problems, either the solution is strictly within the feasible region or else it is on a boundary. In some cases a good procedure is to determine the unconstrained optimum without regard for the constraints and then to test the feasibility of this point. If it is feasible, we are finished; and if it is nonfeasible, we must determine the optimum boundary point. One way to proceed is to move from the unconstrained optimum in the direction of the gradient of the violated constraint. In the event that more than one constraint is violated, we can take our direction as the linear combination of their gradients as determined by the parallelogram rule. This is then an *exterior method*, since we are moving from a nonfeasible point toward the feasible region.

EXAMPLE 7.3 For the problem described by the level curves in Fig. 7.3, it is efficient to ignore the inequality constraint, since any starting point in the shaded region would encounter its boundary unnecessarily.

We now seek an algorithm which will move along the constraint boundary in a more efficient manner than hemstitching. One such algorithm involves forming the vector summation of the gradients of the objective function and the violated constraints. Thus one tends to move along the constraint in the desirable direction. This method of *multiple gradient summation* is defined by the algorithm

$$\mathbf{x}^{p+1} = \mathbf{x}^p + k_p \mathbf{d}^p,$$

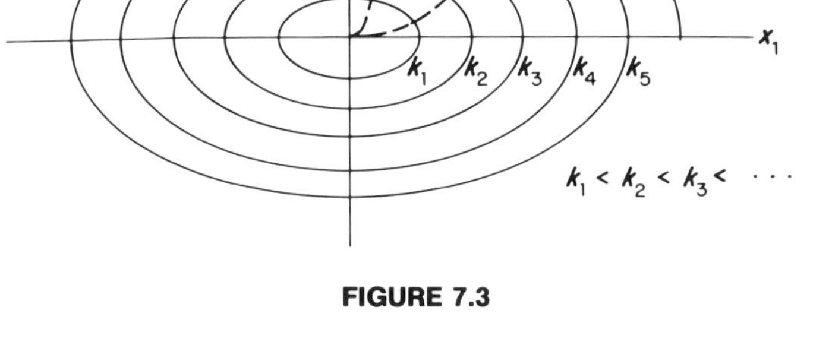

FIGURE 7.3

where

$$\mathbf{d}^p = \nabla f(\mathbf{x}^p)$$

if x^p is inside the feasible region, and

$$\mathbf{d}^p = \frac{\nabla f(\mathbf{x}^p)}{\|\nabla f(\mathbf{x}^p)\|} + \sum_{i=1}^{m} \frac{\nabla g_i(\mathbf{x}^p)}{\|\nabla g_i(\mathbf{x}^p)\|} \tag{7.1}$$

if $\mathbf{x}^p$ is outside the feasible region. Equation (7.1) for $\mathbf{d}^p$ holds for minimization problems where the feasible region is described by

$$g_i(\mathbf{x}) \leqslant 0, \qquad i = 1, 2, \ldots, r,$$

and $k_p < 0$. For other cases the signs of either $\nabla f(\mathbf{x}^p)$ or $\nabla g_i(\mathbf{x}^p)$ must be modified appropriately. The summation in (7.1) is taken only over the set of m constraints which are violated at $\mathbf{x}^p$. This method tends to generate moves into the feasible region, although some difficulty occurs when the objective-function contours are nearly parallel to the constraints. In this case the region of feasible improvement may be too small to find.

EXAMPLE 7.4 For Example 7.1 determine the direction of multiple gradient summation at the point where the discrete gradient path from 1, 4.5 crosses the constraint boundary.

The discrete gradient path from 1, 4.5 is given by

$$x_1 = 1 - 0.111k,$$

$$x_2 = 4.5 - 0.995k.$$

The intersection of this path and the constraint

$$x_2 = 4 - x_1$$

is the point $\hat{\mathbf{x}} = [0.85, 3.15]^T$.

The normalized gradients at this point are

$$\frac{\nabla f(\hat{\mathbf{x}})}{\|\nabla f(\hat{\mathbf{x}})\|} = \frac{1}{\sqrt{2^2(0.85)^2 + 4^2(3.15)^2}} \begin{bmatrix} 2(0.85) \\ 4(3.15) \end{bmatrix} = \begin{bmatrix} 0.134 \\ 0.993 \end{bmatrix},$$

$$\frac{\nabla g(\hat{\mathbf{x}})}{\|\nabla g(\hat{\mathbf{x}})\|} = \begin{bmatrix} 0.707 \\ 0.707 \end{bmatrix}.$$

Since the inequality constraint is greater than or equal to, the sign of the constraint term in Eq. (7.1) is reversed and the direction of multiple

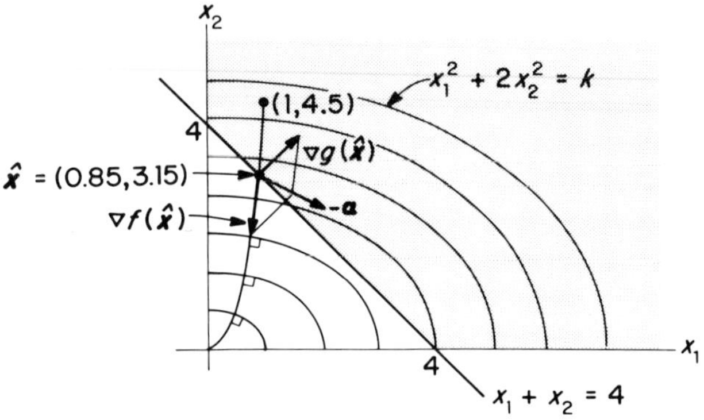

FIGURE 7.4

gradient summation is (taking $k = -1$)

$$-\mathbf{d} = \begin{bmatrix} -0.134 \\ -0.993 \end{bmatrix} - \begin{bmatrix} -0.707 \\ -0.707 \end{bmatrix} = \begin{bmatrix} 0.573 \\ -0.286 \end{bmatrix}.$$

This direction is shown in Fig. 7.4. Note that a move along the boundary would have the components $1, -1$.

We note from Example 7.4 that although the direction of multiple gradient summation points toward the feasible region, it does not follow the constraint boundary as closely as possible. The *method of feasible directions* overcomes this shortcoming by incorporating a linear programming problem into the algorithm at each step to determine the feasible direction which is closest to the gradient of the objective function (and hence closest to the constraint boundary).

Consider the problem

$$\min_{\mathbf{x}} f(\mathbf{x})$$

such that

$$g_i(\mathbf{x}) \leqslant 0, \qquad i = 1, 2, \ldots, r,$$

and expand each nonlinear function in a Taylor series about some point $\mathbf{x}^p$, ignoring terms higher than first order. Then

$$f(\mathbf{x}^p + \mathbf{d}^p) \approx f(\mathbf{x}^p) + (\nabla f(\mathbf{x}^p), \mathbf{d}^p),$$

$$g_i(\mathbf{x}^p + \mathbf{d}^p) \approx g_i(\mathbf{x}^p) + (\nabla g_i(\mathbf{x}^p), \mathbf{d}^p). \tag{7.2}$$

Since we are minimizing, an improvement in $f(\mathbf{x}^p)$ will result if

$$f(\mathbf{x}^p+\mathbf{d}^p)-f(\mathbf{x}^p)\approx(\nabla f(\mathbf{x}^p),\mathbf{d}^p)<0. \tag{7.3}$$

If $\mathbf{x}^p$ is strictly inside the feasible region $[g_i(\mathbf{x}^p)<0]$, then there exists some step size $k_p>0$ such that

$$\mathbf{x}^{p+1}=\mathbf{x}^p+k_p\,\mathbf{d}^p$$

is feasible for any set of directions $\mathbf{d}^p$. On the other hand, if $\mathbf{x}^p$ is on the boundary $g_i(\mathbf{x}^p)=0$, then from Eq. (7.2), $\mathbf{x}^{p+1}$ will be feasible for some $k_p>0$ if

$$g_i(\mathbf{x}^p)+(\nabla g_i(\mathbf{x}^p),\mathbf{d}^p)<0, \qquad i=1,2,\ldots,r. \tag{7.4}$$

If equality is permitted in (7.4), then $\mathbf{x}^{p+1}$ will lie on the tangent hyperplane emanating from $\mathbf{x}^p$, which would be a feasible point only if $g_i(\mathbf{x})$ were linear. In this case, (7.4) becomes

$$(\nabla g_i(\mathbf{x}^p),\mathbf{d}^p)\leqslant 0, \qquad i=1,2,\ldots,r. \tag{7.5}$$

In order to determine the direction $\mathbf{d}^p$ which leads to a feasible point $\mathbf{x}^{p+1}$ and also stay as close as possible to the gradient direction for the objective function, we introduce the slack variable $x_0\geqslant 0$ to convert the strict inequalities (7.3) and (7.4) to normal inequalities, and then solve the following linear programming problem at each point:

$$\max_{\mathbf{d}^p} x_0$$

such that

$$(\nabla f(\mathbf{x}^p),\mathbf{d}^p)+x_0\leqslant 0,$$

$$g_i(\mathbf{x}^p)+(\nabla g_i(\mathbf{x}^p),\mathbf{d}^p)+x_0\leqslant 0, \qquad i=1,2,\ldots,r,$$

or equivalently

$$\max_{\mathbf{d}^p} x_0 \tag{7.6}$$

such that

$$x_0+(\nabla f(\mathbf{x}^p),\mathbf{d}^p)\leqslant 0, \tag{7.7}$$

$$x_0+(\nabla g_i(\mathbf{x}^p),\mathbf{d}^p)\leqslant -g_i(x^p), \qquad i=1,2,\ldots,r. \tag{7.8}$$

When $x_0=0$, no further improvement can be made and the constrained optimum has been found. Thus when $x_0 \geqslant 0$ becomes active, the problem is solved.

Since the elements of $\mathbf{d}^p$ are unconstrained in sign, we must add constraints

$$d_j^p \leqslant b \tag{7.9}$$

for all j in order to avoid unbounded solutions. For the purpose of determing a *direction* $\mathbf{d}^p$, we are interested in relative values of d_j only. Thus b is taken to be an arbitrary positive constant (such as 1).

Summarizing, the algorithm is now:

1. Evaluate $\nabla f(\mathbf{x}^p)$ and $\nabla g_i(\mathbf{x}^p)$, $i=1,2,\ldots,r$.

2. Solve the linear programming problem given by Eqs. (7.6)–(7.9).

3. Stop if $x_0=0$.

4. Otherwise, determine a step size $k_p>0$ such that $f(\mathbf{x}^p+k_p\mathbf{d}^p)$ is minimized over all feasible points.

EXAMPLE 7.5 For Example 7.4, use the method of feasible directions to determine the path from the boundary point $\hat{\mathbf{x}}=(0.85,3.15)$.

Since the constraint is linear, Eq. (7.5) applies. Also the sense of the constraint inequality is reversed from that of the above derivation. Thus the linear programming problem is given by

$$\max_{\mathbf{d}} x_0$$

such that

$$x_0+(\nabla f(\hat{\mathbf{x}}),\mathbf{d}) \leqslant 0,$$

$$(-\nabla g(\hat{\mathbf{x}}),\mathbf{d}) \leqslant 0,$$

$$d_1 \leqslant 1, \qquad d_2 \leqslant 1,$$

where

$$\nabla f(\hat{\mathbf{x}})=\begin{bmatrix}0.134\\0.993\end{bmatrix} \quad \text{and} \quad \nabla g(\hat{\mathbf{x}})=\begin{bmatrix}0.707\\0.707\end{bmatrix}.$$

The constraint equations can be written

$$0.134d_1+0.993d_2 \leqslant -x_0,$$
$$-0.707d_1-0.707d_2 \leqslant 0,$$
$$d_1 \leqslant 1, \qquad d_2 \leqslant 1,$$

or

$$d_2 \leqslant \frac{-1}{0.993}x_0 - \frac{0.134}{0.993}d_1 = -1.008x_0 - 0.135d_1,$$
$$d_2 \geqslant -d_1,$$
$$d_1 \leqslant 1, \qquad d_2 \leqslant 1.$$

The feasible region is shown in Fig. 7.5, and the solution which maximizes x_0 is $d_1=1$, $d_2=-1$, and $x_0=0.993$. This direction leads precisely along the constraint boundary shown in Fig. 7.4.

The minimum occurs at $\mathbf{x}^*=[2,667, 1.333]^T$. At this point the gradients are

$$\nabla g(\mathbf{x}^*)=\begin{bmatrix} 0.707 \\ 0.707 \end{bmatrix}, \qquad \nabla g(\mathbf{x}^*)=\begin{bmatrix} 0.707 \\ 0.707 \end{bmatrix},$$

and the linear programming problem is

$$\max_{\mathbf{d}} x_0$$

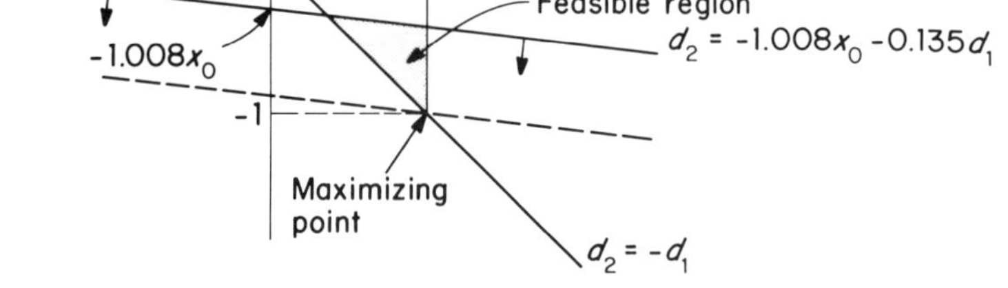

FIGURE 7.5

such that

$$0.707\,d_1 + 0.707\,d_2 \leqslant -x_0,$$

$$-0.707\,d_1 - 0.707\,d_2 \leqslant 0,$$

$$d_1 \leqslant 1, \qquad d_2 \leqslant 1.$$

The feasible region collapses to the line $d_2 = -d_1$ and $\max x_0 = 0$, so the minimum has been reached.

Boundary Following with Equality Constraints

The application of gradient techniques to equality-constrained problems is similar to the inequality-constrained case except that the iterative path must be on the boundary at all times. Thus the feasible region is in fact some boundary surface. Clearly the *initial feasible point* must also be on that surface.

In the case of equality constraints, the problem can be greatly simplified if any of the constraints can be explicitly solved for some variable. The constraint can then be used to eliminate that variable (and the constraint) from the optimization problem. The problem is solved in terms of the remaining variables, and the eliminated constraint is used to determine the optimal value of the eliminated variable. In the case of most nonlinear constraints this technique is rarely feasible; however, the possibility should never be overlooked because of the considerable simplification which results.

EXAMPLES

7.6 Use elimination to solve the problem

$$\min x_1^2 + x_2^2 + x_3^2$$

such that

$$2x_1 + 3x_2 + x_3 = 1.$$

Eliminating x_3, we get

$$x_3 = 1 - 2x_1 - 3x_2,$$

and the new unconstrained problem becomes

$$\min\left[f(\mathbf{x})=x_1^2+x_2^2+(1-2x_1-3x_2)^2\right],$$

$$\frac{\partial f}{\partial x_1}=2x_1+2(1-2x_1-3x_2)(-2)=0,$$

$$\frac{\partial f}{\partial x_2}=2x_2+2(1-2x_1-3x_2)(-3)=0.$$

The solution is

$$x_1^*=0.162, \qquad x_2^*=0.198, \qquad x_3^*=1-2(0.162)-3(0.198)=0.082.$$

7.7 Use elimination to formulate the problem

$$\max x_1^2+2x_2^2+3x_3^2+x_4^2$$

such that

$$x_1+x_2+x_3+x_4=1,$$

$$(x_1-x_2)^2+(x_3-x_4)^2=2.$$

Eliminating x_1, we get

$$x_1=1-x_2-x_3-x_4,$$

and the new problem becomes

$$\max(1-x_2-x_3-x_4)^2+2x_2^2+3x_3^2+x_4^2$$

such that

$$(1-2x_2-x_3-x_4)^2+(x_3-x_4)^2=2.$$

The problem is thus reduced from one having four variables and two constraints to one having three variables and one constraint.

We now seek a general algorithm for utilizing gradient techniques to solve equality-constrained problems. One such algorithm arises from the *gradient projection method*, which is derived below for the case of linear equality constraints. Later a generalization of the algorithm to nonlinear equality constraints will be given.

Consider a problem of the form

$$\min_{\mathbf{x}} f(\mathbf{x})$$

such that

$$\mathbf{Ax}-\mathbf{b}=\mathbf{0}$$

where $\mathbf{x}$ is an n-dimensional vector and $\mathbf{A}$ is an $m \times n$ matrix with $m<n$. Assume that we are at some feasible point $\mathbf{x}^p$ and we want to move to a nearby feasible point $\mathbf{x}^p + d\mathbf{x}^p$ along the path of steepest descent.

Since $\mathbf{x}^p$ is feasible, $\mathbf{Ax}^p - \mathbf{b}=\mathbf{0}$, and we require that

$$\mathbf{A}(\mathbf{x}^p + d\mathbf{x}^p) - \mathbf{b} = \mathbf{0}$$

for the feasibility of $\mathbf{x}^p + d\mathbf{x}^p$. These two relations imply that

$$\mathbf{A}\,d\mathbf{x}^p = \mathbf{0}. \tag{7.10}$$

In the remaining development, we temporarily suppress the superscript p for notational simplicity.

Along some arbitrary path ds, the direction cosines are dx_i/ds, $i=1,2,\ldots,n$, which leads to the constraint

$$\sum_{i=1}^{n} (dx_i)^2 = (ds)^2 \tag{7.11}$$

along this path. We now want to choose the path ds to maximize the (negative) rate of change of the objective function. Using the chain rule, this rate of change is expressed as

$$\frac{df}{ds} = \left(\frac{\partial f}{\partial \mathbf{x}}\right)^T \frac{d\mathbf{x}}{ds}, \tag{7.12}$$

where

$$\frac{\partial f}{\partial \mathbf{x}} = \nabla f(\mathbf{x}),$$

and

$$\frac{d\mathbf{x}}{ds} = \left[\frac{dx_1}{ds}, \frac{dx_2}{ds}, \ldots, \frac{dx_n}{ds}\right]^T.$$

In order to maximize (7.12) subject to the constraints (7.10) and (7.11), we first form the Lagrangian

$$L\left(\frac{d\mathbf{x}}{ds},\boldsymbol{\lambda},\lambda_0\right)=\left(\frac{\partial f}{\partial \mathbf{x}}\right)^T\frac{d\mathbf{x}}{ds}+\boldsymbol{\lambda}^T\mathbf{A}\frac{d\mathbf{x}}{ds}+\lambda_0\left[1-\left(\frac{d\mathbf{x}}{ds}\right)^T\left(\frac{d\mathbf{x}}{ds}\right)\right].$$

The necessary conditions are

$$\boldsymbol{\nabla}_{d\mathbf{x}/ds}L=\frac{\partial f}{\partial \mathbf{x}}-2\lambda_0\frac{d\mathbf{x}}{ds}+\mathbf{A}^T\boldsymbol{\lambda}=\mathbf{0}, \tag{7.13}$$

$$\boldsymbol{\nabla}_{\boldsymbol{\lambda}}L=\mathbf{A}\frac{d\mathbf{x}}{ds}=\mathbf{0}, \tag{7.14}$$

$$\boldsymbol{\nabla}_{\lambda_0}L=1-\left(\frac{d\mathbf{x}}{ds}\right)^T\left(\frac{d\mathbf{x}}{ds}\right)=0. \tag{7.15}$$

Solving (7.13) for $d\mathbf{x}/ds$ yields

$$\frac{d\mathbf{x}}{ds}=\frac{1}{2\lambda_0}\left(\frac{\partial f}{\partial \mathbf{x}}+\mathbf{A}^T\boldsymbol{\lambda}\right). \tag{7.16}$$

Substituting (7.16) into (7.15) and solving for λ_0 gives

$$\lambda_0=\frac{1}{2}\left[\left(\frac{\partial f}{\partial \mathbf{x}}^T+\boldsymbol{\lambda}^T\mathbf{A}\right)\left(\frac{\partial f}{\partial \mathbf{x}}+\mathbf{A}^T\boldsymbol{\lambda}\right)\right]^{1/2}. \tag{7.17}$$

Thus (7.16) can be written as

$$\frac{d\mathbf{x}}{ds}=\frac{\partial f}{\partial \mathbf{x}}+\mathbf{A}^T\boldsymbol{\lambda}, \tag{7.18}$$

where the right-hand side is normalized to a unit vector. Substituting (7.16) into (7.14) and solving for $\boldsymbol{\lambda}$ yields

$$\boldsymbol{\lambda}=-(\mathbf{A}\mathbf{A}^T)^{-1}\mathbf{A}\frac{\partial f}{\partial \mathbf{x}}, \tag{7.19}$$

which has a solution if and only if $\mathbf{A}\mathbf{A}^T$ is nonsingular. Substituting (7.19) into (7.18) gives the direction cosines

$$\frac{d\mathbf{x}}{ds}=\left[\mathbf{I}-\mathbf{A}^T(\mathbf{A}\mathbf{A}^T)^{-1}\mathbf{A}\right]\frac{\partial f}{\partial \mathbf{x}}, \tag{7.20}$$

where $\mathbf{I}$ is an n-dimensional identity matrix. The discrete form of the algorithm arising from (7.20) is

$$\mathbf{x}^{p+1} = \mathbf{x}^p + k_p \mathbf{P} \nabla f(\mathbf{x}^p), \tag{7.21}$$

where $k_p < 0$ is the step size and

$$\mathbf{P} = \left[\mathbf{I} - \mathbf{A}^T (\mathbf{A}\mathbf{A}^T)^{-1} \mathbf{A} \right]$$

is the normalized projection matrix. This matrix projects the gradient onto the linear subspace defined by the constraint set. For minimization problems $k_p < 0$ and for maximization problems $k_p > 0$, as before.

EXAMPLES

7.8 Extend the gradient projection algorithm to problems having nonlinear equality constraints.

In this case the constraints are

$$g_i(\mathbf{x}) = 0, \qquad i = 1, 2, \ldots, m.$$

Linearizing about the feasible point $\mathbf{x}^p$ yields

$$g_i(\mathbf{x}^p + d\mathbf{x}^p) \approx g_i(\mathbf{x}^p) + \left[\nabla g_i(\mathbf{x}^p) \right]^T d\mathbf{x}^p,$$

which must equal zero to ensure the feasibility of $\mathbf{x}^p + d\mathbf{x}^p$. Thus

$$\left[\nabla g_i(\mathbf{x}^p) \right]^T d\mathbf{x}^p = 0, \qquad i = 1, 2, \ldots, m.$$

This constraint is identical to (7.10) where

$$\mathbf{A} = \mathbf{A}(\mathbf{x}^p) = \begin{bmatrix} \dfrac{\partial g_1}{\partial x_1} & \dfrac{\partial g_1}{\partial x_2} & \cdots & \dfrac{\partial g_1}{\partial x_n} \\ \dfrac{\partial g_2}{\partial x_1} & \dfrac{\partial g_2}{\partial x_2} & \cdots & \dfrac{\partial g_2}{\partial x_n} \\ \vdots & \vdots & & \vdots \\ \dfrac{\partial g_m}{\partial x_1} & \dfrac{\partial g_m}{\partial x_m} & \cdots & \dfrac{\partial g_m}{\partial x_n} \end{bmatrix}_{\mathbf{x}^p}.$$

The result is the same as (7.21) except that $\mathbf{A}$ is no longer constant but must be recalculated at each step. Since the projection matrix $\mathbf{P}$ projects the gradient onto the linear subspace defined by $\mathbf{A}(\mathbf{x}^p)$, the constraints are not generally satisfied when we move to the point $\mathbf{x}^{p+1}$. Thus the step size k_p must be kept small, and an additional procedure must be added to each iteration to determine a new feasible point for erecting the supporting hyperplane defined by $\mathbf{A}(\mathbf{x}^{p+1})$. Because of this complication, and the additional computation required to determine $\mathbf{P}$ at each step, the gradient projection method is not convenient for problems with highly nonlinear constraints.

7.9 Extend the gradient projection algorithm to problems having linear inequality constraints

$$\mathbf{Ax}-\mathbf{b}\geqslant \mathbf{0}$$

where $\mathbf{A}$ is an $m\times n$ matrix.

Consider some current point $\mathbf{x}^p$, and let $\mathbf{A}_q$ be the $q^p\times n$ submatrix of coefficients of $\mathbf{A}$ corresponding to constraints which are active at $\mathbf{x}^p$ and also required to be active for the subsequent iteration. The gradient projection algorithm is now

$$\mathbf{x}^{p+1}=\mathbf{x}^p+k_p\left[\mathbf{I}-\mathbf{A}_q^T\left(\mathbf{A}_q\mathbf{A}_q^T\right)^{-1}\mathbf{A}_q\right]\nabla f(\mathbf{x}^p)=\mathbf{x}^p+k_p\mathbf{P}_q\nabla f(\mathbf{x}^p),$$

and $k_p<0$ is chosen to minimize $f(\mathbf{x})$ along the vector $\mathbf{P}_q\nabla f(\mathbf{x}^p)$ without violating any currently satisfied constraints. If a new constraint boundary is reached before $f(\mathbf{x})$ is minimized, its gradient is included as an additional column in the matrix $\mathbf{A}_q$, and $\mathbf{P}_q$ is modified accordingly. If some component of

$$\boldsymbol{\lambda}^q=-\left(\mathbf{A}_q\mathbf{A}_q^T\right)^{-1}\mathbf{A}_q\nabla f(\mathbf{x}^p)$$

is negative, then the constraint corresponding to this component is deleted from $\mathbf{A}q$ and the procedure continues until

$$\mathbf{P}_q\nabla f(\mathbf{x}^p)=\mathbf{0}$$

and

$$\boldsymbol{\lambda}^q\geqslant \mathbf{0}.$$

This procedure arises from the Kuhn-Tucker conditions, which require the nonnegativity of $\boldsymbol{\lambda}$ and the complementary slackness conditions to be satisfied at the optimum. For the active constraints

$$\mathbf{A}^q\mathbf{x}^p-\mathbf{b}=\mathbf{0},\qquad \boldsymbol{\lambda}^q\geqslant \mathbf{0}.$$

And for the inactive constraints, the corresponding components of $\boldsymbol{\lambda}$ are equal to zero.

Solved Problems

7.1 Apply the hemstitching algorithm to the problem

$$\min x_1^2 + x_2^2$$

such that

$$3x_1 + 2x_2 \geqslant 6,$$

$$x_1 \geqslant 0, \qquad x_2 \geqslant 0.$$

Start at the initial feasible point (2,3).

The gradients are

$$\nabla f(\mathbf{x}) = \begin{bmatrix} 2x_1 \\ 2x_2 \end{bmatrix}, \qquad \nabla g(\mathbf{x}) = \begin{bmatrix} 3 \\ 2 \end{bmatrix},$$

and the algorithms are

$$\mathbf{x}^{p+1} = \mathbf{x}^p - \frac{k_p}{\sqrt{x_1^2 + x_2^2}} \begin{bmatrix} x_1 \\ x_2 \end{bmatrix}$$

when $\mathbf{x}^p$ is feasible, and

$$\mathbf{x}^{p+1} = \mathbf{x}^p + \frac{k_p}{13} \begin{bmatrix} 3 \\ 2 \end{bmatrix}$$

when $\mathbf{x}^p$ is nonfeasible. The algorithm converges to the minimum $\frac{18}{13}, \frac{12}{13}$ as shown in Fig. 7.6.

7.2 Consider the application of hemstitching to the problem

$$\max x_1^2 + x_2^2$$

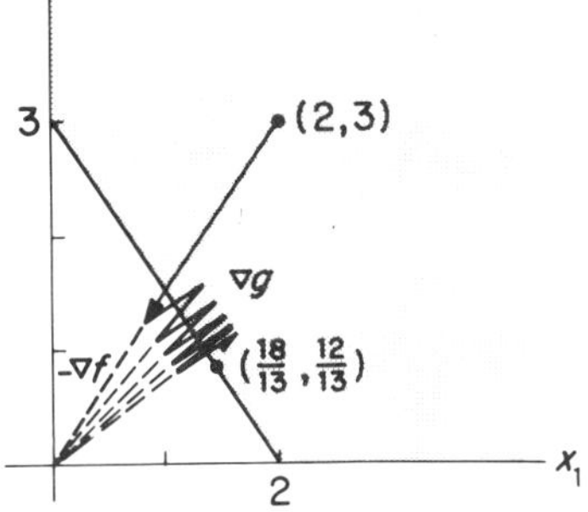

FIGURE 7.6

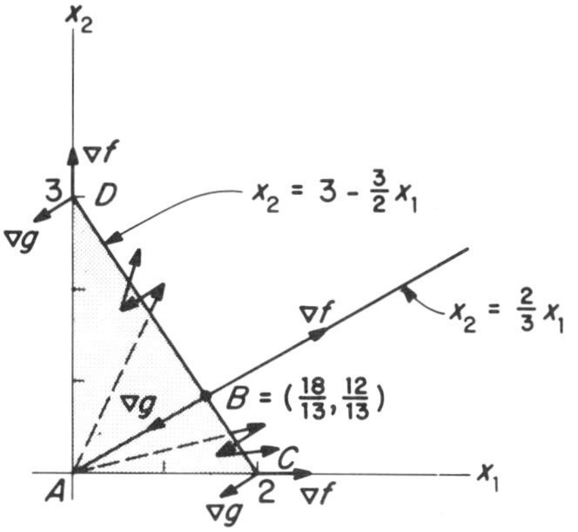

FIGURE 7.7

such that

$$6-3x_1-2x_2\geqslant 0,$$

$$x_1\geqslant 0,\qquad x_2\geqslant 0.$$

Begin at the initial feasible point 0,0. Does this method converge to the maximum?

At the initial feasible point the gradient is $[0,0]^T$ and the algorithm fails to specify a new point, i.e., $\mathbf{x}^{p+1}=\mathbf{x}^p$. Hence the hemstitching method does not converge to the maximum 0,3 shown in Fig. 7.7. The reason for this difficulty is that we are attempting to maximize a convex function over a convex set, and the hemstitching algorithm has been designed to minimize such functions.

If we were to choose an initial feasible point on the line $x_2=\frac{2}{3}x_1$, the algorithm would converge to the point $\frac{18}{13},\frac{12}{13}$, where

$$\nabla f\left(\tfrac{18}{13},\tfrac{12}{13}\right)=\begin{bmatrix}3\\2\end{bmatrix},\qquad \nabla g\left(\tfrac{18}{13},\tfrac{12}{13}\right)=\begin{bmatrix}-3\\-2\end{bmatrix}.$$

This point corresponds to the solution of the minimization problem. If the initial feasible point is in the triangle (A,B,C) in Fig. 7.7, the algorithm converges to the local maximum 2,0; and if the initial feasible point is in the triangle (A,B,D), the algorithm converges to the global maximum 0,3.

7.3 Determine the direction of multiple gradient summation at the point 2,4.5 for the problem

$$\min x_1^2+x_2^2$$

such that

$$g(\mathbf{x})=9-x_1x_2\leqslant 0$$

at the point $\hat{\mathbf{x}}=[2,4.5]^T$.

$$\nabla f(\hat{\mathbf{x}})=\frac{1}{2\sqrt{x_1^2+x_2^2}}\begin{bmatrix}2x_1\\2x_2\end{bmatrix}=\begin{bmatrix}0.405\\0.913\end{bmatrix}$$

and

$$\nabla g(\hat{\mathbf{x}})=\frac{1}{\sqrt{x_1^2+x_2^2}}\begin{bmatrix}-x_2\\-x_1\end{bmatrix}=\begin{bmatrix}-0.918\\-0.408\end{bmatrix}.$$

So the negative gradient direction is given by

$$-\mathbf{d}=\begin{bmatrix}0.513\\-0.505\end{bmatrix}.$$

This vector is shown in Fig. 7.8.

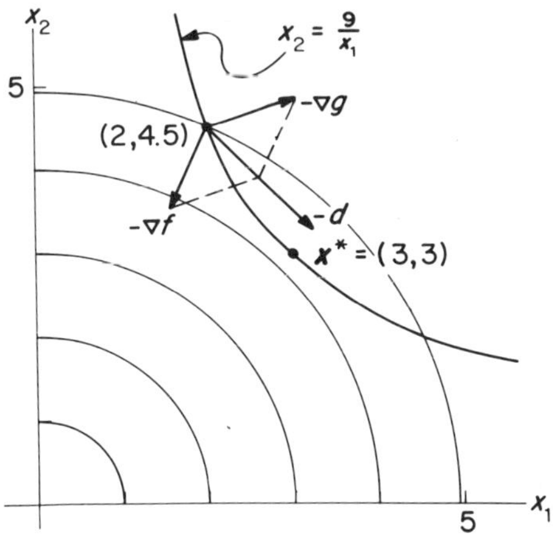

FIGURE 7.8

7.4 Determine the direction of multiple gradient summation at the point 1.67, 2.5 for the problem

$$\max x_1^2 + 2x_2^2$$

such that

$$x_1^2 + x_2^2 \leqslant 9.$$

At the point $\hat{\mathbf{x}} = [1.67, 2.5]^T$,

$$\nabla f(\hat{\mathbf{x}}) = \frac{1}{2\sqrt{x_1^2 + 4x_2^2}} \begin{bmatrix} 2x_1 \\ 4x_2 \end{bmatrix} = \begin{bmatrix} 0.316 \\ 0.947 \end{bmatrix}$$

and

$$\nabla g(\hat{\mathbf{x}}) = \frac{1}{2\sqrt{x_1^2 + x_2^2}} \begin{bmatrix} 2x_1 \\ 2x_2 \end{bmatrix} = \begin{bmatrix} 0.555 \\ 0.830 \end{bmatrix}.$$

For maximization problems,

$$\mathbf{d} = \frac{\nabla f(\mathbf{x})}{\|\nabla f(\mathbf{x})\|} - \frac{\nabla g(\mathbf{x})}{\|\nabla g(\mathbf{x})\|},$$

so

$$\mathbf{d} = \begin{bmatrix} 0.316 \\ 0.947 \end{bmatrix} - \begin{bmatrix} 0.555 \\ 0.830 \end{bmatrix} = \begin{bmatrix} -0.239 \\ 0.117 \end{bmatrix}.$$

This vector is shown in Fig. 7.9.

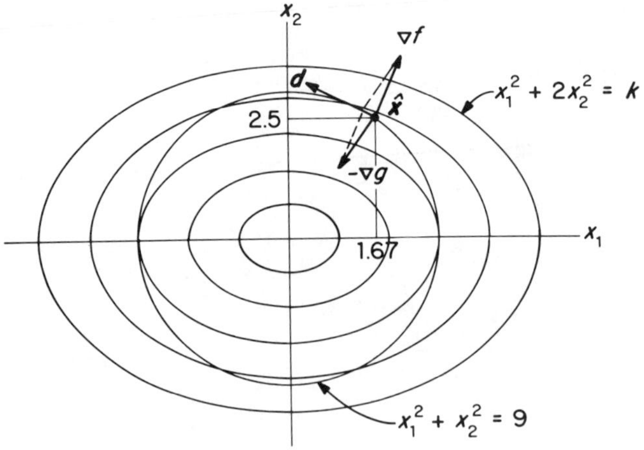

FIGURE 7.9

7.5 Use the feasible directions algorithm to solve the problem

$$\min\left[-\ln(1+x_1)-2\ln(1+x_2)\right]$$

such that

$$g_1(\mathbf{x})=x_1+x_2-2\leqslant 0,$$

$$x_1\geqslant 0, \qquad x_2\geqslant 0.$$

Begin at the initial feasible point $\frac{1}{2}, \frac{1}{2}$.

At the initial point, $\mathbf{x}^0=[\frac{1}{2}, \frac{1}{2}]^T$ and

$$g(\mathbf{x}^0)=\tfrac{1}{2}+\tfrac{1}{2}-2=-1.$$

Thus $\mathbf{x}^0$ is strictly feasible, so

$$\max x_0$$

$$(\nabla f(\mathbf{x}^0), \mathbf{d}^0)\leqslant -x_0$$

indicates that

$$\mathbf{d}^0=-\nabla f(\mathbf{x}^0),$$

i.e., we follow the negative-gradient path until a boundary is encountered:

$$-\nabla f(\mathbf{x})=\begin{bmatrix}\dfrac{1}{1+x_1}\\ \dfrac{2}{1+x_2}\end{bmatrix},$$

and after normalization

$$-\nabla f(\mathbf{x}^0) = \frac{1}{\sqrt{5}}\begin{bmatrix}1\\2\end{bmatrix}.$$

Moving in this direction until a constraint is encountered, we get

$$x_1^1 = \frac{1}{2} + \frac{k_1}{\sqrt{5}},$$

$$x_2^1 = \frac{1}{2} + \frac{2k_1}{\sqrt{5}},$$

$$x_2^1 = 2 - x_1^1,$$

yielding

$$x_1^1 = 0.833,$$

$$x_2^2 = 1.167,$$

$$k_1 = \frac{\sqrt{5}}{3}.$$

This path is shown in Fig. 7.10. To determine $\mathbf{d}^1$ we must solve the linear programming problem

$$\max x_0$$

such that

$$x_0 + (\nabla f(\mathbf{x}^1), \mathbf{d}^1) \leqslant 0,$$

$$(\nabla g(\mathbf{x}^1), \mathbf{d}^1) \leqslant 0,$$

$$d_1^1 \leqslant 1, \qquad d_2^1 \leqslant 1,$$

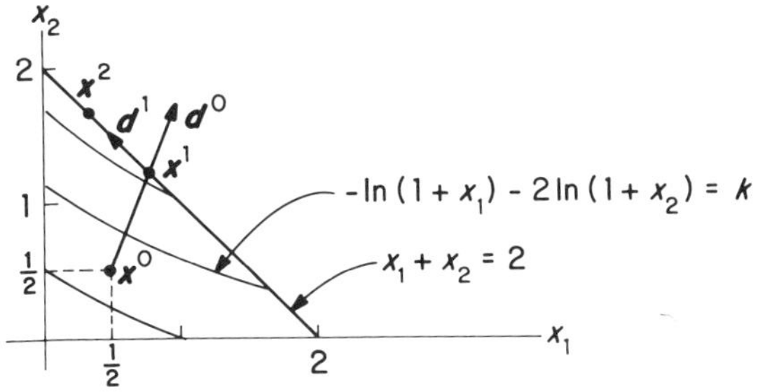

FIGURE 7.10

where

$$\nabla f(\mathbf{x}^1)=\begin{bmatrix} -0.510 \\ -0.826 \end{bmatrix}, \qquad \nabla g(\mathbf{x}^1)=\begin{bmatrix} 0.707 \\ 0.707 \end{bmatrix},$$

or

$$\max x_0$$

such that

$$d_2 \geqslant 1.168x_0 - 0.595\,d_1,$$

$$d_2 \leqslant -d_1,$$

$$d_1 \leqslant 1, \qquad d_2 \leqslant 1.$$

This problem is solved graphically in Fig. 7.11; the solution is

$$\mathbf{d}^1=\begin{bmatrix} -1 \\ 1 \end{bmatrix}$$

and $x_0 = 0.346$.

We now determine the step size by solving the problem

$$\min_k f(\mathbf{x}^1 + k\,\mathbf{d}^1).$$

Normalizing **d**, the solution is given by the necessary condition

$$\frac{\partial}{\partial k}\left[-\ln(1+x_1^1-0.707k)-2\ln(1+x_2^1+0.707k)\right]=0,$$

yielding

$$k=0.678.$$

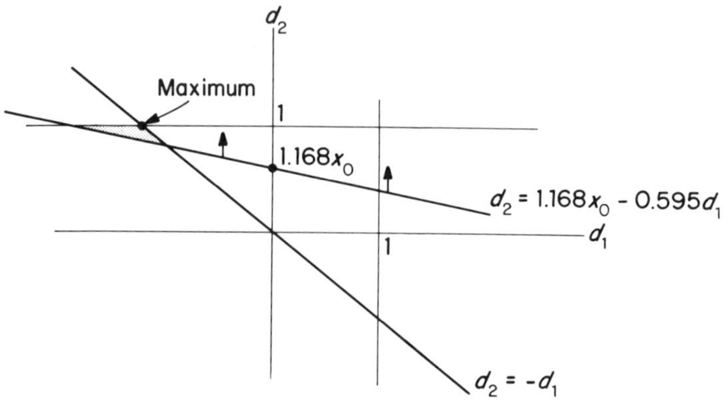

FIGURE 7.11

Now

$$\mathbf{x}^2 = \mathbf{x}^1 + k\mathbf{d}^1,$$

so

$$x_1^2 = 0.832 - 0.707(0.678) = 0.35,$$

$$x_2^2 = 1.168 + 0.707(0.678) = 1.65.$$

Solving the linear programming problem at $\mathbf{x}^2$, we see that $x_0 \approx 0$ so the point $0.35, 1.65$ is the optimal solution.

7.6 For Problem 7.3 determine the direction at the point $2, 4.5$ by the feasible-directions algorithm.

The linear programming problem is

$$\max x_0$$

such that

$$x_0 + 0.405\, d_1 + 0.913\, d_2 \leqslant 0,$$

$$x_0 - 0.918\, d_1 - 0.408\, d_2 \leqslant 0,$$

$$d_1 \leqslant 1, \qquad d_2 \leqslant 1,$$

or

$$d_2 \leqslant -1.095 x_0 - 0.445\, d_1,$$

$$d_2 \geqslant 2.45 x_0 - 2.25\, d_1,$$

$$d_1 \leqslant 1, \qquad d_2 \leqslant 1.$$

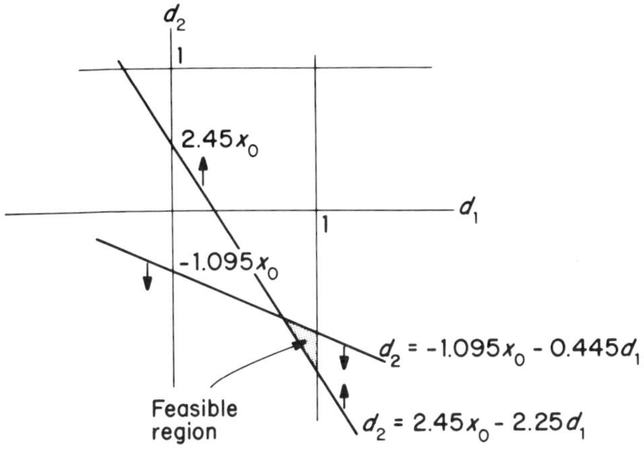

FIGURE 7.12

From Fig. 7.12 it is seen that $\max x_0$ occurs at $d_1 = 1$. Thus

$$d_2 = -1.095x_0 - 0.445,$$

$$d_2 = 2.45x_0 - 2.25,$$

and so

$$d_2 = -1 \quad \text{and} \quad x_0 = 0.507.$$

This direction vector is approximately equal to the vector in Fig. 7.8, which was obtained by the method of multiple gradient summation.

7.7 Use the feasible-directions algorithm to solve the problem

$$\min x_1 + 2x_2 + x_3$$

such that

$$-6x_1 - 4x_2 - 2x_3 + 3x_1^2 + 2x_2^2 + \tfrac{1}{3}x_3^2 + 7.25 \leqslant 0,$$

$$x_1 \geqslant 0, \qquad x_2 \geqslant 0, \qquad x_3 \geqslant 0.$$

Begin at the initial feasible point [1, 1, 3].

The feasible-directions algorithm is

$$\mathbf{x}^{p+1} = \mathbf{x}^p + k_p \mathbf{d}^p,$$

where $\mathbf{d}^p$ is determined from the linear programming problem

$$\max \quad \mathbf{d}^p x_0$$

such that

$$x_0 + (\nabla f(\mathbf{x}^p), \mathbf{d}^p) \leqslant 0,$$

$$x_0 + (\nabla g(\mathbf{x}^p), \mathbf{d}^p) \leqslant -g(\mathbf{x}^p),$$

Table 7.1

ITERATION p	$f(\mathbf{x}^p)$	x_1^p	x_2^p	x_3^p	d_1^p	d_2^p	d_3^p	k_p
0	6	1	1	3	−1	−1	−1	0.37
1	4.51	0.627	0.627	2.627	1	−0.67	−1	0.29
2	4.17	0.918	0.460	2.336	−1	0.71	−1	0.14
3	4.09	0.781	0.557	2.199	1	−0.21	−1	0.12
4	4.05	0.900	0.532	2.080	−1	−0.83	−1	0.07
5	4.02	0.826	0.594	2.006	1	0.10	−1	0.06
9	4.00	0.869	0.625	1.881	1	0	−1	0.01
10	4.00	0.875	0.625	1.875	—	—	—	—
Optimal	4	0.875	0.625	1.875				

and $k_p > 0$. Now

$$\nabla f(\mathbf{x}) = \begin{bmatrix} 1 \\ 2 \\ 1 \end{bmatrix}, \qquad \nabla g(\mathbf{x}) = \begin{bmatrix} -6+6x_1 \\ -4+4x_2 \\ -2+\frac{2}{3}x_3 \end{bmatrix}.$$

The results of applying this algorithm beginning at the point (1, 1, 3) are given in Table 7.1 (Wagner, 1969).

7.8 Determine the projection of the function

$$f(\mathbf{x}) = 5x_1 - 3x_2 + 6x_3$$

onto the x_1, x_2 plane. Sketch the gradient and its projection.

The gradient is

$$\nabla f(\mathbf{x}) = \begin{bmatrix} 5 \\ -3 \\ 6 \end{bmatrix},$$

and the constraint is defined by

$$x_3 = 0.$$

The projection matrix is

$$\mathbf{P} = \left[\mathbf{I} - \mathbf{A}^T(\mathbf{A}\mathbf{A}^T)^{-1}\mathbf{A}\right],$$

where

$$\mathbf{A} = [0, 0, 1],$$

so

$$\mathbf{A}\mathbf{A}^T = [0, 0, 1]\begin{bmatrix} 0 \\ 0 \\ 1 \end{bmatrix} = 1,$$

and

$$(\mathbf{A}\mathbf{A}^T)^{-1} = 1.$$

Thus

$$\mathbf{P} = \begin{bmatrix} 1 & 0 & 0 \\ 0 & 1 & 0 \\ 0 & 0 & 1 \end{bmatrix} - \begin{bmatrix} 0 \\ 0 \\ 1 \end{bmatrix}[0, 0, 1]$$

$$= \begin{bmatrix} 1 & 0 & 0 \\ 0 & 1 & 0 \\ 0 & 0 & 1 \end{bmatrix} - \begin{bmatrix} 0 & 0 & 0 \\ 0 & 0 & 0 \\ 0 & 0 & 1 \end{bmatrix} = \begin{bmatrix} 1 & 0 & 0 \\ 0 & 1 & 0 \\ 0 & 0 & 0 \end{bmatrix},$$

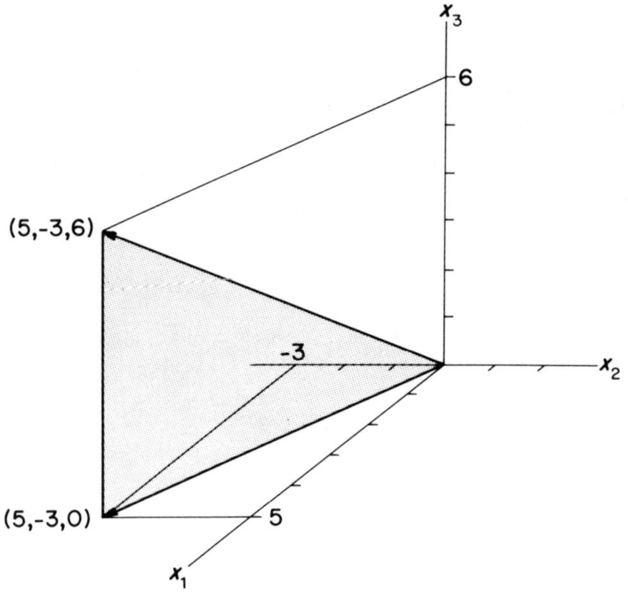

FIGURE 7.13

and the projection is given by

$$\mathbf{P}\nabla f(\mathbf{x})=\begin{bmatrix}1&0&0\\0&1&0\\0&0&0\end{bmatrix}\begin{bmatrix}5\\-3\\6\end{bmatrix}=\begin{bmatrix}5\\-3\\0\end{bmatrix}.$$

The gradient and its projection onto the x_1, x_2 plane are shown in Fig. 7.13.

7.9 Repeat Problem 7.8 for the function

$$f(\mathbf{x})=x_1^2+3x_2^2+2x_3^2$$

evaluated at the point $(1,1,-1)$.

The gradient is

$$\nabla f(1,1,-1)=\begin{bmatrix}2x_1\\6x_2\\4x_3\end{bmatrix}_{1,1,-1}=\begin{bmatrix}2\\6\\-4\end{bmatrix}.$$

The projection matrix is the same as in Problem 7.5. Hence the projection is given

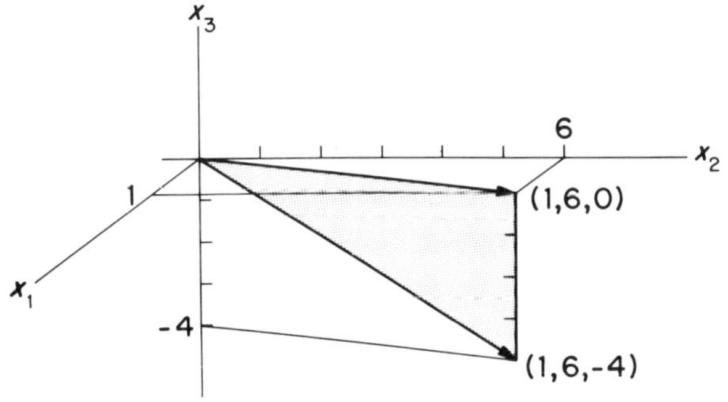

FIGURE 7.14

by

$$\mathbf{P}\nabla f(1,1,-1)=\begin{bmatrix}1&0&0\\0&1&0\\0&0&0\end{bmatrix}\begin{bmatrix}2\\6\\-4\end{bmatrix}=\begin{bmatrix}2\\6\\0\end{bmatrix}.$$

The gradient and its projection onto the x_1, x_2 plane are shown in Fig. 7.14.

7.10 Use the gradient-projection method to compute two steps for the problem

$$\min 2x_1^2+5x_2^2$$

such that

$$x_1+3x_2=6.$$

Start at the initial feasible point 12, -2 and use a step size $k_p=3$.

For this problem, the constraint matrix is

$$\mathbf{A}=[1,3],$$

and

$$\nabla f(\mathbf{x})=\begin{bmatrix}4x_1\\10x_2\end{bmatrix}.$$

The gradient-projection algorithm is

$$\mathbf{x}^{p+1}=\mathbf{x}^p+k_p\left[\mathbf{I}-\mathbf{A}^T(\mathbf{A}\mathbf{A}^T)^{-1}\mathbf{A}\right]\nabla f(\mathbf{x}^p),$$

where

$$\mathbf{A}\mathbf{A}^T=[1,3]\begin{bmatrix}1\\3\end{bmatrix}=10$$

and

$$(\mathbf{AA}^T)^{-1} = 0.1.$$

So the projection matrix is

$$\mathbf{P} = \begin{bmatrix} 1 & 0 \\ 0 & 1 \end{bmatrix} - \begin{bmatrix} 1 \\ 3 \end{bmatrix} 0.1[1,3]$$

$$= \begin{bmatrix} 1 & 0 \\ 0 & 1 \end{bmatrix} - \begin{bmatrix} 0.1 & 0.3 \\ 0.3 & 0.9 \end{bmatrix} = \begin{bmatrix} 0.9 & -0.3 \\ -0.3 & 0.1 \end{bmatrix},$$

and

$$\mathbf{P}\nabla f(\mathbf{x}^0) = \begin{bmatrix} 0.9 & -0.3 \\ -0.3 & 0.2 \end{bmatrix} \begin{bmatrix} 48 \\ -20 \end{bmatrix} = 16.4 \begin{bmatrix} 3 \\ -1 \end{bmatrix}.$$

The normalizing factor is $1/16.4\sqrt{10}$. So

$$\mathbf{x}^1 = \begin{bmatrix} 12 \\ -2 \end{bmatrix} - \frac{3}{\sqrt{10}} \begin{bmatrix} 3 \\ -1 \end{bmatrix} = \begin{bmatrix} 9.15 \\ -1.05 \end{bmatrix}.$$

In a similar manner,

$$\mathbf{x}^2 = \begin{bmatrix} 9.155 \\ -1.052 \end{bmatrix} - \frac{3}{\sqrt{10}} \begin{bmatrix} 3 \\ -1 \end{bmatrix} = \begin{bmatrix} 6.3 \\ -0.1 \end{bmatrix}.$$

Both of these points satisfy the constraint as shown in Fig. 7.15.

7.11 Use the gradient-projection method starting at 0, 1 to solve the problem

$$\min[-2x_1 - x_2]$$

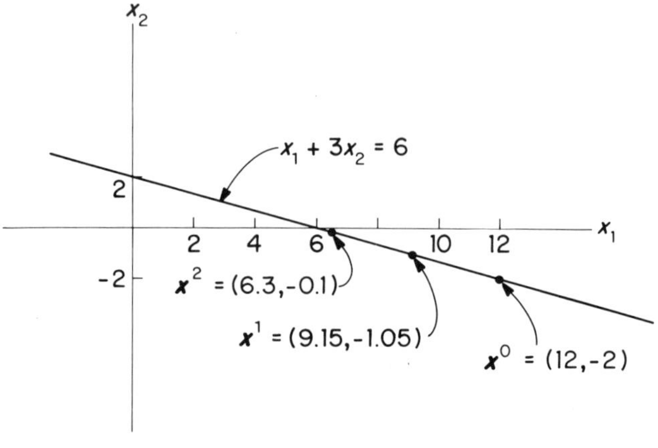

FIGURE 7.15

such that

$$g_1(\mathbf{x}) = -x_2 + 1 \geqslant 0,$$

$$g_2(\mathbf{x}) = -4x_1 - 6x_2 + 7 \geqslant 0,$$

$$g_3(\mathbf{x}) = -10x_1 - 12x_2 + 15 \geqslant 0,$$

$$g_4(\mathbf{x}) = x_2 - 0.5x_1 \geqslant 0,$$

$$x_1 \geqslant 0, \qquad x_2 \geqslant 0.$$

The initial projection matrix is

$$\mathbf{P} = \mathbf{I} - \begin{bmatrix} 1 & 0 \\ 0 & -1 \end{bmatrix} \begin{bmatrix} 1 & 0 \\ 0 & 1 \end{bmatrix}^{-1} \begin{bmatrix} 1 & 0 \\ 0 & -1 \end{bmatrix} = \begin{bmatrix} 0 & 0 \\ 0 & 0 \end{bmatrix},$$

since the constraints $g_1 \geqslant 0$ and $x_1 \geqslant 0$ are active. The multipliers are

$$\boldsymbol{\lambda} = \begin{bmatrix} 1 & 0 \\ 0 & -1 \end{bmatrix} \begin{bmatrix} -2 \\ -1 \end{bmatrix} = \begin{bmatrix} -2 \\ 1 \end{bmatrix}.$$

which means that we can move off of the constraint $x_1 = 0$. Removing this gradient, the projection matrix is

$$\mathbf{P}^0 = \mathbf{I} - \begin{bmatrix} 0 \\ -1 \end{bmatrix} [0, -1] = \begin{bmatrix} 1 & 0 \\ 0 & 0 \end{bmatrix}$$

and

$$\mathbf{d}^0 = \mathbf{P}^0 \nabla f(\mathbf{x}^0) = \begin{bmatrix} 1 & 0 \\ 0 & 0 \end{bmatrix} \begin{bmatrix} -2 \\ -1 \end{bmatrix} = \begin{bmatrix} -2 \\ 0 \end{bmatrix}.$$

We now move along the negative $\mathbf{d}^0$ direction until the constraint $g_2(\mathbf{x}) = 0$ is met at $0.25, 1$. At this point we add the gradient of $g_2(\mathbf{x})$ to $\mathbf{A}$ and compute

$$\boldsymbol{\lambda} = \frac{1}{16} \begin{bmatrix} 52 & -6 \\ -6 & 1 \end{bmatrix} \begin{bmatrix} 0 & -1 \\ -4 & -6 \end{bmatrix} \begin{bmatrix} -2 \\ -1 \end{bmatrix} = \begin{bmatrix} -2 \\ 0.5 \end{bmatrix}.$$

Thus we can leave the constraint $g_1(\mathbf{x}) = 0$ and project along $g_2(\mathbf{x}) = 0$. Our projection matrix is now

$$\mathbf{P}^1 = \mathbf{I} - \begin{bmatrix} -4 \\ -6 \end{bmatrix} \left(\frac{1}{52}\right) (-4, -6) = \frac{1}{13} \begin{bmatrix} 9 & -6 \\ -6 & 4 \end{bmatrix},$$

and the direction is

$$\mathbf{d}^1 = \frac{1}{13} \begin{bmatrix} 9 & -6 \\ -6 & 4 \end{bmatrix} \begin{bmatrix} -2 \\ -1 \end{bmatrix} = \frac{1}{13} \begin{bmatrix} -12 \\ 8 \end{bmatrix}.$$

We move along the negative $\mathbf{d}^1$ direction until the constraint $g_3(\mathbf{x}) = 0$ is met at

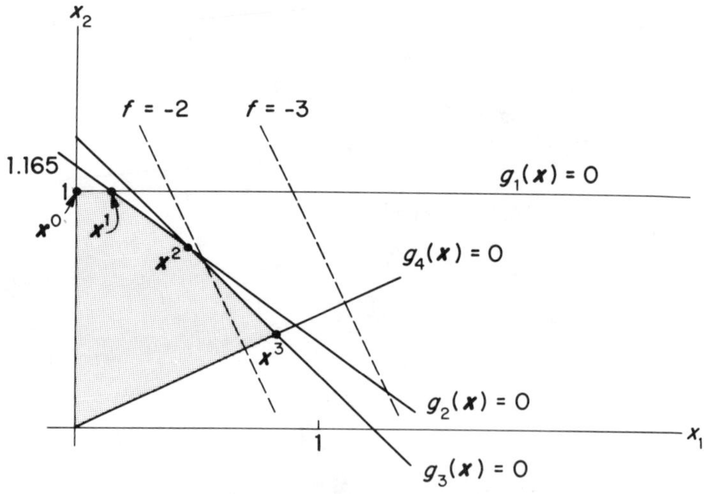

FIGURE 7.16

0.5, 0.834. At this point we add the gradient of $g_3(\mathbf{x})$ to $\mathbf{A}$ and compute

$$\boldsymbol{\lambda}=\frac{1}{36}\begin{bmatrix} 61 & -28 \\ -28 & 13 \end{bmatrix}\begin{bmatrix} -4 & -6 \\ -10 & -12 \end{bmatrix}\begin{bmatrix} -2 \\ -1 \end{bmatrix}=\begin{bmatrix} -1.08 \\ 0.57 \end{bmatrix}$$

Thus $g_2(\mathbf{x})$ can be eliminated from the projection matrix, which becomes

$$\mathbf{P}^2=\mathbf{I}-\begin{bmatrix} -10 \\ -12 \end{bmatrix}\left(\frac{1}{244}\right)[-10,-12]=\frac{1}{61}\begin{bmatrix} 36 & -30 \\ -30 & 25 \end{bmatrix},$$

and the new direction is

$$\mathbf{d}^2=\frac{1}{61}\begin{bmatrix} 36 & -30 \\ -30 & 25 \end{bmatrix}\begin{bmatrix} -2 \\ -1 \end{bmatrix}=\frac{1}{61}\begin{bmatrix} -42 \\ 35 \end{bmatrix}.$$

Projecting along the negative $\mathbf{d}^2$ direction, we encounter the constraint $g_4(\mathbf{x})=0$ at (0.938, 0.469). At this point we add the gradient of $g_4(\mathbf{x})=0$ to the projection matrix and compute

$$\boldsymbol{\lambda}=-\frac{1}{13}\begin{bmatrix} -1.25 & -1 \\ -1 & -136 \end{bmatrix}\begin{bmatrix} -10 & -6 \\ -0.5 & 1 \end{bmatrix}=\begin{bmatrix} 2.69 \\ 2 \end{bmatrix},$$

and since $\boldsymbol{\lambda}\geqslant\mathbf{0}$, the optimum has been reached. The projected gradient path is shown in Fig. 7.16.

7.12 Use the gradient-projection method to solve the problem

$$\max\left[6x_1+4x_2+2x_3-3x_1^2-2x_2^2-\tfrac{1}{3}x_3^2\right]$$

such that

$$x_1+2x_2+x_3-4\leqslant 0.$$

Begin at the initial feasible point (0,0,0).

For this problem,

$$\nabla f(\mathbf{x})=\begin{bmatrix}6-6x_1\\ 4-4x_2\\ 2-\frac{2}{3}x_3\end{bmatrix}$$

and

$$\mathbf{A}=[1,2,1].$$

So

$$\mathbf{AA}^T=6 \quad \text{and} \quad (\mathbf{AA}^T)^{-1}=\tfrac{1}{6}.$$

This

$$\mathbf{P}=\begin{bmatrix}1&0&0\\0&1&0\\0&0&1\end{bmatrix}-\frac{1}{6}\begin{bmatrix}1&2&1\\2&4&2\\1&2&1\end{bmatrix}=\frac{1}{6}\begin{bmatrix}5&-2&-1\\-2&2&-2\\-1&-2&5\end{bmatrix}.$$

The steps taken by the algorithm

$$\mathbf{x}^{p+1}=\mathbf{x}^p+k_p\mathbf{P}\nabla f(\mathbf{x}^p)$$

are given in Table 7.2 (Wagner, 1969).

Table 7.2

Iteration p	$f(\mathbf{x}^p)$	x_1^p	x_2^p	x_3^p	k_p
0	0	0	0	0	0.1981
1	5.547	1.189	0.793	0.396	0.3667
2	6.537	0.774	1.097	1.033	0.2749
3	6.966	1.060	0.817	1.306	0.3075
4	7.137	0.835	0.813	1.539	0.2749
5	7.205	0.949	0.701	1.648	0.3075
6	7.232	0.859	0.700	1.741	0.2749
10	7.249	0.872	0.637	1.854	0.2749
15	7.249	0.876	0.626	1.873	0.3075
Optimal	7.25	0.875	0.625	1.875	—

Supplementary Problems

7.13 Use hemstitching to solve the problem

$$\max 2x_1 + x_2 - x_1^2$$

such that

$$2x_1^2 + 3x_2^2 \leqslant 6,$$

$$x_1 \geqslant 0, \qquad x_2 \geqslant 0.$$

7.14 Use hemstitching to solve the problem

$$\max 3x_1 + 2x_2$$

such that

$$(x_1 - 2)^2 + (x_2 - 1)^2 \leqslant 9,$$

$$x_1 \geqslant 0, \qquad x_2 \geqslant 0.$$

7.15 Use the method of multiple gradient summation to solve Problem 7.13.

7.16 Use the method of multiple gradient summation to solve Problem 7.14.

7.17 Use the method of feasible directions to solve Problem 7.13.

7.18 Use the method of feasible directions to solve Problem 7.14.

7.19 Use the feasible-directions algorithm to solve the problem

$$\max x_1 x_2$$

such that

$$x_2 - x_1 - 1 \leqslant 0,$$

$$2x_2 + x_1 - 6 \leqslant 0,$$

$$x_1 \geqslant 0.$$

7.20 Use the gradient-projection method to solve the linear programming problem

$$\max 0.5x_1 + 2x_2$$

such that

$$x_1 + x_2 \leqslant 6,$$

$$x_1 - x_2 \leqslant 1,$$

$$2x_1 + x_2 \geqslant 6,$$

$$0.5x_1 - x_2 \geqslant -4,$$

$$x_1 \geqslant 1, \qquad x_2 \geqslant 0.$$

Begin with the initial feasible point 3,2.

7.21 Write down the steps in the gradient-projection algorithm for solving problems of the form

$$\min f(\mathbf{x})$$

such that

$$g_i(\mathbf{x}) \leqslant 0, \qquad i = 1, 2, \ldots, r.$$

7.22 Determine the special form of the **P**-matrix in the gradient projection method when constraints of the form

$$x_i \leqslant c, \qquad i = 1, 2, \ldots, r,$$

are encountered.

7.23 Determine the conditions on the $m \times n$ matrix **A** which ensure that the matrix $(\mathbf{A}\mathbf{A}^T)$ in the gradient-projection method is nonsingular.

7.24 Perform four iterations of the gradient-projection algorithm for the problem

$$\min x_1^2 + x_2^2 + 2x_3^2 - 2x_1x_2$$

such that

$$g_1(\mathbf{x}) = 2x_1 + x_2 - 4 = 0,$$

$$g_2(\mathbf{x}) = 5x_1 - x_3 - 8 = 0.$$

7.25 Solve Problem 7.19 using the gradient projection algorithm.

7.26 Perform four iterations of the gradient projection algorithm for the problem

$$\min x_1^2 + 2x_2^2$$

such that

$$3x_1 + x_2 \geqslant 2,$$

$$x_1 + 2x_2 \geqslant 3.$$

Chapter 8

Penalty-Function Methods

In Chapter 7 we discussed two basic approaches to solving constrained optimization problems. The first was to eliminate the (equality) constraint, if possible, and the second was to move along a boundary whenever one was encountered. A third approach is discussed in this chapter and involves the use of penalty functions to convert constrained optimization problems into unconstrained problems. Once the unconstrained problem is formulated, direct methods for solving this type of problem can be employed.

In Chapters 3 and 4, we recall that Lagrange multipliers were also employed to convert constrained problems into unconstrained problems. In this chapter we will show the connection between certain penalty functions used in direct methods for optimization and the Lagrange multipliers used in indirect methods.

A Simple Penalty Function

Consider the equality-constrained problem

$$\min f(\mathbf{x})$$

such that

$$g_i(\mathbf{x}) = 0, \qquad i = 1, 2, \ldots, m.$$

In order to convert this problem into an unconstrained optimization problem, we define the new objective function

$$P(\mathbf{x}; \mathbf{K}) = f(\mathbf{x}) + \sum_{i=1}^{m} K_i [g_i(\mathbf{x})]^2,$$

where the positive constants ($K_i > 0$) are specified parameters. Now as K_i is increased from zero to infinity, more and more weighting is attached to satisfying the ith constraint. When $K_i = 0$, the constraint is ignored, and when $K_i = \infty$, the constraint is satisfied exactly. The positive constants $\mathbf{K} = [K_1, K_2, \ldots, K_m]^T$ are weighting factors specified by the user, depending on how strongly he feels about satisfying each constraint. In most problems, the constraints represent mathematical approximations to physical phenomena which are not known precisely. The value of K_i can be chosen commensurate with the modeling accuracy of the ith constraint. Thus these added weighting terms on the objective function represent penalties for not satisfying the constraints.

EXAMPLE 8.1 The following problem illustrates the use of the penalty constants:

$$\min x_1^2 + x_2^2$$

such that

$$x_2 = 1.$$

We first define the new objective function

$$P(\mathbf{x}; K) = x_1^2 + x_2^2 + K(x_2 - 1)^2.$$

Now to minimize $P(\mathbf{x}; K)$ it is necessary that

$$\frac{\partial P}{\partial x_1} = 2x_1 = 0,$$

$$\frac{\partial P}{\partial x_2} = 2x_2 + 2K(x_2 - 1) = 0.$$

Thus $x_1^* = 0$ and $x_2^* = K/(1+K)$. As K is increased from 0 to ∞, the optimal value of x_2 moves from the unconstrained solution to the constrained solution as shown below:

K	x_2^*
0	0
1	0.5
2	0.67
5	0.83
10	0.91
∞	1

Each of these values represents a solution to the problem as shown in Fig. 8.1. Note that as K increases, the solution moves toward the boundary in a highly nonlinear manner.

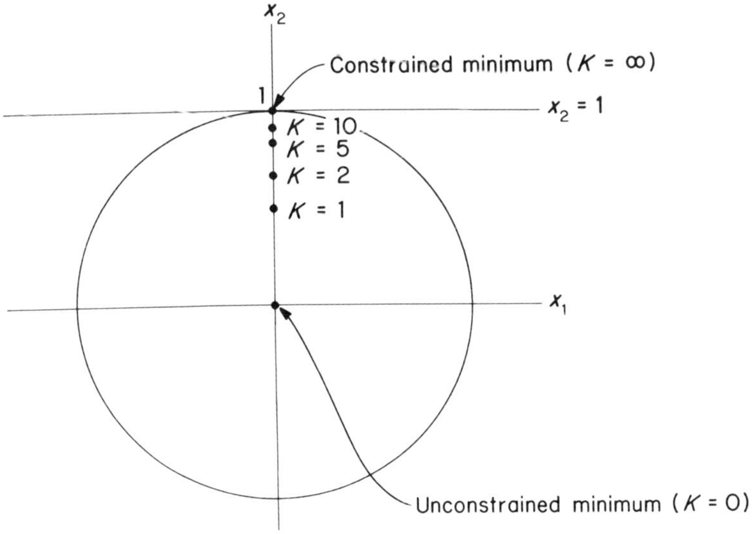

FIGURE 8.1

In Example 8.1, it was possible to obtain the optimal solution analytically as a function of K. In general, this is not possible and the minimum of $P(\mathbf{x};\mathbf{K})$ must be determined by a direct search over $\mathbf{x}$ for each different set of parameter values $\mathbf{K}$. For a given set of constraints, several trials must usually be performed to determine the sensitivity of the solution to changes in $\mathbf{K}$.

To apply this approach to the inequality-constrained problem

$$\min f(\mathbf{x})$$

such that

$$g_i(\mathbf{x}) \leqslant 0, \qquad i = 1, 2, \ldots, r,$$

we define the new objective function

$$P(\mathbf{x};\mathbf{K}) = f(\mathbf{x}) + \sum_{i=1}^{r} K_i[\,g_i(\mathbf{x})\,]^2 u_i(g_i),$$

where

$$u_i(g_i) = \begin{cases} 0 & \text{if} \quad g_i(\mathbf{x}) \leqslant 0, \\ 1 & \text{if} \quad g_i(\mathbf{x}) > 0, \end{cases}$$

and $\mathbf{K}>0$. The step function $u_i(g_i)$ serves to ignore the constraint whenever $\mathbf{x}$ is inside the feasible region and to treat the constraint as an equality whenever $\mathbf{x}$ is outside the feasible region. As before, direct methods can now be applied to minimize the objective function $P(\mathbf{x};\mathbf{K})$ for given values of $\mathbf{K}$.

It is interesting to note the connection between $P(\mathbf{x};\mathbf{K})$ and the Lagrangian

$$L(\mathbf{x},\boldsymbol{\lambda})=f(\mathbf{x})+\sum_{i=1}^{r}\lambda_i g_i(\mathbf{x})$$

for the same problem. If we let

$$\lambda_i=K_i g_i(\mathbf{x})u_i(g_i),$$

then

$$P(\mathbf{x};\mathbf{K})=L(\mathbf{x},\boldsymbol{\lambda}).$$

In addition, it can be shown [Beltrami, 1970] that as K_i approaches infinity, and correspondingly $\mathbf{x}$ approaches some boundary $g_i(\mathbf{x})=0$, then

$$\lim_{K_i\to\infty} 2K_i g_i(\mathbf{x})u_i(g_i)=\lambda_i^*.$$

Since $K_i>0$ and $g_i(\mathbf{x})u_i(g_i)\geqslant 0$ outside the feasible region, the product

$$\lambda_i^*=\lim_{K_i\to\infty} 2K_i g_i(\mathbf{x})u_i(g_i)\geqslant 0,$$

as required (see Problem 8.3). This type of penalty function is included in the class of *exterior methods*, since convergence toward the constraint proceeds from the nonfeasible region.

If we take the above interpretation of λ_i, this penalty-function method will converge to an optimal solution even though the constraint qualification is not satisfied (i.e., even though $\boldsymbol{\lambda}^*$ does not exist at $\mathbf{x}^*$). The reason for this apparent contradiction is that the constraint qualification applies at a single point $(\mathbf{x}^*,\boldsymbol{\lambda}^*)$, while the penalty-function method obtains a sequence of points $(\mathbf{x}^1,\mathbf{x}^2,\ldots;\mathbf{K}^1,\mathbf{K}^2,\ldots)$, which only converges to $(\mathbf{x}^*,\boldsymbol{\lambda}^*)$ in the limit as $\mathbf{K}\to\infty$.

EXAMPLE 8.2

$$\max x_1$$

such that

$$g_1(\mathbf{x}) = (x_2 - 2) + (x_1 - 1)^3 \leqslant 0,$$

$$g_2(\mathbf{x}) = (2 - x_2) + (x_1 - 1)^3 \leqslant 0,$$

$$g_3(\mathbf{x}) = x_1 \geqslant 0, \qquad g_4(\mathbf{x}) = x_2 \geqslant 0.$$

We solved this problem by the Kuhn-Tucker theory in Problem 4.9, page 104, and we saw that the constraint qualification is not satisfied at the optimum $\mathbf{x}^* = [1, 2]^T$. Solving now by penalty-function methods, we define

$$P(\mathbf{x}; \mathbf{K}) = -x_1 + K_1\left[(x_2 - 2) + (x_1 - 1)^3\right]^2 u_1(g_1)$$

$$+ K_2\left[(2 - x_2) + (x_1 - 1)^3\right]^2 u_2(g_2) + K_3 x_1^2 u_3(g_3) + K_4 x_2^2 u_4(g_4).$$

Then

$$\frac{\partial P}{\partial x_1} = -1 + 6K_1\left[(x_2 - 2) + (x_1 - 1)^3\right](x_1 - 1)^2 u_1(g_1)$$

$$+ 6K_2\left[(2 - x_2) + (x_1 - 1)^3\right](x_1 - 1)^2 u_2(g_2) + 2K_3 x_1 u_3(g_3)$$

$$= 0, \tag{8.1}$$

$$\frac{\partial P}{\partial x_2} = 2K_1\left[(x_2 - 2) + (x_1 - 1)^3\right] u_1(g_1)$$

$$- 2K_2\left[(2 - x_2) + (x_1 - 1)^3\right] u_2(g_2) + 2K_4 x_2 u_4(g_4)$$

$$= 0. \tag{8.2}$$

For some arbitrary point in the first quadrant but outside the feasible region,

$$u_1(g_1) = u_2(g_2) = 1 \quad \text{and} \quad u_3(g_3) = u_4(g_4) = 0.$$

Taking $K_1 = K_2 = K$, Eq. (8.2) yields

$$4K(x_2 - 2) = 0,$$

or

$$x_2^* = 2,$$

and Eq. (8.1) yields

$$12K(x_1-1)^5=1.$$

Thus

$$(x_1-1)^5=\lim_{K\to\infty}\frac{1}{12K}=0,$$

so

$$x_1^*=1,$$

which is the correct solution even though the constraint qualification is not satisfied at the point 1,2.

Some care must be taken in the application of this penalty-function method, since the algorithm may converge to a fictitious solution if the problem is not properly posed. An illustration of this defect is given in Example 8.3.

EXAMPLE 8.3 Minimize the distance from the origin subject to the following constraints:

$$g_1(\mathbf{x})=x_1+x_2-1=0,$$

$$g_2(\mathbf{x})=x_1+x_2-2=0.$$

Clearly both of these constraints cannot be satisfied simultaneously, so the problem has no solution. However, ignoring this fact and proceeding with the penalty-function method, we get

$$P(\mathbf{x};K)=x_1^2+x_2^2+K(x_1+x_2-1)^2+K(x_1+x_2-2)^2.$$

The necessary conditions are

$$\frac{\partial P}{\partial x_1}=2x_1+2K(x_1+x_2-1)+2K(x_1+x_2-2)=0,$$

$$\frac{\partial P}{\partial x_2}=2x_2+2K(x_1+x_2-1)+2K(x_1+x_2-2)=0,$$

or

$$x_1+K(2x_1+2x_2-3)=0,$$

$$x_2+K(2x_1+2x_2-3)=0.$$

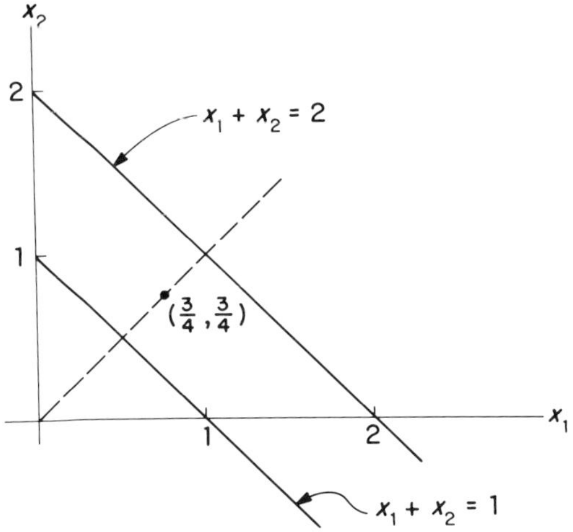

FIGURE 8.2

Thus

$$x_1^* = x_2^*,$$

and

$$x_1^* = \lim_{K \to \infty} \frac{3K}{1+4K} = \frac{3}{4}.$$

Hence the penalty-function method obtains the fictitious solution 0.75, 0.75 which satisfies neither of the equality constraints, as shown in Fig. 8.2.

If the functions $f(\mathbf{x})$ and $g_i(\mathbf{x})$ are convex, then we know that local optimizing procedures will obtain global results, since the constrained surface is unimodal. This result is also obtained for the penalty-function method, since under these convexity conditions $P(\mathbf{x}; \mathbf{K})$ is also convex.

Other Penalty Functions

There is an important class of penalty-function methods called *interior methods* because they proceed toward the constraint boundary from inside the feasible region. Typical of these methods are the sequential unconstrained minimization techniques (SUMT) discussed in this section.

Consider the problem

$$\min f(\mathbf{x})$$

such that

$$g_i(\mathbf{x}) \geqslant 0, \qquad i = 1, 2, \ldots, r,$$

and define the new objective function

$$P(\mathbf{x}; K) = f(\mathbf{x}) + K \sum_{i=1}^{r} \frac{1}{g_i(\mathbf{x})}.$$

If we choose $K_p > 0$ as a positive decreasing sequence, then minimizing $P(\mathbf{x}; K_p)$ yields $\mathbf{x}(K_p)$, and it can be shown that as $p \to \infty$, $K_p \to 0$ and $\mathbf{x}(K_p) \to \mathbf{x}^*$.

Defining

$$R^0 = \{\mathbf{x} | \, g_i(\mathbf{x}) > 0, \, i = 1, 2, \ldots, r\}$$

and

$$R = \{\mathbf{x} | \, g_i(\mathbf{x}) \geqslant 0, \, i = 1, 2, \ldots, r\},$$

we can now state a major result [Fiacco and McCormick, 1968]. The proof of this theorem is outside the scope of this book.

Theorem 8.1 *If*

(1) R^0 *is nonempty,*

(2) $f(\mathbf{x})$ *and* $-g_i(\mathbf{x})$, $i = 1, 2, \ldots, r$, *are convex and twice differentiable,*

(3) *for every finite* k, $\{\mathbf{x} | f(\mathbf{x}) \leqslant k; \; \mathbf{x} \in R\}$ *is a bounded set, and*

(4) *for every* $K > 0$, $P(\mathbf{x}; K)$ *is strictly convex,*

then

(1) *each function* $P(\mathbf{x}; K_p)$ *for* $K_p > 0$ *is minimized over* R^0 *at a unique* $\mathbf{x}(K_p) \in R^0$ *where*

$$\nabla_{\mathbf{x}} P(\mathbf{x}(K_p); K_p) = \mathbf{0},$$

(2) $\lim_{p \to \infty} P(\mathbf{x}(K_p); K_p) = \min_{\mathbf{x} \in R} f(\mathbf{x})$.

Thus under certain conditions, the unconstrained minimization of $P(\mathbf{x}; K)$ will also solve the constrained problem. This theorem leads to the following computational algorithm:

1. Choose $\mathbf{x}(K_0)=\mathbf{x}^0 \in R^0$ (i.e., strictly inside the feasible region), and choose $K_1 > 0$. Set $p=1$.

2. Minimize $P(\mathbf{x}; K_p)$, beginning at the point $\mathbf{x}(K_{p-1})$ and yielding $\mathbf{x}(K_p)$.

3. Stop if the final convergence criterion is satisfied.

4. Otherwise calculate

$$K_{p+1} = \frac{K_p}{c}$$

for $c > 1$, increase p by 1, and return to step 2.

Any suitable algorithm for minimizing (such as the gradient algorithms discussed in Chapter 6) can be used in step 2. Typical convergence criteria for terminating step 2 were given in Chapter 6, page 150. One final convergence criterion for use in step 3 is to stop when

$$\min P(\mathbf{x}; K_p) - \min P(\mathbf{x}; K_{p-1}) \leqslant \varepsilon.$$

Given this iterative procedure over the index p with an imbedded iteration required to minimize $P(\mathbf{x}; K_p)$, it may be computationally advantageous to terminate the imbedded iteration of step 2 with a fairly low degree of accuracy for $\mathbf{x}(K_p)$. This tradeoff arises because the terminal iterations of (for example) the gradient algorithm tend to be less productive than the starting iterations.

The rate of convergence depends heavily upon the values of K_1 and c as well as upon the value of $\mathbf{x}(K_0)$. If K_1 is taken to be too large, many iterations on p will be required to obtain convergence. However, if K_1 is too small, the imbedded minimization in step 2 may converge slowly or not at all. Thus a careful choice of K_1 must be made in order to satisfy both of these requirements (see Problems 8.7 and 8.8).

EXAMPLES

8.4 Use SUMT to solve the problem

$$\min \tfrac{1}{3}(x_1+1)^3 + x_2$$

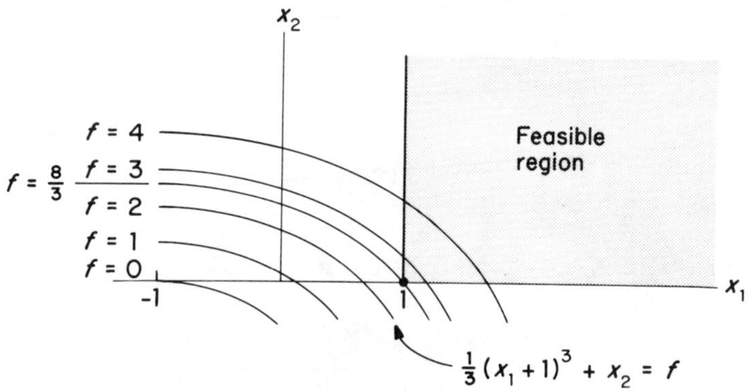

FIGURE 8.3

such that

$$g_1(\mathbf{x}) = x_1 - 1 \geqslant 0,$$

$$g_2(\mathbf{x}) = x_2 \geqslant 0.$$

From Fig. 8.3, we can see that the optimal solution occurs at the point $\mathbf{x}^* = [1,0]^T$ and $f(\mathbf{x}^*) = \frac{8}{3}$. Using SUMT, we define

$$P(\mathbf{x}; K) = \tfrac{1}{3}(x_1+1)^3 + x_2 + \frac{K}{x_1 - 1} + \frac{K}{x_2}.$$

The necessary conditions are

$$\frac{\partial P}{\partial x_1} = (x_1+1)^2 - \frac{K}{(x_1-1)^2} = 0,$$

$$\frac{\partial P}{\partial x_2} = 1 - \frac{K}{x_2^2} = 0.$$

Solving these two equations yields

$$x_1(K) = \sqrt{1 + \sqrt{K}}\ ,$$

$$x_2(K) = \sqrt{K}\ .$$

Thus

$$\mathbf{x}^* = \lim_{K \to 0} \left[\sqrt{1+\sqrt{K}}\ , \sqrt{K}\ \right]^T = [1,0]^T,$$

which is the optimal solution.

8.5 Show the iterative steps obtained by the SUMT algorithm for the problem in Example 8.4. Choose $K_1 = 9$ and $c = 2$.

Since we were able to solve Example 8.4 analytically, we can determine $\mathbf{x}(K_p)$ explicitly as

$$\mathbf{x}(K_p) = \left[\sqrt{1+\sqrt{K_p}}\ , \sqrt{K_p}\ \right]^T$$

rather than obtaining these steps by the application of direct minimizing algorithms. The resulting steps are displayed in Table 8.1 and Fig. 8.4. Thus

$$x_2(K) = \tfrac{1}{64}\left(-1+\sqrt{1+16K}\ \right)^2 + \tfrac{1}{2}K. \tag{8.4}$$

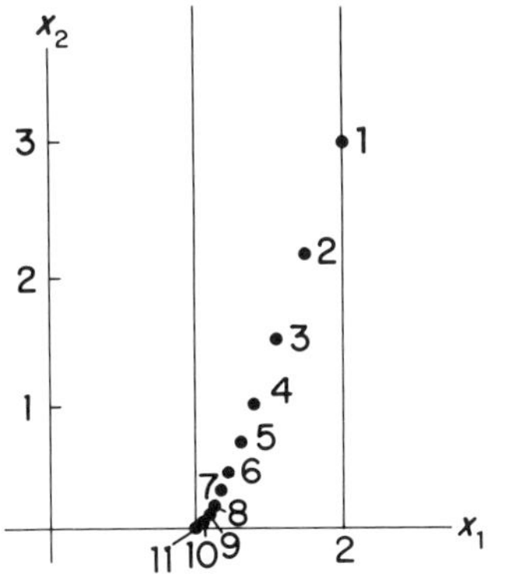

FIGURE 8.4

Table 8.1

p	K_p	$x_1(K_p)$	$x_2(K_p)$
1	9.	2.	3.
2	4.5	1.768	2.12
3	2.25	1.581	1.50
4	1.125	1.439	1.06
5	0.563	1.322	0.752
6	0.282	1.229	0.503
7	0.141	1.170	0.376
8	0.071	1.120	0.257
9	0.036	1.091	0.190
10	0.018	1.062	0.134
11	0.009	1.047	0.095

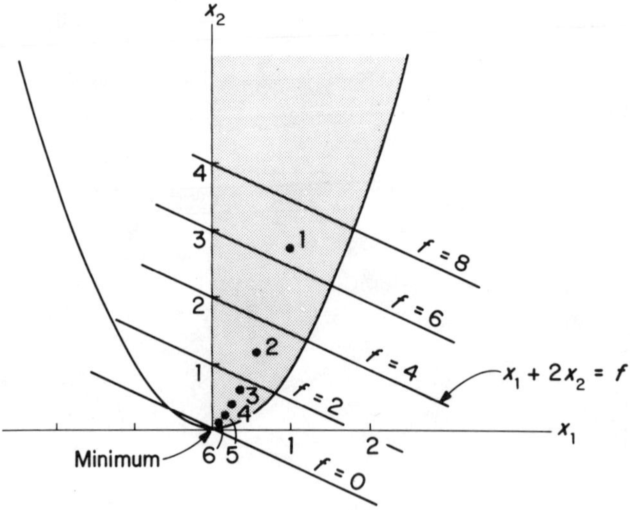

FIGURE 8.5

Then

$$\mathbf{x}^* = \lim_{K\to 0}\left[x_1(K), x_2(K)\right]^T = \left[0,0\right]^T,$$

which is the optimal solution as shown in Fig. 8.5.

Another penalty function in the class of sequential unconstrained minimization techniques is given by

$$P(\mathbf{x}; K) = f(\mathbf{x}) - K\sum_{i=1}^{r} \ln g_i(\mathbf{x})$$

for $K > 0$. This form will be called the *logarithmic penalty function*.

EXAMPLES

8.6 Use the logarithmic penalty function to solve the problem

$$\min x_1 + 2x_2$$

such that

$$g_1(\mathbf{x}) = -x_1^2 + x_2 \geqslant 0,$$

$$g_2(\mathbf{x}) = x_1 \geqslant 0.$$

The objective function is given by

$$P(\mathbf{x}; K) = x_1 + 2x_2 - K\ln(-x_1^2 + x_2) - K\ln(x_1).$$

The necessary conditions are

$$\frac{\partial P}{\partial x_1} = 1 + \frac{2Kx_1}{-x_1^2 + x_2} - \frac{K}{x_1} = 0,$$

$$\frac{\partial P}{\partial x_2} = 2 - \frac{K}{-x_1^2 + x_2} = 0.$$

Solving these two equations yields

$$x_1(K) = \tfrac{1}{8}(-1 + \sqrt{1 + 16K}\,). \tag{8.3}$$

8.7 Show the iteration steps obtained using the logarithmic penalty function in Example 8.6. Choose $K_1 = 4$ and $c = 2$.

These steps are calculated from Eqs. (8.3) and (8.4), and are given in Table 8.2 and Fig. 8.5.

Table 8.2

p	K_p	$x_1(K_p)$	$x_2(K_p)$
1	4.	0.885	2.781
2	2.	0.593	1.350
3	1.	0.391	0.653
4	0.5	0.375	0.390
5	0.25	0.155	0.149
6	0.125	0.090	0.071
7	0.063	0.057	0.035
8	0.032	0.029	0.017
9	0.016	0.016	0.008
10	0.008	0.008	0.004
11	0.004	0.003	0.002

Solved Problems

8.1 Use the penalty function $K[g(\mathbf{x})]^2$ to solve the problem

$$\min\ x_1^2 + 2x_2^2$$

such that

$$g(\mathbf{x}) = 1 - x_1 - x_2 \leqslant 0.$$

The new objective function is

$$P(\mathbf{x};K)=x_1^2+2x_2^2+K(1-x_1-x_2)^2u(g).$$

If $u(g)=0$, the unconstrained solution is 0,0, which is not feasible; hence $u(g)=1$. Then the necessary conditions are

$$\frac{\partial P}{\partial x_1}=2x_1-2K(1-x_1-x_2)=0,$$

$$\frac{\partial P}{\partial x_2}=4x_2-2K(1-x_1-x_2)=0,$$

and

$$x_1=2x_2=\frac{2K}{2+3K}.$$

Thus

$$x_1^*=\lim_{K\to\infty}\frac{2K}{2+3K}=\frac{2}{3},$$

and

$$x_2^*=\tfrac{1}{3}.$$

8.2 Show that the penalty function $K[g(\mathbf{x})]^2$ attains the optimal solution to the equality-constrained problem

$$\min x_1^2+x_2^2$$

such that

$$g(\mathbf{x})=x_2^2-(x_1-1)^3=0,$$

even though the optimal Lagrange multipliers do not exist.

By inspection of Fig. 8.6, the optimal solution is at the point (1,0). From Chapter 3, the Lagrange multipliers exist if the Jacobian of the constraint matrix has rank 1 at 1,0. But

$$\nabla g(1,0)=\begin{bmatrix}-3(x_1-1)^2\\ 2x_2\end{bmatrix}_{1,0}=\begin{bmatrix}0\\0\end{bmatrix},$$

which has rank 0, so the optimal Lagrange multipliers do not exist.

Now by penalty-function methods, the objective function is

$$P(\mathbf{x};K)=x_1^2+x_2^2+K\left[x_2^2-(x_1-1)^3\right]^2,$$

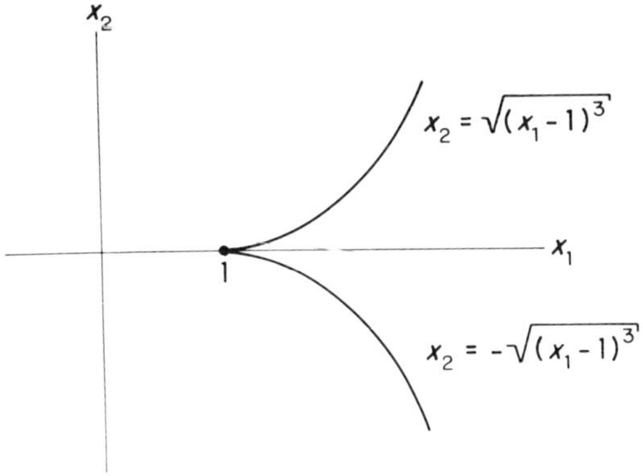

FIGURE 8.6

and the necessary conditions are

$$\frac{\partial P}{\partial x_1} = 2x_1 - 6K(x_1-1)^2\left[x_2^2-(x_1-1)^3\right]=0$$

$$\frac{\partial P}{\partial x_2} = 2x_2 + 4Kx_2\left[x_2^2-(x_1-1)^3\right]=0.$$

Writing these two equations as

$$(x_1-1)^2\left[x_2^2-(x_1-1)^3\right] = \frac{x_1}{3K},$$

$$x_2\left[x_2^2-(x_1-1)^3\right] = -\frac{x_2}{2K},$$

and taking the limit as $K \to \infty$, we see that

$$x_1^* = 1 \quad \text{and} \quad x_2^* = 0,$$

which is the optimal solution.

8.3 Show that

$$\lambda_1^* = \lim_{K_1 \to \infty} 2K_1 g_1(\mathbf{x}) u_1(g_1)$$

for the problem

$$\min x_1^2 + x_2^2$$

such that

$$g_1(\mathbf{x}) = 3x_1 + 2x_2 - 6 \geqslant 0,$$

$$g_2(\mathbf{x}) = x_1 \geqslant 0,$$

$$g_3(\mathbf{x}) = x_2 \geqslant 0.$$

The Lagrangian is given by

$$L(\mathbf{x},\boldsymbol{\lambda}) = x_1^2 + x_2^2 + \lambda_1(3x_1+2x_2-6) + \lambda_2 x_1 + \lambda_3 x_2,$$

and the Kuhn-Tucker conditions are

$$\frac{\partial L}{\partial x_1} = 2x_1 + 3\lambda_1 + \lambda_2 = 0,$$

$$\frac{\partial L}{\partial x_2} = 2x_1 + 2\lambda_1 + \lambda_3 = 0,$$

$$\frac{\partial L}{\partial \lambda_1} = 3x_1 + 2x_2 - 6 \geqslant 0,$$

$$\frac{\partial L}{\partial \lambda_2} = x_1 \geqslant 0, \qquad \frac{\partial L}{\partial \lambda_3} = x_2 \geqslant 0,$$

$$\lambda_1(3x_1+2x_2-6) = 0,$$

$$\lambda_2 x_1 = 0, \qquad \lambda_3 x_2 = 0.$$

The optimal set of $(\mathbf{x},\boldsymbol{\lambda})$ which satisfies these conditions is

$$\mathbf{x}^* = \left[\tfrac{18}{13}, \tfrac{12}{13}\right]^T, \qquad \boldsymbol{\lambda}^* = \left[-\tfrac{12}{13}, 0, 0\right]^T.$$

Using the penalty-function method, the new objective function is given by

$$P(\mathbf{x};\mathbf{K}) = x_1^2 + x_2^2 + K_1(3x_1+2x_2-6)^2 u_1(g_1) + K_2 x_1^2 u_2(g_2) + K_3 x_2^2 u(g_3).$$

For x_1, x_2 in the first quadrant but outside the feasible region,

$$u_1(g_1) = 1 \quad \text{and} \quad u_2(g_2) = u_3(g_3) = 0,$$

and the necessary conditions are

$$\frac{\partial P}{\partial x_1} = 2x_1 + 6K_1(3x_1+2x_2-6) = 0,$$

$$\frac{\partial P}{\partial x_2} = 2x_2 + 4K_1(3x_1+2x_2-6) = 0.$$

Solving these two equations, we get

$$x_1(K_1) = \frac{36K_1}{2+26K_1}, \qquad x_2(K_1) = \frac{24K_1}{2+26K_1},$$

so

$$\mathbf{x}^* = \lim_{K_1\to\infty} \left(x_1(K_1), x_2(K_1)\right) = \left[\tfrac{18}{13}, \tfrac{12}{13}\right]^T$$

and

$$2K_1 g_1(\mathbf{x}) = 2K_1\left[3\left(\frac{36K_1}{2+26K_1}\right)+2\left(\frac{24K_1}{2+26K_1}\right)-6\right] = \frac{-24K_1}{2+26K_1}.$$

So

$$\lambda_1^* = \lim_{K_1\to\infty}\left(\frac{-24K_1}{2+26K_1}\right) = -\frac{12}{13},$$

as expected.

8.4 Use the penalty function $K/g(\mathbf{x})$ to solve the problem

$$\min x$$

such that

$$0 \leqslant x \leqslant 1.$$

We first rewrite the problem as

$$\min x$$

such that

$$x \geqslant 0,$$
$$1 - x \geqslant 0.$$

The new objective function is

$$P(x,K) = x + \frac{K}{x} + \frac{K}{1-x},$$

and the necessary condition is

$$\frac{\partial P}{\partial x} = 1 - \frac{K}{x^2} + \frac{K}{(1-x)^2} = 0,$$

which leads to

$$\frac{x^2(1-x)^2}{1-2x} = K.$$

Thus the limit $K \to 0$ implies that either $x^* = 0$ or $x^* = 1$ (corresponding to the constrained minimum and maximum respectively). In order to distinguish between these two candidate solutions, we examine the behavior of K in the vicinity of each point. For $x = \varepsilon \ll 1$,

$$K = \frac{\varepsilon^2(1-\varepsilon)^2}{1-2\varepsilon} > 0,$$

and for $x=1-\varepsilon$, where $\varepsilon \ll 1$,

$$K=\frac{(1-\varepsilon)^2\varepsilon^2}{2\varepsilon-1}<0.$$

Since we require $K>0$, the minimum is at $x^*=0$.

8.5 Use the penalty function $K/g(\mathbf{x})$ to solve the problem

$$\min x-1$$

such that

$$x \geqslant 0.$$

Choose $K_1=1$, $c=10$, and sketch $P(x;K_p)$ for $p=1$, 2, and 3.

The objective function is

$$P(x;K)=x-1+\frac{K}{x},$$

and the necessary condition is

$$\frac{\partial P}{\partial x}=1-\frac{K}{x^2}=0.$$

Thus

$$x=\pm\sqrt{K},$$

where we choose the positive value to ensure the feasibility of $x(K)$ for $K>0$.

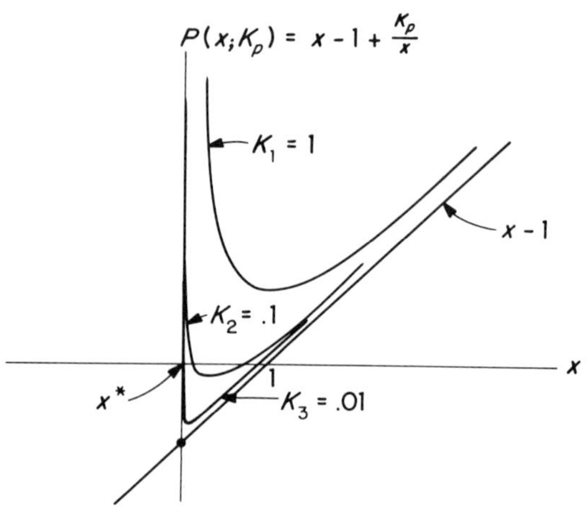

FIGURE 8.7

Clearly

$$x^* = \lim_{K\to 0} \sqrt{K} = 0.$$

The objective functions corresponding to $K=1$, 0.1, and 0.01 are shown in Fig. 8.7. Note that the unconstrained objective function represents a continuous surface strictly within the feasible region. This surface is modified by decreasing K in such a way that its minimum converges to the solution of the constrained problem.

8.6 Devise an appropriate penalty function for use as $K\to-\infty$ and use it to solve Example 8.4 (p. 205).

Since $K\to-\infty$, we need a penalty function which approaches the origin from below. One such function is

$$\frac{1}{\ln g(\mathbf{x})},$$

which yields the objective function

$$P(\mathbf{x};K)=f(\mathbf{x})+K\sum_{i=1}^{r}\frac{1}{\ln g_i(\mathbf{x})}.$$

Applying this penalty function to Example 8.4 yields

$$P(\mathbf{x};K)=\tfrac{1}{3}(x_1+1)^3+x_2+\frac{K}{\ln(x_1-1)}+\frac{K}{\ln x_2}.$$

The necessary conditions are

$$\frac{\partial P}{\partial x_1}=(x_1+1)^2-\frac{K}{(x_1-1)\left[\ln(x_1-1)\right]^2}=0,$$

$$\frac{\partial P}{\partial x_2}=1-\frac{K}{x_2[\ln x_2]^2}=0,$$

which can be written

$$(x_1+1)^2(x_1-1)\left[\ln(x_1-1)\right]^2=K,$$

$$x_2[\ln x_2]^2=K.$$

Then

$$\lim_{K\to-\infty}\left[x_1(K),x_2(K)\right]=[1,0],$$

which is the optimal solution obtained in Example 8.4.

8.7 One estimate of the amount by which a function $f(\mathbf{x})$ evaluated at $\mathbf{x}^0$ exceeds its minimum value is given by

$$\nabla f(\mathbf{x}^0)^T H(\mathbf{x}^0)^{-1} \nabla f(\mathbf{x}^0),$$

where $H(\mathbf{x}^0)$ is the Hessian matrix. Use this estimate to determine K_1 for the penalty function $K/g(\mathbf{x})$ when $\mathbf{x}^0$ is near a constraint boundary.

We let the objective function be

$$P(\mathbf{x}^0; K_1) = f(\mathbf{x}^0) + K_1 p(\mathbf{x}^0),$$

where

$$p(\mathbf{x}^0) = \sum_{i=1}^{r} \frac{1}{g_i(\mathbf{x}^0)}.$$

Then the Hessian matrix [for $P(\mathbf{x}^0, K_1)$] is given by

$$\mathbf{H}_1(\mathbf{x}^0) + K_1 \mathbf{H}_2(\mathbf{x}^0),$$

where $\mathbf{H}_1(\mathbf{x}^0)$ and $\mathbf{H}_2(\mathbf{x}^0)$ are the Hessian matrices for $f(\mathbf{x}^0)$ and $p(\mathbf{x}^0)$ respectively. When these matrices are evaluated near a constraint boundary, we get

$$\mathbf{H}_2(\mathbf{x}^0) \gg \mathbf{H}_1(\mathbf{x}^0)$$

on an element-by-element basis, due to the form of $p(\mathbf{x}^0)$. Ignoring $\mathbf{H}_1(\mathbf{x}^0)$, we can now choose K_1 to minimize

$$\left[\nabla f(\mathbf{x}^0) + K_1 \nabla p(\mathbf{x}^0)\right]^T \left[K_1 \mathbf{H}_2(\mathbf{x}^0)\right]^{-1} \left[\nabla f(\mathbf{x}^0) + K_1 \nabla p(\mathbf{x}^0)\right],$$

obtaining

$$K_1 = \left[\frac{\nabla f(\mathbf{x}^0)^T \mathbf{H}_2^{-1}(\mathbf{x}^0) \nabla f(\mathbf{x}^0)}{\nabla p(\mathbf{x}^0)^T \mathbf{H}_2^{-1}(\mathbf{x}^0) \nabla p(\mathbf{x}^0)}\right]^{1/2}.$$

8.8 Determine a method of estimating K_1 for the penalty function $K/g(\mathbf{x})$.

Since $\nabla_{\mathbf{x}} P(\mathbf{x}; K) = \mathbf{0}$ at the optimum, one approach is to choose K_1 to minimize the magnitude of $\nabla_{\mathbf{x}} P(\mathbf{x}^0; K_1)$. Since

$$\|\nabla_{\mathbf{x}} P(\mathbf{x}^0; K_1)\| = \left[\nabla_{\mathbf{x}} P(\mathbf{x}^0; K_1)\right]^T \left[\nabla_{\mathbf{x}} P(\mathbf{x}^0; K_1)\right],$$

it is necessary that

$$\frac{\partial}{\partial K_1} \left[\nabla f(\mathbf{x}^0) + K_1 \nabla p(\mathbf{x}^0)\right]^T \left[\nabla f(\mathbf{x}^0) + K_1 \nabla p(\mathbf{x}^0)\right] = 0,$$

where

$$p(\mathbf{x}^0)=\sum_{i=1}^{r}\frac{1}{g_i(\mathbf{x}^0)}.$$

Thus

$$K_1=\frac{-\nabla f(\mathbf{x}^0)^T\ \nabla p(\mathbf{x}^0)}{\nabla p(\mathbf{x}^0)^T\ \nabla p(\mathbf{x}^0)}.$$

This estimate of K_1 has been found to work best when $\mathbf{x}^0$ is not close to any constraint boundaries.

Supplementary Problems

8.9 Use the penalty function $K[g(\mathbf{x})]^2$ to solve the problem

$$\min\ 3x_1+4x_2^2$$

such that

$$x_1 \geqslant 0.$$

8.10 Use penalty functions of the form $K_i[g_i(\mathbf{x})]^2$ to solve the problem

$$\min\ (x_1-1)^2+(x_2-2)^2$$

such that

$$x_2-x_1=1,$$

$$x_1+x_2 \leqslant 2,$$

$$x_1 \geqslant 0, \qquad x_2 \geqslant 0.$$

8.11 Show that the penalty function $K[g(\mathbf{x})]^2$ results in the optimal solution to the problem

$$\min\ x_1$$

such that

$$g_1(\mathbf{x})=(x_2-2)+(3-x_1)^2 \leqslant 0,$$

$$g_2(\mathbf{x})=(2-x_2)+(3-x_1)^3 \leqslant 0,$$

even though the constraint qualification is not satisfied.

8.12 Devise an appropriate penalty function which converges as $K\to 1$, and use it to solve Example 8.4 (p. 205).

8.13 Solve Problem 8.9 using the penalty function $K/g(\mathbf{x})$.

8.14 Solve Problem 8.10 using the penalty function $K/g(\mathbf{x})$.

8.15 Solve Problem 8.11 using the penalty function $K/g(\mathbf{x})$.

8.16 Solve Problem 8.9 using the penalty function $\ln g(\mathbf{x})$.

8.17 Solve Problem 8.10 using the penalty function $\ln g(\mathbf{x})$.

8.18 Solve Problem 8.11 using the penalty function $\ln g(\mathbf{x})$.

Chapter 9

Quadratically Convergent Minimization Algorithms

In this chapter we discuss a powerful class of direct methods for function minimization. We denote by $f(\mathbf{x})$ the objective function whose minimum is sought. Since $\max f(\mathbf{x}) = -\min[-f(\mathbf{x})]$, we can restrict our attention to the problem of function minimization without loss of generality. We shall call every algorithm that generates a sequence of points $\mathbf{x}^0, \mathbf{x}^1, \mathbf{x}^2, \ldots$ satisfying $f(\mathbf{x}^0) \geqslant f(\mathbf{x}^1) \geqslant f(\mathbf{x}^2), \ldots$ a *descent method*. In Chapter 6, we encountered an example of a descent method of minimization, called the *steepest-descent method* or the *gradient algorithm*. It is clear that if we want to locate a minimum of a function, then we should use an algorithm that does not increase the function value from one iteration to the next. The descent methods obviously satisfy this requirement.

The descent methods differ from each other in their convergence rates. The rate of convergence of a descent method is measured by the limiting value of the ratio $\ln f(\mathbf{x}^k)/\ln f(\mathbf{x}^{k+1})$ as k tends to infinity, where the index k denotes the number of iterations performed. The limiting value is called the *order of convergence* of the algorithm. The order of convergence of the optimal gradient algorithm is unity. Descent methods whose order of convergence is 2 are said to have *quadratic convergence rates*. However, regardless of their order of convergence, most descent methods cannot locate the minimum of a function in a finite number of steps (see Example 6.4). This is true even when the function is a quadratic form. The methods discussed in this chapter can, however, locate the minimum of a positive definite quadratic function in a finite number of steps.

Thus one measure of convergence for a descent method which is different from the order of convergence introduced earlier is that a descent method is good if it can find the minimum of a positive definite quadratic function in a finite number of steps. Of course, such a measure does not indicate how the descent method will perform with a more general function. However, since we can approximate a twice continuously differentiable function in a neighborhood of its minimum by using a second-order

Taylor series expansion, algorithms which are guaranteed to find the minimum of a quadratic function rapidly will also show rapid convergence to the minimum of more general functions. We shall call an algorithm *quadratically convergent* if it can determine the minimum of a positive definite quadratic function in a finite number of iterations.

Quadratic Functions

A *quadratic function* on E^n is defined as

$$f(x_1,\dots,x_n)=c+\sum_{i=1}^{n} b_i x_i+\frac{1}{2}\sum_{j=1}^{n}\sum_{i=1}^{n} x_j a_{ij} x_i, \tag{9.1}$$

where c, b_i, and a_{ij} are constant scalar quantities. We can rewrite (9.1) in vector-matrix notation as

$$f(\mathbf{x})=c+\mathbf{b}^T\mathbf{x}+\tfrac{1}{2}\mathbf{x}^T\mathbf{A}\mathbf{x} \tag{9.2}$$

where $\mathbf{b}^T=[b_1,\dots,b_n]$ and $\mathbf{A}=[a_{ij}]$ is an $n\times n$ matrix. Without loss of generality we can assume $\mathbf{A}$ to be a symmetric matrix. If $\mathbf{A}$ is positive definite, then we call $f(\mathbf{x})$ in (9.2) a *positive definite quadratic function.*

EXAMPLES

9.1 Consider the quadratic function in two variables

$$f(x_1,x_2)=k+dx_1+ex_2+cx_2^2+bx_1x_2+ax_1^2, \tag{9.3}$$

where a, b, c, d, e, and k are constant real numbers. Find the conditions under which $f(x_1,x_2)$ is positive definite, and determine the shape of its level curves.

We can rewrite (9.3) in vector-matrix notation as

$$f(\mathbf{x})=k+[d,e]\begin{bmatrix}x_1\\x_2\end{bmatrix}+\tfrac{1}{2}[x_1,x_2]\begin{bmatrix}2a & b\\ b & 2c\end{bmatrix}\begin{bmatrix}x_1\\x_2\end{bmatrix}. \tag{9.4}$$

Here the matrix $\mathbf{A}$ is given by

$$\mathbf{A}=\begin{bmatrix}2a & b\\ b & 2c\end{bmatrix}.$$

Now if $a>0$, $c>0$, and $4ac-b^2>0$, then by Sylvester's theorem we know that $\mathbf{A}$ is positive definite. Hence by definition, $f(x_1,x_2)$ is also positive definite. Now by definition, $4ac-b^2$ is the *discriminant* of the quadratic equation given by (9.3). If the discriminant is positive, then the graph in E^2 of $f(x_1,x_2)=k$ (a constant) is an ellipse. Thus the level curves of $f(x_1,x_2)$ are concentric ellipses.

9.2 Consider the quadratic function given by (9.3), where $a=d=k=1$, $e=-1$, $b=0$, and $c=2$. Draw its level curves.

We can write

$$f(x_1,x_2)=1+x_1-x_2+x_1^2+2x_2^2$$

$$=\tfrac{5}{8}+\left(x_1+\tfrac{1}{2}\right)^2+2\left(x_2-\tfrac{1}{4}\right)^2$$

by completing the square. From the above equation it is clear that $f(x_1,x_2)=k$ (a constant) is an ellipse with center at $x_1=-\frac{1}{2}$ and $x_2=\frac{1}{4}$, as shown in Fig. 9.1. Also note that $f(x_1,x_2)\geqslant\frac{5}{8}$ and $f(-\frac{1}{2},\frac{1}{4})=\frac{5}{8}$, i.e., $f(x_1,x_2)$ attains a minimum at the center of the set of concentric ellipses.

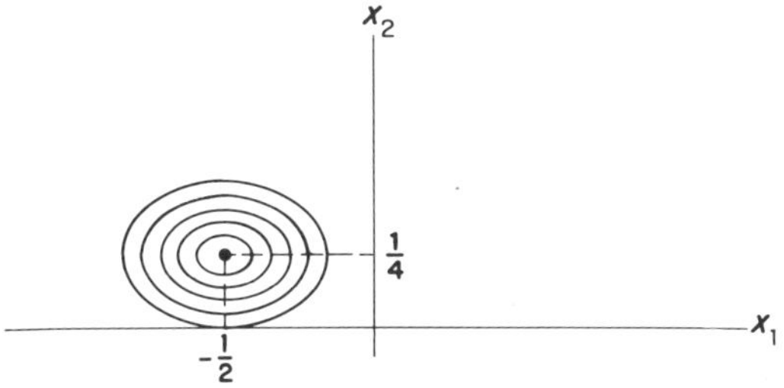

FIGURE 9.1

9.3 Show that the gradient vector of a quadratic function given by Eq. (9.2) is

$$\nabla f(\mathbf{x})=\mathbf{A}\mathbf{x}+\mathbf{b}. \tag{9.5}$$

In order to obtain $\nabla f(\mathbf{x})$ it is necessary to compute $\partial f/\partial x_k$, $k=1,\ldots,n$. Now we can rearrange (9.1) as follows:

$$f(x_1,\ldots,x_k,\ldots,x_n)=c+\sum_i b_ix_i+\frac{1}{2}\left[x_k\sum_j a_{kj}x_j+\sum_{i\neq k}\sum_j x_ia_{ij}x_j\right]$$

$$=c+\sum_i b_ix_i+\frac{1}{2}\left[a_{kk}x_k^2+x_k\sum_{j\neq k}a_{kj}x_j+x_k\sum_{i\neq k}x_ia_{ik}+\sum_{i\neq k}\sum_{j\neq k}x_ia_{ij}x_j\right].$$

Since **A** is symmetric (i.e., $a_{ij} = a_{ji}$), the above equation becomes

$$f(x_1,\ldots,x_k,\ldots,x_n) = c + \sum_i b_i x_i + \frac{1}{2}\left[a_{kk}x_k^2 + 2x_k \sum_{j\neq k} a_{kj}x_j + \sum_{i\neq k}\sum_{j\neq k} x_i a_{ij} x_j\right],$$

from which we obtain

$$\frac{\partial f}{\partial x_k} = b_k + \frac{1}{2}\left[2a_{kk}x_k + 2\sum_{j\neq k} a_{kj}x_j\right] = b_k + \sum_j a_{kj}x_j,$$

which is the required result.

9.4 Show that a positive definite quadratic function attains a unique minimum at $\mathbf{x}^* = -\mathbf{A}^{-1}\mathbf{b}$.

A quadratic function $f(\mathbf{x})$ given by Eq. (9.1) is a twice continuously differentiable function and hence it attains an extremum at *one of* the stationary points given by $\nabla f(\mathbf{x}) = \mathbf{0}$. Since **A** is positive definite, $|\mathbf{A}| \neq 0$, and we conclude from (9.5) that there is only one such stationary point, given by $\mathbf{x}^* = -\mathbf{A}^{-1}\mathbf{b}$. The Hessian matrix of $f(\mathbf{x})$ is **A**, which is positive definite by hypothesis, and hence $f(\mathbf{x})$ attains a unique minimum at $\mathbf{x}^* = -\mathbf{A}^{-1}\mathbf{b}$.

9.5 Consider the quadratic function discussed in Example 9.2. Find its minimum point by using the result of Example 9.4.

We can obtain the matrix **A** from Example 9.1 by substituting the values of a, b, and c given in Example 9.2. Thus we have

$$\mathbf{A} = \begin{bmatrix} 2 & 0 \\ 0 & 4 \end{bmatrix},$$

and we compute

$$\mathbf{A}^{-1} = \begin{bmatrix} -\frac{1}{2} & 0 \\ 0 & \frac{1}{4} \end{bmatrix}$$

In a similar way we obtain $\mathbf{b} = [1, -1]^T$. Now using the result of Example 9.4 we obtain

$$\mathbf{x}^* = -\mathbf{A}^{-1}\mathbf{b} = -\begin{bmatrix} \frac{1}{2} & 0 \\ 0 & \frac{1}{4} \end{bmatrix}\begin{bmatrix} 1 \\ -1 \end{bmatrix} = \left[-\tfrac{1}{2}, \tfrac{1}{4}\right]^T$$

As pointed out in Example 9.2, $f(x_1, x_2)$ attains a minimum at $x_1^* = -\frac{1}{2}$ and $x_2^* = \frac{1}{4}$.

Conjugate Directions in E^n

Let **u** and **v** denote two vectors in E^n. As noted earlier, **u** and **v** are said to be *mutually orthogonal* if their scalar product equals zero, i.e., $\mathbf{u}^T\mathbf{v}=(\mathbf{u},\mathbf{v})=0$. We shall frequently use the convenient notation $(\mathbf{u},\mathbf{v})$ for the scalar product in this chapter. Now suppose we have an $n\times n$, symmetric positive definite matrix **A**. Then **u** and **v** are said to be *mutually conjugate* with respect to **A** if **u** and **Av** are mutually orthogonal, i.e., if $\mathbf{u}^T\mathbf{A}\mathbf{v}=(\mathbf{u},\mathbf{A}\mathbf{v})=0$. Clearly if **u** and **v** are mutually conjugate with respect to the identity matrix, then they are mutually orthogonal. Thus the concept of mutual orthogonality can be thought of as a special case of mutual conjugacy of vectors.

A square matrix **A** is said to have an eigenvalue λ and an associated eigenvector **x** if $\mathbf{x}\neq 0$ and if $\mathbf{A}\mathbf{x}=\lambda\mathbf{x}$. A result from linear algebra states that if **A** is a symmetric positive definite matrix and **x** and **y** are eigenvectors of **A**, then **x** and **y** are mutually orthogonal, i.e., $(\mathbf{x},\mathbf{y})=0$. Since $(\mathbf{y},\mathbf{A}\mathbf{x})=(\mathbf{y},\lambda\mathbf{x})=\lambda(\mathbf{y},\mathbf{x})=0$, we can say that the eigenvectors of a positive definite matrix **A** are mutually conjugate with respect to **A**. Hence given a positive definite matrix, we can be sure of the existence of at least one set of vectors mutually conjugate with respect to the given matrix. The algorithms we are going to discuss use a set of mutually conjugate vectors in E^n for one-dimensional search directions in order to locate the minimum of a given function. However, if a set of mutually conjugate vectors in E^n does not span E^n, then we could be searching in a proper subspace of E^n not containing the minimum. In the next example we show that this is not the case, since a set of mutually conjugate vectors in E^n constitutes a basis and therefore spans E^n.

EXAMPLE 9.6 Let $\mathbf{u}^i$, $i=1,\ldots,n$, denote a set of fectors in E^n mutually conjugate with respect to a given symmetric positive definite matrix **A**. Show that for every vector $\mathbf{x}\in E^n$, $\mathbf{x}=\sum_{i=1}^n \alpha_i\mathbf{u}^i$, where $\alpha_i = (\mathbf{u}^i,\mathbf{A}\mathbf{x})/(\mathbf{u}^i,\mathbf{A}\mathbf{u}^i)$.

Let us consider a linear combination $\sum_{i=1}^n \alpha_i\mathbf{u}^i$ such that

$$\sum_{i=1}^{n} \alpha_i\mathbf{u}^i=0.$$

Then

$$\mathbf{A}\left[\sum_{i=1}^{n} \alpha\mathbf{u}^i\right]=\sum_{i=1}^{n} \alpha_i\mathbf{A}\mathbf{u}^i=0.$$

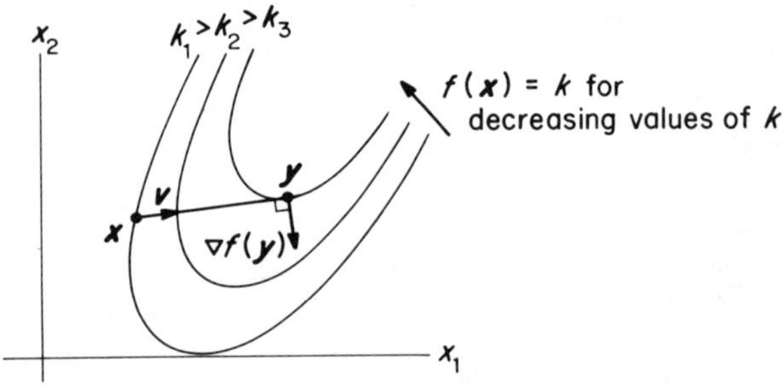

FIGURE 9.2

Therefore, by mutual conjugacy of the given set of vectors,

$$\left(\mathbf{u}^k, \mathbf{A}\left[\sum_{i=1}^{n} \alpha_i \mathbf{u}^i\right]\right) = \sum_{i=1}^{n} \alpha_i (\mathbf{u}^k, \mathbf{A}\mathbf{u}^i) = \alpha_k (\mathbf{u}^k, \mathbf{A}\mathbf{u}^k) = 0.$$

Since $\mathbf{A}$ is positive definite and $\mathbf{u}^k \neq \mathbf{0}$, we have $(\mathbf{u}^k, \mathbf{A}\mathbf{u}^k) \neq 0$ and therefore $\alpha_k = 0$, $k = 1, \ldots, n$. Hence $\{u^i, i = 1, \ldots, n\}$ is a set of linearly independent vectors in E^n which can be used to form a basis. Thus for every $\mathbf{x} \in E^n$ there exist unique α_i, $i = 1, \ldots, n$, such that

$$\mathbf{x} = \sum_{i=1}^{n} \alpha_i \mathbf{u}^i.$$

Since $\mathbf{u}^i$ are mutually conjugate with respect to $\mathbf{A}$, it follows that $(\mathbf{u}^i, \mathbf{A}\mathbf{x}) = \alpha_i (\mathbf{u}^i, \mathbf{A}\mathbf{u}^i)$, i.e., $\alpha_i = (\mathbf{u}^i, \mathbf{A}\mathbf{x}) / (\mathbf{u}^i, \mathbf{A}\mathbf{u}^i)$.

In concluding this section we remind the reader of the geometric aspects of one-dimensional search techniques discussed in Chapter 6. Let $\mathbf{x}$ be any vector in E^n, and let $\mathbf{v}$ be the direction of search starting from $\mathbf{x}$. We want to find that λ^* which minimizes $f(\mathbf{x} + \lambda \mathbf{v})$ with respect to λ. Figure 9.2 shows the geometric description of the problem. Letting $\mathbf{y} = \mathbf{x} + \lambda^* \mathbf{v}$, we showed in Problem 6.7 that $(\mathbf{v}, \nabla f(\mathbf{y})) = 0$. This result is called the *condition for optimal descent*, and it is used in the later sections.

Minimization with Conjugate Directions of Descent

We now consider the principal advantage of using mutually conjugate directions for any descent algorithm for minimization. The main result is given by the following:

Theorem 9.1 *Every descent method of minimization, using mutually conjugate directions of descent, is quadratically convergent.*

Proof: Let the positive definite quadratic function be given by Eq. (9.2), and let $\mathbf{x}^0 \in E^n$ denote an arbitrary initial approximation to $\mathbf{x}^* = -\mathbf{A}^{-1}\mathbf{b}$. Let $\mathbf{u}^i$, $i = 1, \ldots, n$, denote the directions of descent which we assume to be mutually conjugate with respect to $\mathbf{A}$. Furthermore, let $\mathbf{x}^i$ denote the ith approximation to $\mathbf{x}^*$, i.e.,

$$\mathbf{x}^i = \mathbf{x}^0 + \sum_{k=1}^{i} \lambda_k \mathbf{u}^k.$$

We arrive at $\mathbf{x}^{i+1} = \mathbf{x}^i + \lambda_{i+1}\mathbf{u}^{i+1}$ from $\mathbf{x}^i$ by moving along the direction $\mathbf{u}^{i+1}$ an amount λ_{i+1}, where λ_{i+1} minimizes $f(\mathbf{x}^i + \lambda \mathbf{u}^{i+1})$ with respect to λ. Here we are using a one-dimensional search for a local minimum along $\mathbf{u}^{i+1}$ as explained in Chapter 6. From the condition for optimal descent we have

$$\left(\mathbf{u}^{i+1}, \nabla f\left(\mathbf{x}^i + \lambda_{i+1}\mathbf{u}^{i+1}\right)\right) = 0. \tag{9.6}$$

Using (9.5) for $\nabla f(\mathbf{x})$ and the mutual conjugacy of the given vectors, we compute the optimal value of λ_{i+1} from (9.6) to be

$$\lambda_{i+1} = -\frac{(\mathbf{u}^{i+1}, \mathbf{A}\mathbf{x}^0 + \mathbf{b})}{(\mathbf{u}^{i+1}, \mathbf{A}\mathbf{u}^{i+1})}. \tag{9.7}$$

Hence

$$\begin{aligned}
\mathbf{x}^n &= \mathbf{x}^0 + \sum_{k=1}^{n} \lambda_k \mathbf{u}^k \\
&= \mathbf{x}^0 - \sum_{k=1}^{n} \frac{(\mathbf{u}^k, \mathbf{A}\mathbf{x}^0 + \mathbf{b})}{(\mathbf{u}^k, \mathbf{A}\mathbf{u}^k)} \mathbf{u}^k \\
&= \mathbf{x}^0 - \sum_{k=1}^{n} \frac{(\mathbf{u}^k, \mathbf{A}\mathbf{x}^0)}{(\mathbf{u}^k, \mathbf{A}\mathbf{u}^k)} \mathbf{u}^k - \sum_{k=1}^{n} \frac{(\mathbf{u}^k, \mathbf{A}[\mathbf{A}^{-1}\mathbf{b}])}{(\mathbf{u}^k, \mathbf{A}\mathbf{u}^k)} \mathbf{u}^k \\
&= \mathbf{x}^0 - \mathbf{x}^0 - \mathbf{A}^{-1}\mathbf{b} = -\mathbf{A}^{-1}\mathbf{b} = \mathbf{x}^*.
\end{aligned}$$

Note that the result obtained in Example 9.6 is used to identify the terms $\mathbf{x}^0$ and $\mathbf{A}^{-1}\mathbf{b}$ in the above derivation.

We observe from Theorem 9.1 that quadratic convergence results from any arbitrary initial approximation $\mathbf{x}^0$. The maximum number of iterations required is n, the number of variables in $f(\mathbf{x})$. Also, quadratic convergence can be obtained regardless of the actual sequence in which the directions $\mathbf{u}^k$ are used in practice.

EXAMPLE 9.7 Consider the quadratic function discussed in Examples 9.2 and 9.5. Find the minimum of this function by using Theorem 9.1.

From Example 9.5 we have

$$\mathbf{A}=\begin{bmatrix} 2 & 0 \\ 0 & 4 \end{bmatrix}.$$

Let $\mathbf{u}^1=[1,0]^T$ and $\mathbf{u}^2=[0,1]^T$. Then

$$(\mathbf{u}^1,\mathbf{A}\mathbf{u}^2)=[1,0]\begin{bmatrix} 2 & 0 \\ 0 & 4 \end{bmatrix}\begin{bmatrix} 0 \\ 1 \end{bmatrix}=[1,0]\begin{bmatrix} 0 \\ 4 \end{bmatrix}=0.$$

Hence $\mathbf{u}^1$ and $\mathbf{u}^2$ are mutually conjugate with respect to $\mathbf{A}$. As an initial approximation we take $\mathbf{x}^0=[0,0]^T$. Now using (9.7), we obtain $\lambda_1=-\frac{1}{2}$ and $\lambda_2=\frac{1}{4}$. Hence

$$\mathbf{x}^2=\mathbf{x}^0+\lambda_1\mathbf{u}^1+\lambda_2\mathbf{u}^2=\left[-\tfrac{1}{2},\tfrac{1}{4}\right]^T=\mathbf{x}^*.$$

Conjugate-Direction Algorithms

We can modify many well-known descent methods for minimization, such as the gradient method discussed in Chapter 6, to take advantage of mutually conjugate directions of descent. For such modifications we must develop methods for generating mutually conjugate directions. Computationally convenient algorithms should use only function values and gradient vectors for generating mutually conjugate directions of search, since these quantities can usually be readily computed. In particular, such algorithms should not require explicit knowledge of the matrix $\mathbf{A}$ (second partial derivatives) in order to generate mutually conjugate vectors with respect to $\mathbf{A}$. One such method of minimization is the Fletcher-Reeves *conjugate gradient algorithm*. This algorithm, described below, is a modification of the gradient algroithm discussed in Chapter 6.

CONJUGATE-GRADIENT ALGORITHM

1. Let $\mathbf{x}^0$ denote the first approximation to $\mathbf{x}^*$. Compute $\nabla f(\mathbf{x}^0)$, and define $\mathbf{v}^0=-\nabla f(\mathbf{x}^0)$.

2. For $i=1,\ldots,n-1$,
   (i) set $\mathbf{x}^i=\mathbf{x}^{i-1}+\lambda_{i-1}\mathbf{v}^{i-1}$, where λ_{i-1} minimizes $f(\mathbf{x}^{i-1}+\lambda\mathbf{v}^{i-1})$ with respect to λ.
   (ii) compute $\nabla f(\mathbf{x}^i)$.
   (iii) when $i<n$, define

$$\mathbf{v}^i=-\nabla f(\mathbf{x}^i)+\frac{\|\nabla f(\mathbf{x}^i)\|^2}{\|\nabla f(\mathbf{x}^{i-1})\|^2}\mathbf{v}^{i-1}. \tag{9.8}$$

3. Replace $\mathbf{x}^0$ by $\mathbf{x}^n$ and go to 1 unless the stopping rule is satisfied.

To start the algorithm, the initial approximation $\mathbf{x}^0$ can be chosen in the same way as in the case of the gradient method. Also, the stopping rules for the gradient method discussed in Chapter 6 can be used to terminate the iterations. In the gradient method, we move from $\mathbf{x}^i$ to $\mathbf{x}^{i+1}$ along $\mathbf{v}^i=-\nabla f(\mathbf{x}^i)$ for function minimization. In the conjugate-gradient method we modify the gradient direction by adding

$$\frac{\|\nabla f(\mathbf{x}^i)\|^2}{\|\nabla f(\mathbf{x}^{i-1})\|^2}\mathbf{v}^{i-1}.$$

When $f(\mathbf{x})$ is a positive definite quadratic function, this modification results in a set of mutually conjugate vectors $\mathbf{v}^i$, $i=1,\ldots,n$. A detailed analysis of the conjugate-gradient algorithm is given in Problem 9.8.

EXAMPLES

9.8 Consider a positive definite quadratic form $f(\mathbf{x})=\frac{1}{2}\mathbf{x}^T\mathbf{A}\mathbf{x}$, where

$$\mathbf{A}=\begin{bmatrix}1 & 1\\ 1 & 2\end{bmatrix}.$$

Find the minimum of this function by the Fletcher-Reeves conjugate-gradient method.

The minimum of $f(\mathbf{x})$ is located at $[0,0]^T$ in E^2. We have learned in Chapter 6, Problem 6.4, that the optimal-gradient method cannot locate this minimum point in a finite number of iterations unless the initial point is chosen along one of the coordinate axes. Since $f(\mathbf{x})$ is a quadratic form, the convergence of the conjugate-gradient method is not affected by the choice of the initial point. So we choose arbitrarily $\mathbf{x}^0=[10,-5]^T$.

Now $\nabla f(\mathbf{x})=\mathbf{A}\mathbf{x}=[x_1+x_2,\ x_1+2x_2]^T$. According to step 1 of the conjugate-gradient algorithm, we define $\mathbf{v}^0=-\nabla f(\mathbf{x}^0=-[5,0]^T$. Then

$\mathbf{x}^0+\lambda\mathbf{v}^0=[10-5\lambda,\ -5]^T$ and

$$f(\mathbf{x}^0+\lambda\mathbf{v}^0)=\tfrac{1}{2}[10-5\lambda,\ -5]\begin{bmatrix}1 & 1\\ 1 & 2\end{bmatrix}\begin{bmatrix}10-5\lambda\\ -5\end{bmatrix}$$
$$=\tfrac{1}{2}[50-75\lambda+50\lambda^2].$$

Since $d^2f(\mathbf{x}^0+\lambda\mathbf{v}^0)/d\lambda^2=50>0$, $f(\mathbf{x}^0+\lambda\mathbf{v}^0)$ attains a minimum with respect to λ at $\lambda_0=0.75$, where $df(\mathbf{x}^0+\lambda\mathbf{v}^0)/d\lambda=0$. Therefore, $\mathbf{x}^1=\mathbf{x}^0+\lambda_0\mathbf{v}^0=[6.25,\ -5]^T$ and $\nabla f(\mathbf{x}^1)=[1.25,\ -3.75]^T$.

Now in order to use (9.8) we need $\|\nabla f(\mathbf{x}^1)\|^2/\|\nabla f(\mathbf{x}^0)\|^2$. In this case $\|\nabla f(\mathbf{x}^0)\|^2=5^2+0^2=25$ and $\|\nabla f(\mathbf{x}^1)\|^2=(1.25)^2+(-3.75)^2=15.55$. Then according to (9.8) we have $\mathbf{v}^1=-[1.25,\ -3.75]^T+(15.55/25)[-5,0]^T=[-4.36,3.75]^T$. As before, we compute $f(\mathbf{x}^1+\lambda\mathbf{v}^1)=\frac{1}{2}(26.55-38.95\lambda+14.45\lambda^2)$, which attains a minimum at $\lambda_1=1.34$.

Then we have $\mathbf{x}^2=\mathbf{x}^1+\lambda_1\mathbf{v}^1=[0.4,0.01]^T$. In theory we should have obtained $\mathbf{x}^2=[0,0]^T$, since the conjugate-gradient method can find the minimum of a positive definite quadratic form in at most n steps and here $n=2$. Our numerical solution $\mathbf{x}^2=[0.4,0.01]^T$ is close to $[0,0]^T$, and the difference is due to the effect of the computational roundoff errors. For a better answer we can now replace $\mathbf{x}^0$ by $\mathbf{x}^2$ and repeat the iterations.

9.9 Let $f(x_1,x_2)=x_1^2+2x_2^2$, which can be rewritten as $f(x)=\frac{1}{2}\mathbf{x}^T\mathbf{A}\mathbf{x}$, where

$$\mathbf{A}=\begin{bmatrix}2 & 0\\ 0 & 4\end{bmatrix}.$$

Apply the conjugate-gradient method for locating the minimum of $f(\mathbf{x})$ and show two iterations.

Since $f(\mathbf{x})$ is a positive definite quadratic form, the convergence characteristics of the conjugate-gradient method are not affected by our choice of the initial point. So we arbitrarily choose $\mathbf{x}^0=[5,5]^T$. Then we compute $\nabla f(\mathbf{x}^0)=\mathbf{A}\mathbf{x}^0=[2x_1^0,4x_2^0]^T=[10,20]^T$, and following step 1 of the conjugate-gradient method set $\mathbf{v}^0=-\nabla f(\mathbf{x}^0)=[-10,\ -20]^T$. Then $\mathbf{x}^0+\lambda\mathbf{v}^0=[5-10\lambda,5-20\lambda]^T$ and $f(\mathbf{x}^0+\lambda\mathbf{v}^0)=\frac{1}{2}(150-1000\lambda+1800\lambda^2)$. Since $d^2f[\mathbf{x}^0+\lambda\mathbf{v}^0)/d\lambda^2=1800>0$, $f(\mathbf{x}^0+\lambda\mathbf{v}^0)$ attains a minimum at $\lambda^0=0.28$, where $df(\mathbf{x}^0+\lambda\mathbf{v}^0)/d\lambda=0$. Thus we obtain $\mathbf{x}^1=\mathbf{x}^0+\lambda_0\mathbf{v}^0=[2.2,\ -0.6]^T$.

Next we compute $\nabla f(\mathbf{x}^1)=[4.4,\ -2.4]^T$, $\|\nabla f(x^1)\|^2=25.4$, and $\|\nabla f(x^0)\|^2=500$. Following step 2 of the conjugate-gradient method, we set

$$\mathbf{v}^1=-\nabla f(\mathbf{x}^1)+\frac{\|\nabla f(\mathbf{x}^1)\|^2}{\|\nabla f(\mathbf{x}^0)\|^2}\mathbf{v}^0$$

$$=-[4.4,\ -2.4]^T+\frac{25.4}{500}[-10,\ -20]^T=[-4.9,1.4]^T.$$

Then $\mathbf{x}^1+\lambda\mathbf{v}^1=(2.2-4.9\lambda,\ -0.6+1.4\lambda]$, and minimizing $f(\mathbf{x}^1+\lambda\mathbf{v}^1)$ with respect to λ we get $\lambda_1=0.44$. Hence $\mathbf{x}^2=\mathbf{x}^1+\lambda_1\mathbf{v}^1=[0.05, 0.15]^T$, and we can now repeat the iterations by replacing $\mathbf{x}^0$ by $\mathbf{x}^2$ for higher accuracy.

Generation of Conjugate Directions by One-Dimensional Search

The conjugate-gradient algorithm requires the computation of $\nabla f(\mathbf{x})$ at each iteration to generate the direction of descent. Since each component of $\nabla f(\mathbf{x})$ is about as complex as $f(\mathbf{x})$ itself, this amounts to computing $n+1$ function values at each iteration. Powell has developed an alternative method of generating conjugate directions by one-dimensional searches in each iteration [Powell, 1964]. If $\mathbf{x}^{i+1}$ is a local minimum obtained by one-dimensional search from $\mathbf{x}^i$ in the direction $\mathbf{v}^i$, then using the condition for optimal descent defined earlier, we obtain $(\mathbf{v}^i, \nabla f(\mathbf{x}^{i+1}))=0$. Powell's method is based on the observation (see Fig. 9.3) that if $\mathbf{x}^1$ and $\mathbf{x}^2$ are obtained by one-dimensional searches in the same direction $\mathbf{v}$ but from different base points, then in the case of a quadratic function, the vector $\mathbf{x}^2-\mathbf{x}^1$ is mutually conjugate to $\mathbf{v}$. We have $(\mathbf{v}, \mathbf{A}(\mathbf{x}^2-\mathbf{x}^1))=(\mathbf{v}, (\mathbf{g}^2-\mathbf{g}^1))=0$, where $\mathbf{g}^i=\nabla f(\mathbf{x}^i)$, $i=1,2$. We use this observation to generate conjugate directions by means of one-dimensional searches.

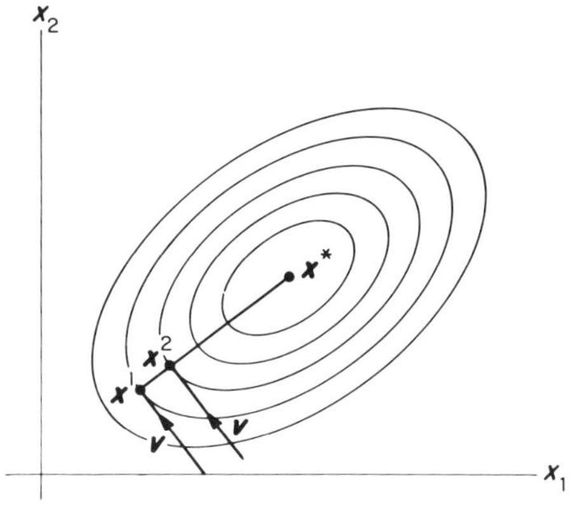

FIGURE 9.3

Powell's Algorithm

Let $\{\mathbf{v}^i,\ i=1,\ldots,n\}$ be a given set of linearly independent vectors in E^n, and let $\mathbf{x}^0$ be an initial approximation to $\mathbf{x}^*$. Then Powell's algorithm consists of the following five steps:

1. For $i=1,2,\ldots,n$, find λ_i that minimized $f(\mathbf{x}^{i-1}+\lambda\mathbf{v}^i)$ with respect to λ, and define
$$\mathbf{x}^i=\mathbf{x}^{i-1}+\lambda_i\mathbf{v}^i.$$
2. For $i=1,\ldots,n-1$, replace $\mathbf{v}^i$ by $\mathbf{v}^{i+1}$.
3. Replace $\mathbf{v}^n$ by $\mathbf{x}^n-\mathbf{x}^0$.
4. Find λ_n that minimizes $f(\mathbf{x}^n+\lambda(\mathbf{x}^n-\mathbf{x}^0))$ with respect to λ, and replace $\mathbf{x}^0$ by $\mathbf{x}^0+\lambda(\mathbf{x}^n-\mathbf{x}^0)$.
5. Go to 1 unless stopping rule is satisfied.

To start the algorithm, the coordinate vectors in E^n can be used in place of $\{\mathbf{v}^i, i=1,\ldots,n\}$ for the first cycle of iterations. Any of the stopping rules given in Chapter 6 can be used to terminate the algorithm.

We can think of Powell's algorithm as an improvement on the one-dimensional search technique. In the one-dimensional search technique, we vary one variable at a time to find a local minimum of the given function. This is equivalent to a sequential one-dimensional search along the coordinate directions in E^n. Powell's improvement consists of introducing mutually conjugate directions of search in place of the coordinate directions.

EXAMPLES

9.10 Let $f(x_1,x_2)$ be a positive definite quadratic function of two variables with circular level curves centered at $\mathbf{x}^*$ as shown in Fig. 9.4. Illustrate graphically the iterations of Powell's algorithm applied to this function starting from $\mathbf{x}^0$ with $\mathbf{v}^1$ and $\mathbf{v}^2$ as the initial directions of search as shown in Fig. 9.4.

We start a one-dimensional search for the local minimum from $\mathbf{x}^0$ in the direction $\mathbf{v}^1$. Since the level curves of $f(x_1,x_2)$ are circles, this local minimum is attained at $\mathbf{x}^1$, where the line segment $\mathbf{x}^*-\mathbf{x}^1$ is orthogonal to the direction $\mathbf{v}^1$.

From $\mathbf{x}^1$ we carry on another one-dimensional search in the direction $\mathbf{v}^2$. Using arguments like those above, we conclude that this local minimum is attained at $\mathbf{x}^2$. Now, following Powell's algorithm, we replace $\mathbf{v}^1$ by $\mathbf{v}^2$ and $\mathbf{v}^2$ by $\mathbf{u}^1=\mathbf{x}^2-\mathbf{x}^0$.

From $\mathbf{x}^2$ we start a one-dimensional search along $\mathbf{u}^1$, which terminates at $\mathbf{x}^3$. Now, following Powell's algorithm, we apply a one-dimensional search

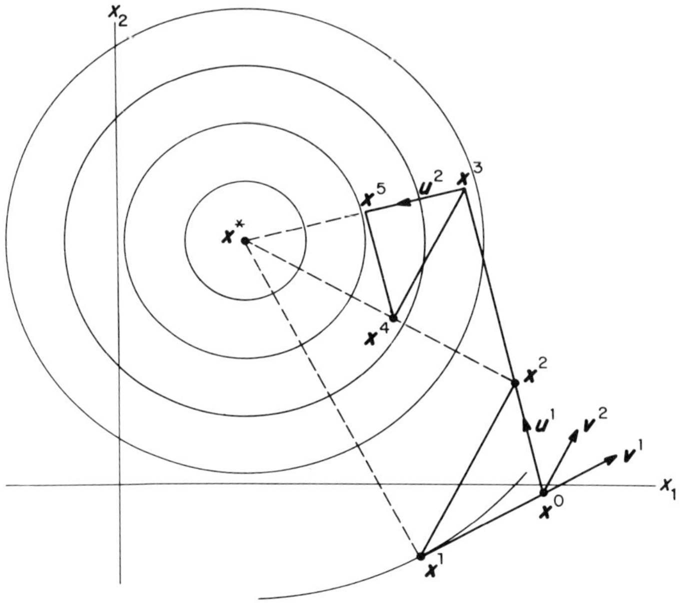

FIGURE 9.4

from $\mathbf{x}^3$ in the direction $\mathbf{v}^2$, which terminates at $\mathbf{x}^4$. From $\mathbf{x}^4$ we arrive at $\mathbf{x}^5$ by a one-dimensional search along $\mathbf{u}^1$. Now in Fig. 9.4 we observe that $\mathbf{x}^5$ and $\mathbf{x}^3$ have been obtained from $\mathbf{x}^4$ and $\mathbf{x}^2$ respectively by one-dimensional searches along $\mathbf{u}^1$. Hence according to our previous observations, $\mathbf{u}^2=\mathbf{x}^5-\mathbf{x}^3$ and $\mathbf{u}^1$ are mutually conjugate directions. Following Powell's algorithm we replace $\mathbf{v}^2$ by $\mathbf{u}^1$ and $\mathbf{u}^1$ by $\mathbf{u}^2$.

Now we have two mutually conjugate search directions given by $\mathbf{u}^1$ and $\mathbf{u}^2$. So we should be able to locate $\mathbf{x}^*$ with one-dimensional searches along $\mathbf{u}^1$ and $\mathbf{u}^2$. If we consider $\mathbf{x}^4$ as a starting point, then we have already arrived at $\mathbf{x}^5$ by a one-dimensional search along $\mathbf{u}^1$. Hence a one-dimensional search along $\mathbf{u}^2$ from $\mathbf{x}^5$ leads us to $\mathbf{x}^*$.

9.11 Consider the objective function $f(\mathbf{x})$ used in Example 9.9, and let $\mathbf{v}^1=[1,-1]^T$ and $\mathbf{v}^2=[1,1]^T$. Apply Powell's method of one-dimensional search from $\mathbf{x}^0=[20,20]^T$ and show two iterations.

Following Powell's method we construct $\mathbf{x}^0+\lambda\mathbf{v}^1=[20+\lambda,20-\lambda]^T$ and compute $f(\mathbf{x}^0+\lambda\mathbf{v}^1)=\frac{1}{2}[2400-80\lambda+6\lambda^2]$. Since $d^2f(\mathbf{x}^0+\lambda\mathbf{v}^1)/d\lambda^2=6>0$, $f(\mathbf{x}^0+\lambda\mathbf{v}^1)$ attains a minimum with respect to λ at $\lambda_1=6.66$, where $df(\mathbf{x}^0+\lambda\mathbf{v}^1)/d\lambda=0$. Then we get $\mathbf{x}^1=\mathbf{x}^0+\lambda_1\mathbf{v}^1=[26.66,13.34]^T$. The remaining one-dimensional searches are done in a similar manner, and we summarize the results below:

$$\lambda_2 = -17.8, \qquad \mathbf{x}^2 = \mathbf{x}^1 + \lambda_2 \mathbf{v}^2 = [8186, -4.46]^T,$$

$$\mathbf{u}^1 = \mathbf{x}^2 - \mathbf{x}^0 = [-11.14, -24.46]^T,$$

$$\lambda_3 = -0.15, \qquad \mathbf{x}^3 = \mathbf{x}^2 + \lambda_3 \mathbf{u}^1 = [10.48, -0.92]^T,$$

$$\lambda_4 = -3.46, \qquad \mathbf{x}^4 = \mathbf{x}^3 + \lambda_4 \mathbf{v}^2 = [7.02, -4.18]^T,$$

$$\lambda_5 = -0.15, \qquad \mathbf{x}^5 = \mathbf{x}^4 + \lambda_5 \mathbf{u}^1 = [8.68, -0.64]^T,$$

$$\mathbf{u}^2 = \mathbf{x}^5 - \mathbf{x}^3 = [-1.8, 0.28]^T.$$

We now check to see if $\mathbf{u}^1$ and $\mathbf{u}^2$ are mutually conjugate with respect to A:

$$(\mathbf{u}^1, \mathbf{A}\mathbf{u}^2) = [-11.14, -24.46] \begin{bmatrix} 2 & 0 \\ 0 & 4 \end{bmatrix} \begin{bmatrix} -1.8 \\ 0.28 \end{bmatrix}$$

$$= 3$$

Since $(\mathbf{u}^1, \mathbf{A}\mathbf{u}^2) \neq 0$, $\mathbf{u}^1$ and $\mathbf{u}^2$ are not mutually conjugate with respect to $\mathbf{A}$. This is due to the effect of the computational roundoff errors. Next we use a one-dimensional search from $\mathbf{x}^5$ along $\mathbf{u}^2$ to obtain $\mathbf{x}^6 = \mathbf{x}^5 + \lambda_6 \mathbf{u}^2 = [0.03, 0.7]^T$. We should have $\mathbf{x}^6 = [0, 0]^T$; the discrepancy can be attributed again to the computational errors. For greater accuracy we can replace $\mathbf{x}^0$ by $\mathbf{x}^6$ and repeat the iterations.

Variable-Metric Algorithms

An advantage of a conjugate-direction algorithm over the optimal-gradient algorithm is a faster convergence rate near the optimum point. If the initial approximation $\mathbf{x}^0$ is close to the optimum point $\mathbf{x}^*$, then (as shown in Fig. 9.5) faster convergence will result with a one-dimensional search along $\mathbf{x}^* - \mathbf{x}^0$, since $\nabla f(\mathbf{x}^0)$ is almost orthogonal to $\mathbf{x}^* - \mathbf{x}^0$. Another class of quadratically convergent algorithms, called *variable-metric algorithms*, seeks to determine $\mathbf{x}^* - \mathbf{x}^0$ from $\nabla f(\mathbf{x}^0)$ in order to avoid convergence difficulties near the optimum. This is done by approximating $f(\mathbf{x})$ near $\mathbf{x}^*$ by a sequence of quadratic functions.

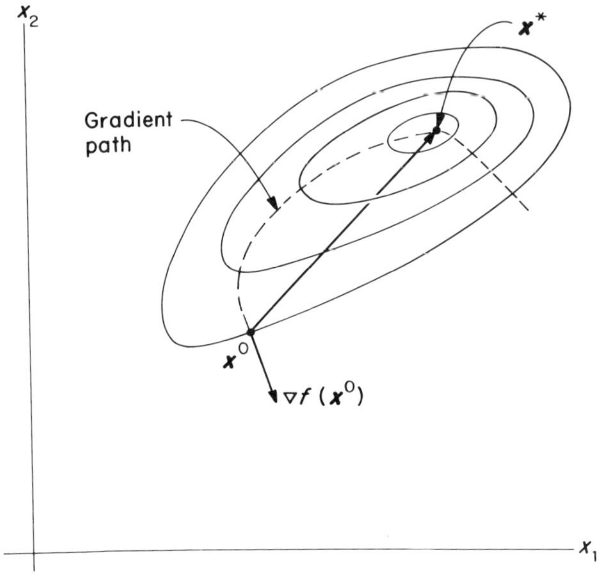

FIGURE 9.5

EXAMPLE 9.12 Let $f(\mathbf{x})$ be a twice continuously differentiable function with a global minimum at $\mathbf{x}^*$. Let $\mathbf{x}^0$ be a point in a sufficiently small neighborhood of $\mathbf{x}^*$. Determine the relation between $\nabla f(\mathbf{x}^0)$ and $\mathbf{x}^0-\mathbf{x}^*$.

Consider the vector-valued function $\nabla f(\mathbf{x})$ in the neighborhood of $\mathbf{x}^*$. Using a first-order Taylor series expansion, we can write approximately

$$\nabla f(\mathbf{x}^0)=\nabla f(\mathbf{x}^*)+\mathbf{H}(\mathbf{x}^*)(\mathbf{x}^0-\mathbf{x}^*), \tag{9.9}$$

where $\mathbf{H}(\mathbf{x}^*)$ is the Hessian matrix of $f(\mathbf{x})$ at $\mathbf{x}^*$. Since $f(\mathbf{x})$ attains a minimum at $\mathbf{x}^*$, $\nabla f(\mathbf{x}^*)=\mathbf{0}$ and $\mathbf{H}(\mathbf{x}^*)$ is positive definite, i.e., $|\mathbf{H}(\mathbf{x}^*)|\neq 0$. Hence we can rewrite (9.9) as

$$\mathbf{x}^0-\mathbf{x}^*=\mathbf{H}(\mathbf{x}^*)^{-1}\nabla f(\mathbf{x}^0),$$

or equivalently as

$$\mathbf{x}^*=\mathbf{x}^0-\mathbf{H}(\mathbf{x}^*)^{-1}\nabla f(\mathbf{x}^0). \tag{9.10}$$

If $f(\mathbf{x})$ is a quadratic function given by (9.2), then $\mathbf{H}(\mathbf{x})=\mathbf{A}$, and Eq. (9.10) becomes

$$\mathbf{x}^*=\mathbf{x}^0-\mathbf{A}^{-1}\nabla f(\mathbf{x}^0), \tag{9.11}$$

where $\nabla f(\mathbf{x}^0)=\mathbf{A}\mathbf{x}^0+\mathbf{b}$ and $\mathbf{x}^*=-\mathbf{A}^{-1}\mathbf{b}$. Equation (9.11) implies that if $f(\mathbf{x})$ is a quadratic function, then starting from any arbitrary point $\mathbf{x}^0$, we can find $\mathbf{x}^*=-\mathbf{A}^{-1}\mathbf{b}$ without a one-dimensional search by taking a step length of unity along the modified gradient direction given by $-\mathbf{A}^{-1}\nabla f(\mathbf{x}^0)$. Equation (9.10) has a similar meaning. However, (9.10) is only an approximate result, since we used an approximate Taylor series expansion of $\nabla f(\mathbf{x})$. These two equations motivate us to look for algorithms which use modified gradient directions of the form $-\mathbf{P}\nabla f(\mathbf{x})$, where $\mathbf{P}$ is a symmetric positive definite matrix. One possible choice for $\mathbf{P}$ is of course the inverse of the Hessian matrix of $f(\mathbf{x})$.

In practice, the Hessian matrix of $f(\mathbf{x})$ is either not available or very expensive to compute. Furthermore, numerical inversion of an $n\times n$ matrix can be time-consuming, since the number of multiplications to be performed in the process increases approximately as n^3. So a practical approach is to start with an approximation of $\mathbf{H}(\mathbf{x})^{-1}$ and then update it sequentially with each iteration. Note that $\mathbf{A}^{-1}\nabla f(\mathbf{x})$ is the gradient of the quadratic function $f(\mathbf{x})$ given by (9.2) when we use $(\mathbf{x},\mathbf{A}\mathbf{x})^{1/2}$ as the norm or *metric* (distance measure) in E^n. A similar meaning can be given to $\mathbf{H}(\mathbf{x})^{-1}\nabla f(\mathbf{x})$ when $f(\mathbf{x})$ is not a quadratic function. Thus this type of modified gradient algorithm can be thought of as a gradient algorithm with a modified metric on E^n. Since $\mathbf{H}(\mathbf{x})^{-1}$ is computed sequentially, such algorithms are called *variable-metric algorithms*. An important algorithm of this type was developed by Fletcher, Powell, and Davidon. [See Fletcher and Powell, 1963; Davidon, 1959.]

Fletcher-Powell Variable-Metric Algorithm

Let $\mathbf{x}^i$ and $\mathbf{H}_i$ denote the approximations to $\mathbf{x}^*$ and the inverse of the Hessian matrix respectively at the ith iteration. For notational convenience we denote $\mathbf{g}^i=\nabla f(\mathbf{x}^i)$. The Fletcher-Powell algorithm consists of the following four steps:

1. Choose $\mathbf{x}^0$ and $\mathbf{H}_0$.
2. For $i=0,1,\dots,n-1$, define $\mathbf{x}^{i+1}=\mathbf{x}^i+\lambda_i\mathbf{v}^i$, where $\mathbf{v}^i=-\mathbf{H}_i\mathbf{g}^i$ and λ_i minimizes $f(\mathbf{x}^i+\lambda\mathbf{v}^i)$ with respect to λ.
3. For $i=0,1,\dots,n-1$, let $\mathbf{u}^i=\lambda_i\mathbf{v}^i$ and $\mathbf{y}^i=\mathbf{g}^{i+1}-\mathbf{g}^i$. Define

$$\mathbf{H}_{i+1}=\mathbf{H}_i+\mathbf{A}_i+\mathbf{B}_i, \tag{9.12}$$

where

$$\mathbf{A}_i = \frac{\mathbf{u}^i(\mathbf{u}^i)^T}{(\mathbf{u}^i)^T\mathbf{y}^i} \quad \text{and} \quad \mathbf{B}_i = -\frac{\mathbf{H}_i\mathbf{y}^i(\mathbf{H}_i\mathbf{y}^i)^T}{(\mathbf{y}^i)^T\mathbf{H}_i\mathbf{y}^i}.$$

4. If the stopping criterion is not satisfied, let $\mathbf{x}^0 = \mathbf{x}^n$ and go to 2.

The iterations can be started from any arbitrary $\mathbf{x}^0$ in E^n. Methods for choosing a suitable $\mathbf{x}^0$ are given in Chapter 6. We can use any symmetric positive definite matrix $\mathbf{H}_0$ as our initial approximation to the inverse of the Hessian matrix. A convenient choice is a diagonal matrix with positive scalars along the diagonal. Stopping rules discussed in Chapter 6 can be used for terminating iterations. To minimize the effect of roundoff errors in computation, a good strategy is to discard $\mathbf{H}_n$ after every n iterations and start with a new approximation. Fletcher and Powell have demonstrated by numerical computations that this algorithm is superior both to the PARTAN method mentioned in Chapter 6 and to Powell's algorithm discussed earlier. Note that when the iterations are terminated, an estimate of the inverse of the Hessian matrix of $f(\mathbf{x})$ is automatically available for fitting a quadratic surface to $f(\mathbf{x})$ as explained in Chapter 6. Further discussions of this algorithm can be found in Problems 9.5 and 9.6.

EXAMPLES

9.13 Consider a positive definite quadratic form $f(\mathbf{x}) = \frac{1}{2}\mathbf{x}^T\mathbf{A}\mathbf{x}$, where

$$\mathbf{A} = \begin{bmatrix} 3 & -1 \\ -1 & 1 \end{bmatrix}.$$

Find the minimum point by using the Fletcher-Powell variable-metric algorithm.

We make an arbitrary choice $\mathbf{x}^0 = [10, 10]^T$ as our initial estimate of the minimum point. The initial estimate of the inverse of the Hessian matrix, denoted by $\mathbf{H}_0$, can be taken to be any symmetric positive definite matrix. For convenience we choose

$$\mathbf{H}_0 = \begin{bmatrix} 1 & 0 \\ 0 & 1 \end{bmatrix}.$$

Now the gradient vector $\nabla f(\mathbf{x}) = \mathbf{A}\mathbf{x} = [3x_1 - x_2, -x_1 + x_2]^T$. By direct substitution of x_1^0 and x_2^0 we have $\nabla f(\mathbf{x}^0) = [20, 0]^T$. Following step 2 of the Fletcher-Powell algorithm, we set $\mathbf{v}^0 = -\mathbf{H}_0\nabla f(\mathbf{x}^0) = -[20, 0]^T$ and obtain

$\mathbf{x}^0+\lambda\mathbf{v}^0=[10-20\lambda, 10]^T$. Hence

$$f(\mathbf{x}^0+\lambda\mathbf{v}^0)=\tfrac{1}{2}[10-20\lambda, 10]\begin{bmatrix} 3 & -1 \\ -1 & 1 \end{bmatrix}\begin{bmatrix} 10-20\lambda \\ 10 \end{bmatrix}$$

$$=100-400\lambda+600\lambda^2,$$

which attains a minimum with respect to λ at $\lambda_0=0.33$, and so $\mathbf{x}^1=\mathbf{x}^0+\lambda_0\mathbf{v}^0=[3.34, 10]^T$.

Following step 3 of the Fletcher-Powell algorithm, we compute $\nabla f(\mathbf{x}^1)=[0.02, 6.66]^T$, $\mathbf{u}^0=\lambda_0\mathbf{v}^0=-[6.6, 0]^T$, and $\mathbf{y}^0=\nabla f(\mathbf{x}^1)-\nabla f(\mathbf{x}^0)=-[20, -6.66]^T$. Then $(\mathbf{u}^0,\mathbf{y}^0)=132$ and $(\mathbf{y}^0,\mathbf{H}_0\mathbf{y}^0)=(\mathbf{y}^0,\mathbf{y}^0)=443.6$. By direct computation we obtain

$$\mathbf{A}_0=\begin{bmatrix} 0.33 & 0 \\ 0 & 0 \end{bmatrix} \quad \text{and} \quad \mathbf{B}_0=-\begin{bmatrix} 0.9 & -0.3 \\ -0.3 & 0.01 \end{bmatrix}.$$

Hence according to (9.12) we have

$$\mathbf{H}_1=\mathbf{H}_0+\mathbf{A}_0+\mathbf{B}_0=\begin{bmatrix} 0.43 & 0.3 \\ 0.3 & 0.99 \end{bmatrix}.$$

Now we start the second cycle of iterations from step 2. We set $\mathbf{v}^1=-\mathbf{H}_1\nabla f(\mathbf{x}^1)=-[2.22, 6.65]^T$ and obtain $\mathbf{x}^1+\lambda\mathbf{v}^1=[3.34-2.22\lambda, 10-6.65\lambda]^T$. Then $f(\mathbf{x}^1+\lambda\mathbf{v}^1)$ attains a minimum at $\lambda_1=1.5$. So $\mathbf{x}^2=\mathbf{x}^1+\lambda_1\mathbf{v}^1=[-0.01, 0.15]^T$. In theory we should have $\mathbf{x}^2=[0,0]^T$, which is the minimum point being sought; however, $\mathbf{x}^2\neq\mathbf{0}$, due to computational roundoff errors.

Let us now start with step 3 to compute $\mathbf{H}_2$. We have

$$\mathbf{u}^1=\lambda_1\mathbf{v}^1=-[3.33, 9.85]^T,$$

$$\nabla f(\mathbf{x}^2)=[-0.18, 0.16]^T,$$

$$\mathbf{y}^1=\nabla f(\mathbf{x}^2)-\nabla f(\mathbf{x}^1)=-[0.2, 6.5]^T,$$

$$\mathbf{H}_1\mathbf{y}^1=-[2.036, 6.56], \qquad (\mathbf{u}^1,\mathbf{y}^1)=64.66, \qquad (\mathbf{y}^1,\mathbf{H}_1\mathbf{y}^1)=42.9.$$

Then by direct computation we get

$$\mathbf{A}_1=\begin{bmatrix} 0.17 & 0.51 \\ 0.51 & 1.5 \end{bmatrix} \quad \text{and} \quad \mathbf{B}_1=-\begin{bmatrix} 0.1 & 0.31 \\ 0.31 & 0.99 \end{bmatrix}.$$

Hence according to (9.12) we have

$$\mathbf{H}_2 = \mathbf{H}_1 + \mathbf{A}_1 + \mathbf{B}_1 = \begin{bmatrix} 0.5 & 0.5 \\ 0.5 & 1.5 \end{bmatrix}.$$

We observe that $\mathbf{H}_2$ is the exact inverse of $\mathbf{A}$, the Hessian matrix.

9.14 Consider the objective function $f(x_1, x_2) = \frac{1}{2}(x_1^2 + 2x_2^2)$, which can be rewritten as $f(\mathbf{x}) = \frac{1}{2}\mathbf{x}^T\mathbf{A}\mathbf{x}$, where

$$\mathbf{A} = \begin{bmatrix} 1 & 0 \\ 0 & 2 \end{bmatrix}.$$

Apply the variable-metric algorithm for minimization starting from $\mathbf{x}^0 = [4,4]^T$ and show two iterations.

We choose $\mathbf{H}_0$, the initial estimate of the inverse of the Hessian matrix $\mathbf{A}$, to be

$$\mathbf{H}_0 = \begin{bmatrix} 1 & 0 \\ 0 & 1 \end{bmatrix}.$$

Following step 2 of the variable-metric algorithm, we compute $\nabla f(\mathbf{x}^0) = \mathbf{A}\mathbf{x}^0 = [4,8]^T$ and set $\mathbf{v}^0 = -\mathbf{H}_0 \nabla f(\mathbf{x}^0) = -[4,8]^T$. Then $f(\mathbf{x}^0 + \lambda\mathbf{v}^0) = \frac{1}{2}(48 - 160\lambda + 144\lambda^2)$, which attains a minimum at $\lambda_0 = 0.56$, and hence $\mathbf{x}^1 = \mathbf{x}^0 + \lambda_0\mathbf{v}^0 = [1.76, -0.48]^T$.

Following step 3 of the variable-metric algorithm, we compute

$$\mathbf{u}^0 = \lambda_0\mathbf{v}^0 = [-2.24, -4.48]^T,$$

$$\nabla f(\mathbf{x}^1) = [1.76, -0.96],$$

$$\mathbf{y}^0 = \nabla f(\mathbf{x}^1) - \nabla f(\mathbf{x}^0) = [-2.24, -8.96]^T, \qquad (\mathbf{u}^0, \mathbf{y}^0) = 45, \quad (\mathbf{y}^0, \mathbf{H}_0\mathbf{y}^0) = 85.$$

Next, using the formula for $\mathbf{A}_0$ and $\mathbf{B}_0$ given in step 3, we obtain

$$\mathbf{A}_0 = \begin{bmatrix} 0.11 & 0.22 \\ 0.22 & 0.44 \end{bmatrix}, \qquad \mathbf{B}_0 = \begin{bmatrix} -0.059 & -0.235 \\ -0.235 & -0.94 \end{bmatrix}.$$

Hence using (9.12) we get

$$\mathbf{H}_1 = \mathbf{H}_0 + \mathbf{A}_0 + \mathbf{B}_0 = \begin{bmatrix} 1.04 & -0.02 \\ -0.02 & 0.5 \end{bmatrix}.$$

Now we start the second cycle of iterations from $\mathbf{x}^1=[1.76,-0.48]^T$. Following step 2, we set $\mathbf{v}^1=-\mathbf{H}_1\nabla f(\mathbf{x}^1)=[-1.85,0.51]^T$. Then $f(\mathbf{x}^1+\lambda\mathbf{v}^1)$ attains a minimum at $\lambda_1=0.94$ and we have $\mathbf{x}^2=\mathbf{x}^1+\lambda_1\mathbf{v}^1=[0.02,-0.001]^T$. Following step 3, we compute

$$\mathbf{u}^1=\lambda_1\mathbf{v}^1=[-1.78,0.48]^T,$$

$$\nabla f(\mathbf{x}^2)=[0.02,-0.002]^T,$$

$$\mathbf{y}^1=\nabla f(\mathbf{x}^2)-\nabla f(\mathbf{x}^1)=[-1.46,0.69]^T, \qquad \mathbf{H}_1\mathbf{y}^1=[-1.86,0.52]^T,$$

$$(\mathbf{u}^1,\mathbf{y}^1)=3.51, \qquad (\mathbf{y}^1,\mathbf{H}_1\mathbf{y}^1)=3.78.$$

Then as before, $\mathbf{A}_1$ and $\mathbf{B}_1$ are computed to be

$$\mathbf{A}_1=\begin{bmatrix} 0.87 & -0.236 \\ -0.236 & 0.065 \end{bmatrix}, \qquad \mathbf{B}_1=\begin{bmatrix} -0.89 & 0.256 \\ 0.256 & -0.066 \end{bmatrix},$$

and so by (9.12),

$$\mathbf{H}_2=\mathbf{H}_1+\mathbf{A}_1+\mathbf{B}_1=\begin{bmatrix} 1.02 & 0 \\ 0 & 0.499 \end{bmatrix}.$$

As expected, $\mathbf{H}_2$ is the approximate inverse of the Hessian matrix $\mathbf{A}$.

Solved Problems

9.1 Show that the quadratic function $f(\mathbf{x})$ given by (9.2) has an absolute or global minimum at $\mathbf{x}^*=-\mathbf{A}^{-1}\mathbf{b}$ in E^n.

The definition of an absolute or global minimum was given in Chapter 2. First we note that $f(\mathbf{x}^*)=f(-\mathbf{A}^{-1}\mathbf{b})=c-\frac{1}{2}\mathbf{b}^T\mathbf{A}^{-1}\mathbf{b}$. Next we define $\phi(x)=\frac{1}{2}(\mathbf{x}+\mathbf{A}^{-1}\mathbf{b})^T\mathbf{A}(\mathbf{x}+\mathbf{A}^{-1}\mathbf{b})$. Since $\mathbf{A}$ is positive definite, $\phi(\mathbf{x})>0$ for all $\mathbf{x}\neq-\mathbf{A}^{-1}\mathbf{b}$ and $\phi(\mathbf{x}^*)=0$. Furthermore we observe that

$$\begin{aligned}\phi(\mathbf{x})&=\tfrac{1}{2}\mathbf{x}^T\mathbf{A}\mathbf{x}+\mathbf{b}^T\mathbf{x}+\tfrac{1}{2}\mathbf{b}^T\mathbf{A}^{-1}\mathbf{b}\\ &=f(\mathbf{x})-c+\tfrac{1}{2}\mathbf{b}^T\mathbf{A}^{-1}\mathbf{b}\\ &=f(\mathbf{x})-f(\mathbf{x}^*).\end{aligned}$$

Now using the properties of $\phi(\mathbf{x})$ given above, we conclude that $f(\mathbf{x})>f(\mathbf{x}^*)$ for all $\mathbf{x}\neq\mathbf{x}^*$. Hence by the definition given in Chapter 2, $f(\mathbf{x})$ has a global minimum at $\mathbf{x}=\mathbf{x}^*$.

9.2 Let u^i, $i=1,\ldots,n$, be mutually conjugate vectors with respect to an $n\times n$ symmetric positive definite matrix $\mathbf{A}$. Show that

$$\mathbf{A}^{-1}=\sum_{i=1}^{n}\frac{\mathbf{u}^i(\mathbf{u}^i)^T}{(\mathbf{u}^i)^T\mathbf{A}\mathbf{u}^i}.$$

Let $\mathbf{x}$ be an arbitrary vector in E^n, and define

$$\mathbf{B}=\sum_{i=1}^{n}\frac{\mathbf{u}^i(\mathbf{u}^i)^T}{(\mathbf{u}^i)^T\mathbf{A}\mathbf{u}^i}.$$

Then using the result of Example 9.6, we obtain

$$\mathbf{BAx}=\left[\sum_{i=1}^{n}\frac{\mathbf{u}^i(\mathbf{u}^i)^T}{(\mathbf{u}^i)^T\mathbf{A}\mathbf{u}^i}\right]\mathbf{Ax}$$

$$=\sum_{i=1}^{n}\frac{(\mathbf{u}^i)^T\mathbf{Ax}}{(\mathbf{u}^i)^T\mathbf{A}\mathbf{u}^i}\mathbf{u}^i$$

$$=\mathbf{x}.$$

Since $\mathbf{BAx}=\mathbf{x}$ for every $\mathbf{x}$ in E^n, then $\mathbf{BA}=\mathbf{I}$, where $\mathbf{I}$ denotes the $n\times n$ identity matrix and hence $\mathbf{B}=\mathbf{A}^{-1}$.

9.3 Let $\mathbf{u}^k$, $k=1,\ldots,n$, denote mutually conjugate directions along an optimal descent path used on a positive definite quadratic function $f(\mathbf{x})$. Show that $(\mathbf{u}^k,\nabla f(\mathbf{x}^{i+1}))=0$, $k=1,\ldots,i$, for all i, $1\leqslant i\leqslant n$.

In this problem we point out an important global characteristic of descent algorithms using conjugate directions of descent. Note that for every descent algorithm using one-dimensional search in each iteration, the local orthogonality condition of $(\mathbf{u}^i,\nabla f(\mathbf{x}^{i+1}))=0$ is always satisfied. However, $(\mathbf{u}^k,\nabla f(\mathbf{x}^{i+1}))=0$ is not necessarily true for every descent algorithm. The vectors $\mathbf{u}^k$, $k=1,\ldots,n$ are linearly independent, as shown in Example 9.6; it follows that $(\mathbf{u}^ik,\nabla f(\mathbf{x}^n))=0$, $k=1,\ldots,n$, implies $\nabla f(\mathbf{x}^n)=0$ and hence $\mathbf{x}^n=\mathbf{x}^*$.

In an optimal-descent method, $\mathbf{x}^{k+1}$ is obtained from $\mathbf{x}^k$ by a one-dimensional search along $\mathbf{u}^k$. It follows from Eq. (9.5) that at $\mathbf{x}^{i+1}=\mathbf{x}^{k+1}+\sum_{j=k+1}^{i+1}\lambda_j u^j$,

$$\nabla f(\mathbf{x}^{i+1})=\mathbf{A}\mathbf{x}^{i+1}+\mathbf{b}$$

$$=\mathbf{A}\left[\mathbf{x}^{k+1}+\sum_{j=k+1}^{i+1}\lambda_j\mathbf{u}^j\right]+\mathbf{b}$$

$$=\nabla f(\mathbf{x}^{k+1})+\sum_{j=k+1}^{i+1}\lambda_j\mathbf{A}\mathbf{u}^j.$$

Hence

$$(\mathbf{u}^k, \nabla f(\mathbf{x}^{i+1})) = (\mathbf{u}^k, \nabla f(\mathbf{x}^{k+1})) + \left(\mathbf{u}^k, \sum_{j=k+1}^{i+1} \lambda_j \mathbf{A}\mathbf{u}^u\right).$$

Using the condition for optimal descent, we have $(\mathbf{u}^k, \nabla f(\mathbf{x}^{k+1})) = 0$, and furthermore, by mutual conjugacy of the given vectors,

$$\left(\mathbf{u}^k, \sum_{j=k+1}^{i+1} \lambda_j \mathbf{A}\mathbf{u}^j\right) = \sum_{j=k+1}^{n} \lambda_j (\mathbf{u}^k, \mathbf{A}\mathbf{u}^j) = 0.$$

Hence $(\mathbf{u}^k, \nabla f(\mathbf{x}^{i+1})) = 0$ for $k = 1, \ldots, i$ and for all i, $1 \leqslant i \leqslant n$.

9.4 Show that if $\mathbf{H}_i$ is a positive definite matrix and $\mathbf{H}_{i+1}$ is computed by Eq. (9.12), then $\mathbf{H}_{i+1}$ is positive definite.

Since $\mathbf{H}_i$ is positive definite, there exists a unique positive definite matrix $\mathbf{G}_i$ such that $(\mathbf{G}_i)^2 = \mathbf{H}_i$. Then from Eq. (9.12) we obtain

$$(\mathbf{x}, \mathbf{H}_{i+1}\mathbf{x}) = (\mathbf{x}, \mathbf{H}_i\mathbf{x}) + \frac{(\mathbf{u}^i, \mathbf{x})^2}{(\mathbf{y}^i, \mathbf{u}^i)} - \frac{(\mathbf{x}, \mathbf{H}_i\mathbf{y}^i)^2}{(\mathbf{y}^i, \mathbf{H}_i\mathbf{y}^i)}$$

$$= \frac{(\mathbf{p}^i, \mathbf{p}^i)(\mathbf{q}^i, \mathbf{q}^i) - (\mathbf{p}^i, \mathbf{q}^i)^2}{(\mathbf{q}^i, \mathbf{q}^i)} + \frac{(\mathbf{u}^i, \mathbf{x})^2}{(\mathbf{y}^i, \mathbf{u}^i)},$$

where $\mathbf{p}^i = \mathbf{G}_i\mathbf{x}$ and $\mathbf{q}^i = \mathbf{G}_i\mathbf{y}^i$. If $\mathbf{x} \neq \theta\mathbf{y}^i$ for some scalar θ, then the first factor in the above equation is strictly positive due to the Schwartz inequality. So let $\mathbf{x} = \theta\mathbf{y}^i$; then since $\mathbf{y}^i = \mathbf{g}^{i+1} - \mathbf{g}^i = \mathbf{A}(\mathbf{x}^{i+1} - \mathbf{x}^i) = \mathbf{A}\mathbf{u}^i$, we have $\mathbf{x} = \theta\mathbf{A}\mathbf{u}^i$. Then

$$(\mathbf{x}, \mathbf{H}_{i+1}\mathbf{x}) = \theta^2 \frac{(\mathbf{u}^i, \mathbf{A}\mathbf{u}^i)^2}{(\mathbf{u}^i, \mathbf{A}\mathbf{u}^i)} = \theta^2 (\mathbf{u}^i, \mathbf{A}\mathbf{u}^i),$$

which is strictly positive, since $\mathbf{A}$ is positive definite. Hence $\mathbf{H}_{i+1}$ is positive definite.

9.5 For the Fletcher-Powell algorithm, verify that

$$(\mathbf{u}^i, \mathbf{A}\mathbf{u}^j) = 0, \qquad 0 \leqslant i < j < k, \tag{9.13}$$

and

$$\mathbf{H}_k\mathbf{A}\mathbf{u}^i = \mathbf{u}^i, \qquad 0 \leqslant i < k, \tag{9.14}$$

where $k = 0, \ldots, n-1$.

The proof is by induction. The most important part of the proof is the induction step which we consider here. We assume that (9.13) and (9.14) are valid for k. Then it is necessary to show that they are also valid for $k+1$. By Eq. (9.5) with $\mathbf{g}^k = \nabla f(\mathbf{x}^k)$,

$$\mathbf{g}^k = \mathbf{b} + \mathbf{A}\mathbf{x}^k = \mathbf{b} + \mathbf{A}\mathbf{x}^{i+1} + \mathbf{A}\sum_{j=i+1}^{k-1} \mathbf{u}^j = \mathbf{g}^{i+1} + \mathbf{A}\sum_{j=i+1}^{k-1} \mathbf{u}^j.$$

Then from the induction hypothesis and the condition for optimal descent in each iteration, we have

$$\begin{aligned}(\mathbf{u}^i, \mathbf{g}^k) &= (\mathbf{u}^i, \mathbf{g}^{i+1}) \\ &= 0, \qquad 0 \leqslant i < k. \end{aligned} \tag{9.15}$$

Hence $(\mathbf{u}^i, \mathbf{A}\mathbf{H}_k\mathbf{g}^k) = (\mathbf{H}_k\mathbf{A}\mathbf{u}^i, \mathbf{g}^k) = (\mathbf{u}^i, \mathbf{g}^k) = 0$; and since $\lambda_k \mathbf{H}_k \mathbf{g}^k = \mathbf{u}^k$, we have $(\mathbf{u}^i, \mathbf{A}\mathbf{u}^k)/\lambda_k = 0$, and $\lambda_k > 0$ implies $(\mathbf{u}^i, \mathbf{A}\mathbf{u}^k) = 0$, $0 \leqslant i < k$. Also $(\mathbf{y}^k, \mathbf{H}_k\mathbf{A}\mathbf{u}^i) = (\mathbf{y}^k, \mathbf{u}^i) = (\mathbf{u}^i\mathrm{k}, \mathbf{A}\mathbf{u}^i) = 0$, $0 \leqslant i < k$. Therefore $\mathbf{H}_{k+1}\mathbf{A}\mathbf{u}^i = \mathbf{H}_k\mathbf{A}\mathbf{u}^i = \mathbf{u}^i$, $0 \leqslant i < k$, where $k = 0, 1, \ldots, n-1$.

We know from (9.13) that the vectors $\mathbf{u}^k$, $k = 0, \ldots, n-1$, are mutually conjugate with respect to $\mathbf{A}$ and hence are linearly independent. By Eq. (9.15), $\mathbf{g}^n$ is orthogonal to all $\mathbf{u}^k$. Thus $\mathbf{g}^n = \mathbf{0}$ and the Fletcher-Powell algorithm is quadratically convergent according to our definition.

9.6 In the Fletcher-Powell algorithm, show that

$$\mathbf{A}^{-1} = \sum_{i=0}^{n-1} \mathbf{A}_i.$$

We know from (9.13) that the $\mathbf{u}^k$ are mutually conjugate with respect to $\mathbf{A}$. Also

$$\mathbf{A}_i = \frac{\mathbf{u}^i(\mathbf{u}^i)^T}{(\mathbf{u}^i)^T\mathbf{y}^i} = \frac{\mathbf{u}^i(\mathbf{u}^i)^T}{(\mathbf{u}^i)^T\mathbf{A}\mathbf{u}^i}.$$

Letting

$$\mathbf{B} = \sum_{i=0}^{n-1} \mathbf{A}^i = \sum_{i=0}^{n-1} \frac{\mathbf{u}^i(\mathbf{u}^i)^T}{(\mathbf{u}^i)^T\mathbf{A}\mathbf{u}^i},$$

we have

$$\mathbf{B}\mathbf{A}\mathbf{x} = \sum_{i=0}^{n-1} \frac{(\mathbf{u}^i)^T\mathbf{A}\mathbf{x}}{(\mathbf{u}^i)^T\mathbf{A}\mathbf{u}^i}\mathbf{u}^i = \mathbf{x}$$

for arbitrary $\mathbf{x} \in E^n$, as shown in Problem 9.2. Therefore $\mathbf{B}\mathbf{A} = \mathbf{I}$, where $\mathbf{I}$ is the $n \times n$ identity matrix, and so $\mathbf{B} = \mathbf{A}^{-1}$.

9.7 Let $f(\mathbf{x})$ be a positive definite quadratic function given by (9.2). Show that the modified gradient direction given by (9.8) generates a set of vectors $\mathbf{v}^i$, $i=1,\ldots,n$, mutually conjugate with respect to $\mathbf{A}$.

The proof is by induction, and we discuss the induction step, since it is the most important part of the proof. The induction hypothesis is that the vectors $\mathbf{v}^k$, $k=1,\ldots,i-1$, are mutually conjugate with respect to $\mathbf{A}$. Letting $\mathbf{g}^i=\nabla f(\mathbf{x}^i)$ and $\mathbf{v}^i=-\mathbf{g}^i+\alpha_i\mathbf{v}^{i-1}$, where α_i is an unknown parameter, and using the results of Problem 9.3, we obtain $(\mathbf{v}^k,\mathbf{g}^i)=0$ for $k=1,\ldots,i-1$, and $i=1,\ldots,n-1$. Since $\mathbf{g}^i$ is orthogonal to $\mathbf{v}^{k-1}$ for $k<i-1$, we have $-(\mathbf{g}^k,\mathbf{g}^i)=(-\mathbf{g}^k+\alpha_k\mathbf{v}^{k-1},\mathbf{g}^i)=(\mathbf{v}^k,\mathbf{g}^i)=0$, where in the last step $\mathbf{v}^k$ was obtained by setting $i=k$ in the definition of $\mathbf{v}^i$ given above. Recalling that $\mathbf{x}^{k+1}$ is obtained from $\mathbf{x}^k$ by moving an amount λ_k in the direction $\mathbf{v}^k$, we can write $\mathbf{v}^k=(1/\lambda_k)(\mathbf{x}^{k+1}-\mathbf{x}^k)$. Using this result we obtain

$$-(\mathbf{v}^k,\mathbf{A}\mathbf{v}^i)=(\mathbf{v}^k,\mathbf{A}\mathbf{g}^i)=\frac{1}{\lambda_k}(\mathbf{x}^{k+1}-\mathbf{x}^k,\mathbf{A}\mathbf{g}^i)$$

$$=\frac{1}{\lambda_k}(\mathbf{A}(\mathbf{x}^{k+1}-\mathbf{x}^i),\mathbf{g}^i)=\frac{1}{\lambda_k}(\mathbf{g}^{k+1}-\mathbf{g}^i,\mathbf{g}^i)=0$$

for $k<i-1$, i.e., $\mathbf{v}^i$ and $\mathbf{v}^k$, $k=1,\ldots,i-2$, are mutually conjugate with respect to $\mathbf{A}$.

However, we still have to show that $\mathbf{v}^i$ and $\mathbf{v}^{i-1}$ are mutually conjugate with respect to $\mathbf{A}$. This is achieved by the proper choice of the unknown parameter α_i. We have $\mathbf{v}^{i-1}=\mathbf{g}^{i-1}+\alpha_{i-1}\mathbf{v}^{i-2}$, where $\alpha_{i-1}=\|\mathbf{g}^{i-1}\|^2/\|\mathbf{g}^{i-2}\|^2$ by the induction hypothesis and (9.8). By direct computation we have $-(\mathbf{v}^{i-1},\mathbf{g}^{i-1})=\|\mathbf{g}^{i-1}\|^2$, since $(\mathbf{v}^{i-2},\mathbf{g}^{i-1})=0$ by Problem 9.3. We observe that $\mathbf{v}^i$ and $\mathbf{v}^{i-1}$ are mutually conjugate with respect to $\mathbf{A}$ [i.e., $(\mathbf{v}^i,\mathbf{A}\mathbf{v}^{i-1})=0$] if we choose $\alpha_i=(\mathbf{g}^i,\mathbf{A}\mathbf{v}^{i-1})/(\mathbf{v}^{i-1},\mathbf{A}\mathbf{v}^{i-1})$. To obtain a simpler expression for α_i we notice that $(\mathbf{g}^i,\mathbf{A}\mathbf{v}^{i-1})=(1/\lambda_i)\|\mathbf{g}^i\|^2$, where we have used $\mathbf{v}^{i-1}=(1/\lambda_i)(\mathbf{x}^i-\mathbf{x}^{i-1})$ and $(\mathbf{v}^{i-1},\mathbf{A}\mathbf{v}^{i-1})=(1/\lambda_i)\|\mathbf{g}^{i-1}\|^2$.

Hence we have an alternative expression for $\alpha_i\| g^i\|^2/\| g^{i-1}\|^2$, which is the value of α_i used in (9.8). Thus the modified gradient directions generated by (9.8) are mutually conjugate with respect to $\mathbf{A}$.

Supplementary Problems

9.8 For $f(x_1,x_2)=\frac{1}{2}[(x_1^2/a)+(x_2^2/b)]$, $a,b>0$, consider the level curve $C=\{(x_1,x_2)|f(x_1,x_2)=1\}$, which is an ellipse. Show that every $(x_1,x_2)\in C$ is mutually conjugate to the vector tangent to C at (x_1,x_2) with respect to the positive definite matrix

$$\mathbf{A}=\begin{bmatrix}1/a & 0\\ 0 & u/b\end{bmatrix}.$$

9.9 Consider a positive definite quadratic form $f(\mathbf{x})=\frac{1}{2}(\mathbf{x},\mathbf{A}\mathbf{x})$, where $\mathbf{A}$ is a 2×2 matrix. Let $\mathbf{x}^0\neq\mathbf{0}$ be any arbitrary vector in E^2. Assume $\mathbf{x}^1$ is obtained from $\mathbf{x}^0$, and $\mathbf{x}^2$ from $\mathbf{x}^1$, by the optimal-gradient algorithm. Show that the line joining $\mathbf{x}^0$ and $\mathbf{x}^2$ passes through the origin. Discuss the possible usefulness of this result for accelerating the optimal gradient algorithm.

9.10 Let $f(\mathbf{x})=\frac{1}{2}(\mathbf{x},\mathbf{Ax})$ be a positive definite quadratic form, and consider the sequence of vectors

$$\mathbf{x}^{k+1}=\mathbf{x}^k-\lambda_k\mathbf{v}^k,$$

where $\mathbf{v}^k=\nabla f(\mathbf{x}^k)$. Show that for $\lambda_k=1/\|\mathbf{A}\|$, the sequence $\{\mathbf{x}^k\}$ converges to the minimum of $f(\mathbf{x})$. Discuss the rate of convergence for three possible choices of the norm of $\mathbf{A}$; namely,

(a) $\|\mathbf{A}\|=(\Sigma_{i,j}|a_{ij}|^2)^{1/2}$,
(b) $\|\mathbf{A}\|=\max_i(\Sigma_j|a_{ij}|)$,
(c) $\|\mathbf{A}\|=\max_j(\Sigma_i|a_{ij}|)$.

9.11 Consider a positive definite quadratic function $f(\mathbf{x})$ given by (9.2) with $c=0$, $\mathbf{b}=[1,-2]^T$, and

$$\mathbf{A}=\begin{bmatrix}1 & -1\\ -1 & 2\end{bmatrix}.$$

Find the minimum point by the Fletcher-Powell variable-metric algorithm as shown in Example 9.11.

9.12 Consider $f(x_1,x_2)=2x_1^2+2x_1x_2+5x_2^2$. Apply four iterations of the conjugate-gradient method. Compare the result with that of Problem 6.10.

9.13 Consider the positive definite quadratic form given in Problem 9.12. Find the minimum point by the Fletcher-Reeves conjugate-gradient method as shown in Example 9.8.

Chapter 10

Quadratic Programming

The theory of quadratic programming is concerned with problems of constrained minimization where the constraint functions are linear and the objective function is a positive definite quadratic function. As such it represents a natural transition from the theory of linear programming (see Appendix) to the theory of nonlinear programming. However, there are some important differences between solutions of linear programming problems and those of quadratic programming problems. An optimal solution to a linear program, if one exists, is either an extreme point of the set of all feasible solutions or a convex combination of such extreme points (see Fig. 10.1). On the other hand, an optimal solution to a quadratic program can be an interior point or a boundary point which is not necessarily an extreme point (see Figs. 10.2 and 10.3). In Figs. 10.1, 10.2, and 10.3, $\mathbf{x}^*$ denotes the optimal solution.

Equality Constraints

The simplest problem of quadratic programming is to find the constrained minimum of a positive definite quadratic function with linear equality constraints. Positive definite quadratic functions have been discussed in detail in Chapter 9. Their level surfaces are hyperellipsoids, while each linear equality constraint defines a hyperplane. We can see intuitively that the constrained minimum is located at the point of tangency between the boundary of the feasible set and the minimizing hyperellipsoid. The situation in E^2 is shown in Fig. 10.4, where the straight line passing through the points $0,a$ and $b,0$ represents the constraint hyperplane, and $\mathbf{x}^0$ and $\mathbf{x}^*$ denote respectively the unconstrained and constrained minima.

Feasible Solutions

There are three equivalent ways of defining the set of all feasible solutions. Let $\mathbf{B}$ be an $m \times n$ matrix and $\mathbf{c}$ a vector in E^m. We denote the set of all

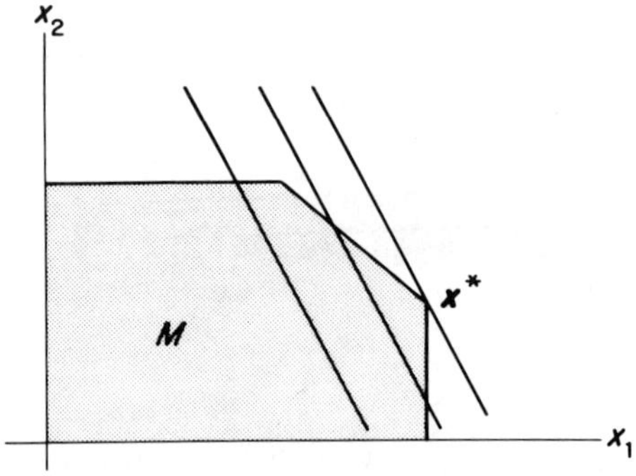

FIGURE 10.1

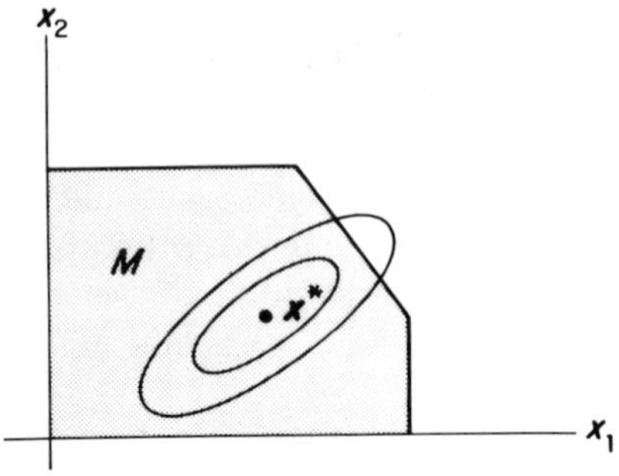

FIGURE 10.2

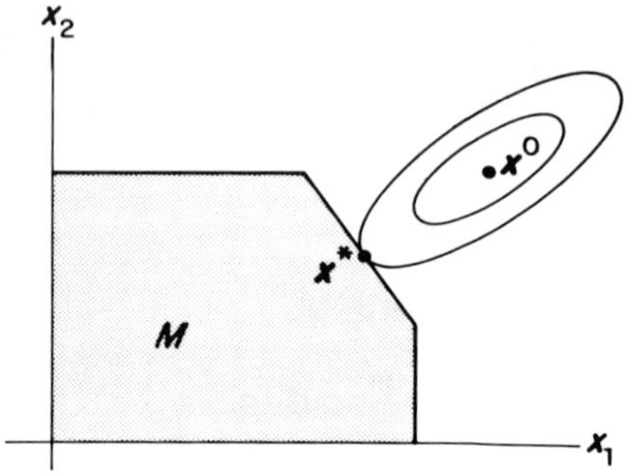

FIGURE 10.3

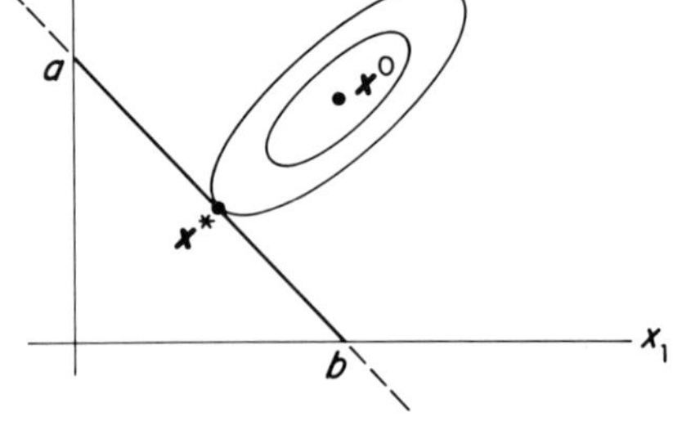

FIGURE 10.4

feasible solutions by M. Then we can define M as follows:

$$M=\{\mathbf{x}|\mathbf{x}\in E^n, \mathbf{Bx}\leqslant \mathbf{c}, \mathbf{x}\geqslant \mathbf{0}\}. \tag{10.1}$$

We can replace the inequalities $\mathbf{Bx}\leqslant\mathbf{c}$ in (10.1) by equalities with the introduction of slack variables. Let $\mathbf{u}=\mathbf{c}-\mathbf{Bx}$, $y=[\mathbf{x}^T,\mathbf{u}^T]^T$, and $\overline{\mathbf{B}}=[\mathbf{B}|\mathbf{I}]$. Then the system $\mathbf{Bx}\leqslant\mathbf{c}$, $\mathbf{x}\geqslant\mathbf{0}$ is equivalent to $\overline{\mathbf{B}}\mathbf{y}=\mathbf{c}$, $\mathbf{y}\geqslant\mathbf{0}$. For convenience we use the notation $\mathbf{B}$ for $\overline{\mathbf{B}}$ and $\mathbf{x}$ for $\mathbf{y}$, and obtain the following equivalent definition of M:

$$M=\{\mathbf{x}|\mathbf{x}\in E^n, \mathbf{Bx}=\mathbf{c}, \mathbf{x}\geqslant \mathbf{0}\}. \tag{10.2}$$

Since the constraints $\mathbf{x}\geqslant\mathbf{0}$ in (10.1) are linear in $\mathbf{x}$, we can include these in the constraints $\mathbf{Bx}\leqslant\mathbf{c}$ by suitably augmenting the matrix $\mathbf{B}$ and the vector $\mathbf{c}$. This leads to another equivalent definition of M,

$$M=\{\mathbf{x}|\mathbf{x}\in E^n, \mathbf{Bx}\leqslant \mathbf{c}\}. \tag{10.3}$$

The geometric properties of the set M defined by (10.1) are discussed in the Appendix. We assume that the positive definite quadratic function $f(\mathbf{x})$ to be minimized is given by

$$f(\mathbf{x})=\mathbf{b}^T\mathbf{x}+\tfrac{1}{2}\mathbf{x}^T\mathbf{Ax}. \tag{10.4}$$

Kuhn-Tucker Conditions

In Chapter 4, the Kuhn-Tucker theorem was proved for a general class of constrained minimization problems. The hypothesis of the theorem requires that both the constraint functions and the objective function be convex in $\mathbf{x}$. The constraint functions in this case are linear in $\mathbf{x}$, and hence they are convex. Furthermore, since $\mathbf{A}$, the Hessian matrix of $f(\mathbf{x})$ in (10.1), is positive definite, $f(\mathbf{x})$ is convex in $\mathbf{x}$ and we can apply the Kuhn-Tucker theorem to the quadratic programming problem.

First we consider the case where M is defined by (10.1). We introduce vectors $\boldsymbol{\lambda}\in E^m$ and $\boldsymbol{\mu}\in E^n$ as Lagrange multipliers. Let $\mathbf{x}^*$, $\boldsymbol{\lambda}^*$, and $\boldsymbol{\mu}^*$ denote the optimum values of the corresponding variables $\mathbf{x}$, $\boldsymbol{\lambda}$, and $\boldsymbol{\mu}$, and let $\mathbf{y}^*=\mathbf{c}-\mathbf{Bx}^*$. Then the following equations represent the Kuhn-Tucker conditions for this problem:

$$\mathbf{B}\mathbf{x}^* + \mathbf{y}^* = \mathbf{c}, \tag{10.5a}$$

$$\mathbf{A}\mathbf{x}^* + \mathbf{B}^T\boldsymbol{\lambda}^* - \boldsymbol{\mu}^* = -\mathbf{b}, \tag{10.5b}$$

$$(\mathbf{x}^*, \boldsymbol{\mu}^*) + (\boldsymbol{\lambda}^*, \mathbf{y}^*) = 0, \tag{10.5c}$$

$$\mathbf{x}^*, \mathbf{y}^*, \boldsymbol{\lambda}^*, \boldsymbol{\mu}^* \geqslant \mathbf{0}. \tag{10.5d}$$

Next suppose M is given by (10.2). Then the corresponding Kuhn-Tucker conditions are

$$\mathbf{B}\mathbf{x}^* = \mathbf{c}, \tag{10.6a}$$

$$\mathbf{A}\mathbf{x}^* + \mathbf{B}^T\boldsymbol{\lambda}^* - \boldsymbol{\mu}^* = -\mathbf{b}, \tag{10.6b}$$

$$(\mathbf{x}^*, \boldsymbol{\mu}^*) = 0, \tag{10.6c}$$

$$\mathbf{x}^*, \boldsymbol{\mu}^* \geqslant \mathbf{0}. \tag{10.6d}$$

Finally, suppose M is given by (10.3). The corresponding Kuhn-Tucker conditions become

$$\mathbf{B}\mathbf{x}^* + \mathbf{y}^* = \mathbf{c}, \tag{10.7a}$$

$$\mathbf{A}\mathbf{x}^* + \mathbf{B}^T\boldsymbol{\lambda}^* = -\mathbf{b}, \tag{10.7b}$$

$$(\boldsymbol{\lambda}^*, \mathbf{y}^*) = 0, \tag{10.7c}$$

$$\boldsymbol{\lambda}^*, \mathbf{y}^* \geqslant \mathbf{0}. \tag{10.7d}$$

Influence of the Constraints on the Minimization Problem

In Chapter 9, we discussed the problem of unconstrained minimization of $f(\mathbf{x})$ given by (10.4). We now investigate the influence of the linear constraints on the solution of this problem. We assume that M is defined by (10.3) and let $\mathbf{x}^0$ denote the unconstrained minimum of $f(\mathbf{x})$ given by (10.4). Then from Chapter 9 we know that $\mathbf{x}^0 = -\mathbf{A}^{-1}\mathbf{b}$. Let $\mathbf{x}^*$ denote the constrained minimum of $f(\mathbf{x})$ over the set M given by (10.3). If M is nonempty, then $\mathbf{x}^*$ is unique, since $f(\mathbf{x})$ is strictly convex in $\mathbf{x}$. Letting $\mathbf{b}^j$ denote the transpose of the jth row vector of $\mathbf{B}$, $j = 1, \ldots, m$, we can write the constraint inequalities as $(\mathbf{b}^j, \mathbf{x}) \leqslant c_j$, $j = 1, \ldots, m$. We often refer to the constraint $(\mathbf{b}^j, \mathbf{x}) \leqslant c_j$ as the jth constraint. The jth constraint is *active* at $\mathbf{x}$ if $(\mathbf{b}^j, \mathbf{x}) = c_j$. It is *inactive* if $(\mathbf{b}^j, \mathbf{x}) < c_j$, and $\mathbf{x}$ violates the jth constraint if $(\mathbf{b}^j, \mathbf{x}) > c_j$.

Let S^* denote a subset of the set of integers $\{1, \ldots, m\}$ such that for all $j \in S^*$ the jth constraint is active at $\mathbf{x}^*$. Then solve the problem of

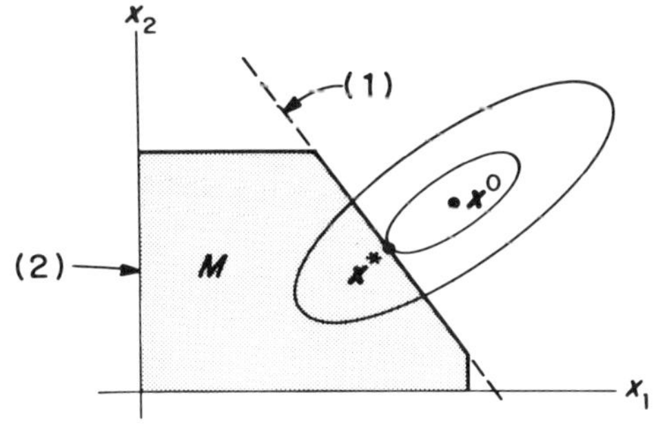

FIGURE 10.5

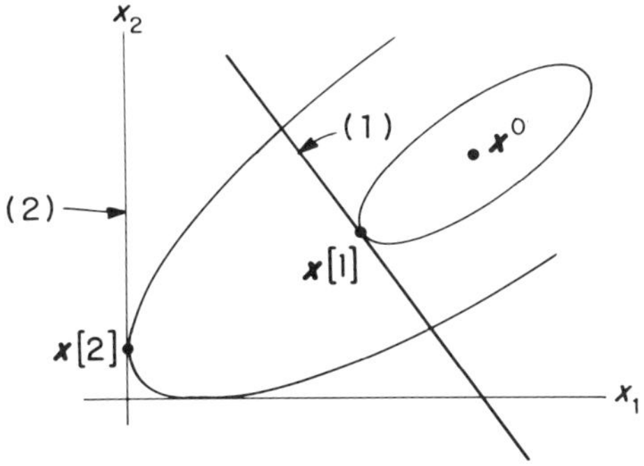

FIGURE 10.6

minimizing $f(\mathbf{x})$ subject to the constraints in S^*, and denote this solution by $\mathbf{x}^* = \mathbf{x}[S^*]$. (See Figs. 10.5 and 10.6.) In Fig. 10.5, the constraint denoted by (1) is active at $\mathbf{x}^*$. Figure 10.6 shows the solution of the minimization problem with only constraint (1) as an equality constraint. In this case the other constraints have been ignored and $\mathbf{x}[1]$ denotes the corresponding solution. Then it is clear from Figs. 10.5 and 10.6 that $\mathbf{x}[1] = \mathbf{x}^*$.

Now let S be a subset of the integers $\{1, \ldots, m\}$, and as before, let $\mathbf{x}[S]$ be the minimum point of $f(\mathbf{x})$ with the constraints in S active. Assuming that $\mathbf{x}[S] \in M$, we arbitrarily select a constraint from the set S and drop it from further consideration. Next we minimize $f(\mathbf{x})$ subject to the remaining constraints. If the solution to this problem violates the constraint dropped from consideration, then $\mathbf{x}[S] = \mathbf{x}^*$. Letting $S = \{1\}$, we see from Figs. 10.5 and 10.6 that $\mathbf{x}[1] \in M$. If we drop constraint (1) from S, then S contains no constraint and we are left with an unconstrained minimization problem. The solution to this problem clearly violates constraint (1) (see Fig. 10.6), and hence $\mathbf{x}[1] = \mathbf{x}^*$. On the other hand, suppose $S = \{2\}$, where constraint (2) is shown in Figs. 10.5 and 10.6. Then $\mathbf{x}[2]$ shown in Fig. 10.6

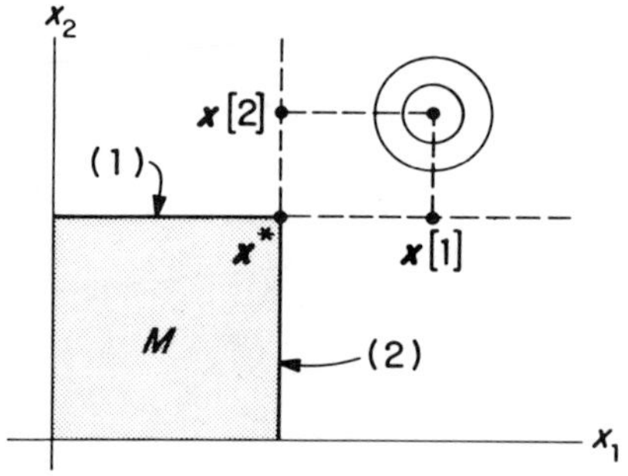

FIGURE 10.7

minimizes $f(\mathbf{x})$ with (2) active. However, as is clear from these figures, the unconstrained minimum does not violate constraint (2), and hence $\mathbf{x}[2] \neq \mathbf{x}^*$. Now let $\mathbf{x}^* = \mathbf{x}[S]$, where S is some nonempty subset of the integers $\{1,\ldots,m\}$. Let S_1 be a subset of S, and let $\mathbf{x}[S_1]$ minimize $f(\mathbf{x})$ with the constraints in S_1 as active constraints. Then $\mathbf{x}[S_1]$ violates at least one constraint in $S - S_1$. In Fig. 10.7, $S = \{1,2\}$ and $\mathbf{x}^* = \mathbf{x}[S]$, and we remove constraint (2) from S, i.e., we take $S_1 = \{2\}$. Then $\mathbf{x}[1]$ clearly violates constraint (2).

Figures 10.5, 10.6, and 10.7 indicate the influence of the linear constraints on the solution of a quadratic programming problem. General results of this nature are proved in Problems 10.3, 10.4, and 10.5.

Method of Theil and Van de Panne

In the previous section, we have discussed the influence of the linear constraints on the solution of a quadratic programming problem. It is clear that if the active constraints at $\mathbf{x}^*$ are known, then we can find $\mathbf{x}^*$ by minimizing $f(\mathbf{x})$ with respect to these constraints. However, the active constraints at $\mathbf{x}^*$ are generally not known in advance. The method of Theil and Van de Panne finds these active constraints by an iterative algorithm [Theil and Van de Panne, 1961]. The basic steps of this algorithm are obtained from the observations in the last section:

1. Compute $\mathbf{x}^0 = -\mathbf{A}^{-1}\mathbf{b}$ and $(\mathbf{b}^j, \mathbf{x}^0)$ for all j, $1 \leqslant j \leqslant m$. Obtain $V[\mathbf{x}^0] = \{j | j \in (1,\ldots,m), (\mathbf{b}^j, \mathbf{x}^0) > c_j\}$. If $V[\mathbf{x}^0]$ is empty, then $\mathbf{x}^0 = \mathbf{x}^*$ and the iterations are stopped, Otherwise go to 2.

2. Let k denote the iteration index and C_k denote the collection of subsets S_k of the integers $\{1,\ldots,m\}$ for which $V[\mathbf{x}[S_k]]$ is nonempty.

Considering all sets of the form $S_{k+1} = S_k \cup \{k\}$, where $k \in V[\mathbf{x}[S_k]]$, compute $\mathbf{x}[S_{k+1}]$, obtain $V[\mathbf{x}[S_{k+1}]]$, and go to 3.

3. If $V[\mathbf{x}[S_{k+1}]]$ is empty, then drop from consideration each constraint in turn whose index is in S_{k+1}, and minimize $f(\mathbf{x})$ subject to the remaining constraints in S_{k+1} as equality constraints. If the solutions of these problems violate the constraint dropped from S_{k+1}, then $S_{k+1} = S^*$, $\mathbf{x}[S_{k+1}] = \mathbf{x}^*$, and the iterations are stopped. Otherwise, obtain C_{k+1} as the collection of all sets S_{k+1} for which $V[\mathbf{x}[S_{k+1}]]$ is nonempty, and go to 2.

EXAMPLES

10.1 Consider a quadratic programming problem with M as shown in Fig. 10.8, where the level curves of this positive definite quadratic function are assumed to be circles with center at $\mathbf{x}^0$. The problem is to find the constrained minimum $\mathbf{x}^*$.

There are five constraints, denoted by (1),...,(5), in Fig. 10.8. Let $V[\mathbf{x}]$ denote the set of constraints violated by $\mathbf{x}$. Then from Fig. 10.8, $V[\mathbf{x}^0] = \{1,2,3\}$. Let $S_1 = \{1\}$; then $\mathbf{x}[S_1] = \mathbf{a}$ and $V[\mathbf{a}] = \{2,3\}$. Now let $S_1 = \{2\}$; then $\mathbf{x}[S_1] = \mathbf{b}$ and $V[\mathbf{b}] = \{3\}$. Similarly when $S_1 = \{3\}$, we have $\mathbf{x}[S_1] = \mathbf{c}$ and $V[\mathbf{c}] = \{1,2\}$. Since $V[\mathbf{x}[S_1]]$ is nonempty for every choice of S_1, by step 1 the optimal solution has not yet been found.

Now S_2 can be either $\{1,2\}$, $\{1,3\}$, or $\{2,3\}$. When $S_2 = \{1,2\}$, $\mathbf{x}[S_2] = \mathbf{u}$; and from Fig. 10.8, $V[\mathbf{u}]$ is empty. Similarly when $S_2 = \{2,3\}$, $\mathbf{x}[S_2] = \mathbf{v}$ and $V[\mathbf{v}]$ is empty. However, when $S_2 = \{1,3\}$, $\mathbf{x}[S_2] = \mathbf{w}$, and from Fig. 10.8, $V[\mathbf{w}]$ is nonempty.

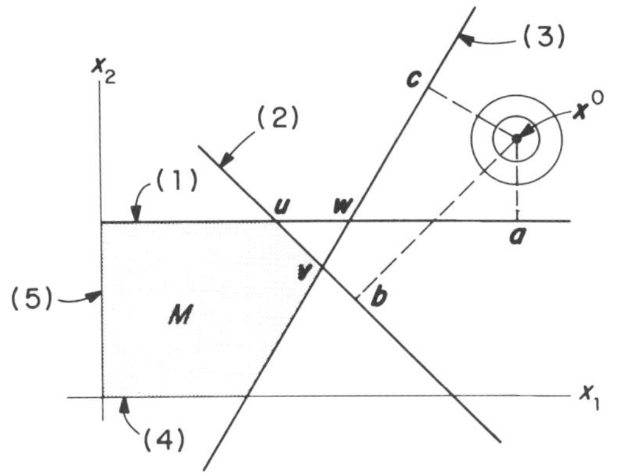

FIGURE 10.8

Since $V[\mathbf{u}]$ and $V[\mathbf{v}]$ are both empty, we must test **u** and **v** for optimality. First we consider the test for **u**. Here $S_2=\{1,2\}$, so let $S=S_2-\{1\}=\{2\}$. Then $\mathbf{x}[S]=\mathbf{b}$, and from Fig. 10.8, **b** does not violate constraint (1); so **u** cannot be the optimal solution.

Next we consider **v** where $S_2=\{2,3\}$ and $S=S_2-\{2\}=\{3\}$. Then $\mathbf{x}[S]=\mathbf{c}$, which violates constraint (2). Now letting $S=S_2-\{3\}=\{2\}$, we have $\mathbf{x}[S]=\mathbf{b}$, which also violates constraint (3). Thus **v** satisfies the test for optimality, and by step 3 we have $\mathbf{x}^*=\mathbf{v}$.

10.2 Consider the quadratic programming problem where M is defined by the following inequalities: $x_1,x_2\geqslant 0$, $x_1+x_2\leqslant 3$, and $2x_1-x_2\leqslant 4$, as shown in Fig. 10.9. The objective function is $f(\mathbf{x})=\frac{1}{2}\mathbf{x}^T\mathbf{A}\mathbf{x}+\mathbf{b}^T\mathbf{x}$, where

$$\mathbf{b}=[-2,1]^T \quad \text{and} \quad \mathbf{A}=\begin{bmatrix} 1 & -1 \\ -1 & 2 \end{bmatrix}.$$

Find the constrained minimum $\mathbf{x}^*$ by the method of Theil and Van de Panne.

First we find the unconstrained minimum of $f(\mathbf{x})$, which is

$$\mathbf{x}^0=-\mathbf{A}^{-1}\mathbf{b}=-\begin{bmatrix} 2 & 1 \\ 1 & 1 \end{bmatrix}\begin{bmatrix} -2 \\ 1 \end{bmatrix}=[3,1]^T.$$

The location of $\mathbf{x}^0$ is shown in Fig. 10.9 and clearly violates both constraints $x_1+x_2\leqslant 3$ and $2x_1-x_2\leqslant 4$. If we call $x_1+x_2\leqslant 3$ constraint (1) and $2x_1-x_2\leqslant 4$ constraint (2), then $V[\mathbf{x}^0]=\{1,2\}$ and we take $S_1=\{1\}$. To find $\mathbf{x}[S_1]$ we form the Lagrangian $L(\mathbf{x},\lambda)=f(\mathbf{x})+\lambda(x_1+x_2-3)$ and obtain

$$\frac{\partial L}{\partial x_1}=x_1-x_2-2+\lambda,$$

$$\frac{\partial L}{\partial x_2}=-x_1+2x_2+1+\lambda.$$

Setting $\partial L/\partial x_1=\partial L/\partial x_2=0$, we have $x_1=3-3\lambda$ and $x_2=1-2\lambda$. Since constraint (1) is now active, we should choose λ such that $x_1+x_2=3$. Thus $\lambda=0.2$ and $\mathbf{x}[S_1]=[2.4,0.6]^T$, which is shown as point **a** in Fig. 10.9. Clearly point **a** violates constraint (2).

Taking $S_1=\{2\}$ and using the Lagrange multiplier technique, we obtain $\mathbf{x}[S_1]=[2.4,0.8]^T$, which is shown as point **b** in Fig. 10.9. Point **b** similarly violates constraint (1).

Since $V[\mathbf{a}]=\{2\}$ and $V[\mathbf{b}]=\{1\}$, neither **a** nor **b** is the optimal solution. So we form $S_2=\{1,2\}$ and proceed to compute $\mathbf{x}[S_2]$. Now the constraints in S_2 are active, and from Fig. 10.9 we see that **c** is the only point which

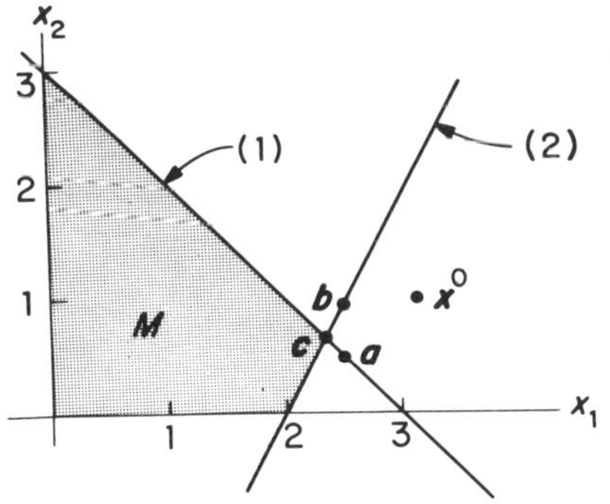

FIGURE 10.9

satisfies both constraint (1) and constraint (2), and hence $\mathbf{x}[S_2]=\mathbf{c}=[2.3, 0.7]^T$.

Since **c** does not violate any constraint, it must be tested for optimality. Letting $S=S_2-\{1\}=\{2\}$, we find $\mathbf{x}[S]=\mathbf{b}$, which violates constraint (1). Similarly letting $S=S_2-\{2\}=\{1\}$, we find $\mathbf{x}[S]=\mathbf{a}$, which violates constraint (2). Hence according to step 3, $\mathbf{x}^*=\mathbf{c}=[2.3, 0.7]^T$.

Dual Problem

We now consider the problem of minimizing $f(\mathbf{x})$ described by (10.4) over the set M given by (10.3). In this section we show that the solution to this problem can be obtained by solving a much simpler quadratic programming problem. The Kuhn-Tucker conditions for the original problem are given by (10.7a–d), where we assumed that **A** is strictly positive definite. We then solve (10.7b) to obtain

$$\mathbf{x}^* = -\mathbf{A}^{-1}(\mathbf{B}^T\boldsymbol{\lambda}^* + \mathbf{b}). \tag{10.8}$$

Substituting the expression for $\mathbf{x}^*$ in (10.7a), we obtain

$$-\mathbf{B}\mathbf{A}^{-1}(\mathbf{B}^T\boldsymbol{\lambda}^* + \mathbf{b}) + \mathbf{y}^* = \mathbf{c},$$

which can be rewritten as

$$\mathbf{D}\boldsymbol{\lambda}^* - \mathbf{y}^* = -\mathbf{d}, \tag{10.9}$$

where $\mathbf{D}=\mathbf{B}\mathbf{A}^{-1}\mathbf{B}^T$ and $\mathbf{d}=\mathbf{c}+\mathbf{B}\mathbf{A}^{-1}\mathbf{b}$. Now (10.7c), (10.7d), and (10.9) are the Kuhn-Tucker conditions for the constrained minimization problem given by

$$\min_{\boldsymbol{\lambda} \geqslant 0} \left(\mathbf{d}^T\boldsymbol{\lambda} + \tfrac{1}{2}\boldsymbol{\lambda}^T\mathbf{D}\boldsymbol{\lambda}\right).$$

This quadratic programming problem is simpler to solve than the original one, since the only constraints are of the form $\boldsymbol{\lambda} \geqslant \mathbf{0}$. It is called the dual problem and is discussed in Problems 10.9–10.11. If $\boldsymbol{\lambda}^*$ is the solution to this constrained minimization problem, then the solution $\mathbf{x}^*$ to the original problem can be obtained from (10.8).

Method of Hildreth and D'Espo

This method is based on solving the constrained minimization problem

$$\min_{\boldsymbol{\lambda} \geqslant 0} \mathbf{d}^T\boldsymbol{\lambda} + \tfrac{1}{2}\boldsymbol{\lambda}^T\mathbf{D}\boldsymbol{\lambda}$$

and then using (10.8) to obtain $\mathbf{x}^*$. If the matrix $\mathbf{B}$ in (10.3) has full row rank, then $\mathbf{D}$ is strictly positive definite. Otherwise $\mathbf{D}$ is nonnegative definite. In either case the diagonal elements of $\mathbf{D}$ are positive. The method of Hildreth and D'Espo solves the above constrained minimization problem by using a one-dimensional search technique starting from $\boldsymbol{\lambda}=0$:

1. Computing $\mathbf{D}=\mathbf{B}\mathbf{A}^{-1}\mathbf{B}^T$ and $\mathbf{d}=\mathbf{c}+\mathbf{B}\mathbf{A}^{-1}\mathbf{b}$, define $\mathbf{g}(\boldsymbol{\lambda})=\mathbf{d}^T\boldsymbol{\lambda} + \tfrac{1}{2}\boldsymbol{\lambda}^T\mathbf{D}\boldsymbol{\lambda}$, and using $\nabla g(\boldsymbol{\lambda})=0$, obtain the following linear equations in λ_j, $j=1,\ldots,m$:

$$d_i + \sum_{j=1}^{m} d_{ij}\lambda_j = 0. \tag{10.10}$$

   In (10.10), $i=1,\ldots,m$, d_i is the ith component of $\mathbf{d}$, and d_{ij} is the ijth element of $\mathbf{D}$.

2. Letting k denote the iteration index, solve (10.10) for $\hat{\lambda}_i$ by substituting $\lambda_j=\lambda_j^k$, $j=1,\ldots,m$, $j\neq i$, and obtain the new value of λ_i from $\lambda_i^{k+1}=\max(0,\hat{\lambda}_i)$.

3. If $\boldsymbol{\lambda}^k \neq \boldsymbol{\lambda}^{k+1}$, then go to 2. Otherwise the iterations are stopped and we have $\boldsymbol{\lambda}^k=\boldsymbol{\lambda}^{k+1}=\boldsymbol{\lambda}^*$, and by (10.8), $\mathbf{x}^*=-\mathbf{A}^{-1}(\mathbf{B}^T\boldsymbol{\lambda}^*+\mathbf{b})$.

EXAMPLES

10.3 Solve the following quadratic programming problem where the set M is defined by the inequalities $0 \leqslant x_1 \leqslant 1$ and $0 \leqslant x_2 \leqslant 1$:

$$\min_{\mathbf{x}\in M} f(\mathbf{x}),$$

where

$$f(x_1,x_2)=\tfrac{1}{2}\left[(x_1-2)^2+(x_2-2)^2\right].$$

The problem is depicted graphically in Fig. 10.10. The inequalities defining M can be written as $\mathbf{Bx}\leqslant\mathbf{c}$, where $\mathbf{c}=[1,1,0,0]^T$ and

$$\mathbf{B}^T=\begin{bmatrix}1 & 0 & -1 & 0\\ 0 & 1 & 0 & -1\end{bmatrix}.$$

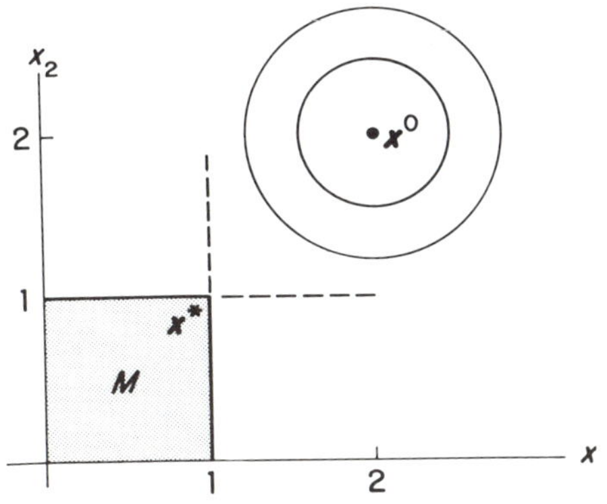

FIGURE 10.10

The objective function $f(\mathbf{x})$ can be written as $f(\mathbf{x})=\frac{1}{2}\mathbf{x}^T\mathbf{Ax}+\mathbf{b}^T\mathbf{x}$, where

$$\mathbf{b}=[-2,-2]^T \quad\text{and}\quad \mathbf{A}=\begin{bmatrix}1 & 0\\ 0 & 1\end{bmatrix}.$$

By direct computation we obtain

$$\mathbf{D}=\mathbf{BA}^{-1}\mathbf{B}^T=\begin{bmatrix}1 & 0\\ 0 & 1\\ -1 & 0\\ 0 & -1\end{bmatrix}\begin{bmatrix}1 & 0\\ 0 & 1\end{bmatrix}\begin{bmatrix}1 & 0 & -1 & 0\\ 0 & 1 & 0 & -1\end{bmatrix}$$

$$=\begin{bmatrix}1 & 0 & -1 & 0\\ 0 & 1 & 0 & -1\\ -1 & 0 & 1 & 0\\ 0 & -1 & 0 & 1\end{bmatrix},$$

and

$$\mathbf{d} = \mathbf{c} + \mathbf{B}\mathbf{A}^{-1}\mathbf{b}$$

$$= \begin{bmatrix} 1 \\ 1 \\ 0 \\ 0 \end{bmatrix} + \begin{bmatrix} 1 & 0 \\ 0 & 1 \\ -1 & 0 \\ 0 & -1 \end{bmatrix} \begin{bmatrix} 1 & 0 \\ 0 & 1 \end{bmatrix} \begin{bmatrix} -2 \\ -2 \end{bmatrix}$$

$$= [-1, -1, 2, 2]^T.$$

We start the one-dimensional search with $\boldsymbol{\lambda}^0 = [0,0,0,0]^T$, and for $i = 1$ we obtain from (10.10)

$$-1 + \lambda_1 - \lambda_3 = 0.$$

Solving this equation for $\hat{\lambda}_1$ with $\lambda_3 = \lambda_3^0 = 0$, we have $\lambda_1 = 1$ and hence $\lambda_1^1 = \max(0, \hat{\lambda}_1) = \max(0, 1) = 1$. Using (10.10) in a similar manner for $i = 2$, 3, and 4, we have $\lambda_2^1 = 1$, $\lambda_3^1 = 0$, and $\lambda_4^1 = 0$, respectively, and hence $\boldsymbol{\lambda}^1 = [1,1,0,0]^T$. Now we use $\boldsymbol{\lambda}^1 = [1,1,0,0]^T$ in (10.10) and employ the same procedure to obtain $\boldsymbol{\lambda}^2 = [1,1,0,0]^T$.

Since $\boldsymbol{\lambda}^1 = \boldsymbol{\lambda}^2$, the one-dimensional search technique has converged, and we set $\boldsymbol{\lambda}^* = [1,1,0,0]^T$. Now using (10.8), we obtain

$$\mathbf{x}^* = -\mathbf{A}^{-1}(\mathbf{B}^T\boldsymbol{\lambda}^* + \mathbf{b})$$

$$= -\begin{bmatrix} 1 & 0 \\ 0 & 1 \end{bmatrix} \left\{ \begin{bmatrix} 1 & 0 & -1 & 0 \\ 0 & 1 & 0 & -1 \end{bmatrix} \begin{bmatrix} 1 \\ 1 \\ 0 \\ 0 \end{bmatrix} + \begin{bmatrix} -2 \\ -2 \end{bmatrix} \right\}$$

$$= [1, 1]^T.$$

In Fig. 10.10, the level curves $f(\mathbf{x})$ are circles, one of which passes through the boundary point $[1,1]^T$ of the set M. From geometric considerations we know that the minimum of $f(\mathbf{x})$ on M is the point $[1,1]^T$, which is also the answer we obtained by using the method of Hildreth and D'Espo.

10.4 Let M be defined by the inequalities

$$x_1, x_2 \geqslant 0 \quad \text{and} \quad 3x_1 + 2x_2 \leqslant 6; \text{ and let } \quad f(\mathbf{x}) = \tfrac{1}{2}\mathbf{x}^T\mathbf{A}\mathbf{x} + \mathbf{b}^T\mathbf{x},$$

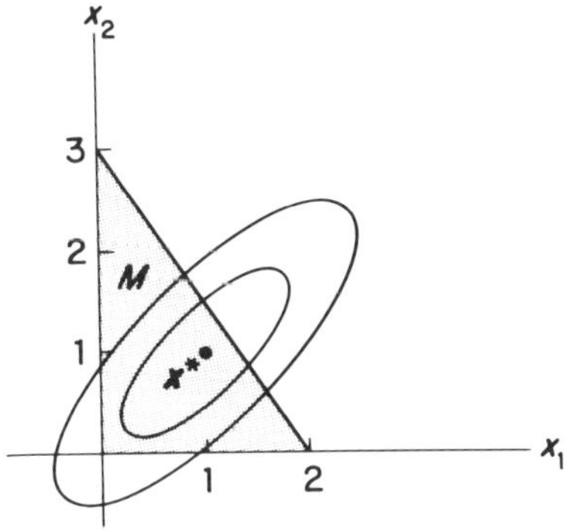

FIGURE 10.11

where

$$\mathbf{b}=[-1,0]^T \quad \text{and} \quad \mathbf{A}=\begin{bmatrix} 2 & -1 \\ -1 & 1 \end{bmatrix}.$$

Solve the problem $\min_{\mathbf{x}\in M} f(\mathbf{x})$.

The objective function $f(\mathbf{x})$ has a global minimum at

$$\mathbf{x}^*=-\mathbf{A}^{-1}\mathbf{b}=-\begin{bmatrix} 1 & 1 \\ 1 & 2 \end{bmatrix}\begin{bmatrix} -1 \\ 0 \end{bmatrix}=[1,1]^T.$$

We note that $3x_1^*+2x_2^*=3+2=5\leqslant 6$; hence $\mathbf{x}^*$ is feasible, since it is an interior point of M (see Fig. 10.11). Thus the constrained minimum of $f(\mathbf{x})$ is the same as the global minimum.

Solving now by the method of Hildreth and D'Espo, the inequalities defining M can first be written as $\mathbf{Bx}\leqslant\mathbf{c}$, where

$$\mathbf{c}=[6,0,0]^T \quad \text{and} \quad \mathbf{B}^T=\begin{bmatrix} 3 & -1 & 0 \\ 2 & 0 & 1 \end{bmatrix}.$$

Then by direct computation

$$\mathbf{D}=\mathbf{BA}^{-1}\mathbf{B}^T=\begin{bmatrix} 3 & 2 \\ -1 & 0 \\ 0 & -1 \end{bmatrix}\begin{bmatrix} 1 & 1 \\ 1 & 2 \end{bmatrix}\begin{bmatrix} 3 & -1 & 0 \\ 2 & 0 & -1 \end{bmatrix}$$

$$=\begin{bmatrix} 29 & -5 & -7 \\ -5 & 1 & 1 \\ -7 & 1 & 2 \end{bmatrix}$$

and

$$\mathbf{d}=\mathbf{c}+\mathbf{B}\mathbf{A}^{-1}\mathbf{b}=\begin{bmatrix}6\\0\\0\end{bmatrix}+\begin{bmatrix}3 & 2\\-1 & 0\\0 & -1\end{bmatrix}\begin{bmatrix}1 & 1\\1 & 2\end{bmatrix}\begin{bmatrix}-1\\0\end{bmatrix}=[1,1,1]^T.$$

Starting the one-dimensional search from $\boldsymbol{\lambda}^0=[0,0,0]^T$, for $i=1$ we obtain from (10.10)

$$1+29\lambda_1-5\lambda_2-7\lambda_3=0.$$

Solving this equation for $\hat{\lambda}_1$with $\lambda_2=\lambda_3=0$, we have $\hat{\lambda}_1=-0.035$ and hence $\lambda_1^1=\max(0,\hat{\lambda}_1)=\max(0,-0.035)=0$. Using (10.10) in a similar manner for $i=2$ and 3, we obtain $\lambda_2^1=0$ and $\lambda_3^1=0$, and hence $\boldsymbol{\lambda}^1=[0,0,0]^T$. Since $\boldsymbol{\lambda}^0=\boldsymbol{\lambda}^1$, the one-dimensional search technique has converged, and we set $\boldsymbol{\lambda}^*=[0,0,0]^T$. Now using (10.8), we get

$$\begin{aligned}\mathbf{x}^*&=-\mathbf{A}^{-1}(\mathbf{B}^T\boldsymbol{\lambda}^*+\mathbf{b})\\&=-\begin{bmatrix}1 & 1\\1 & 2\end{bmatrix}\left(\begin{bmatrix}3 & -1 & 0\\2 & 0 & -1\end{bmatrix}\begin{bmatrix}0\\0\\0\end{bmatrix}+\begin{bmatrix}-1\\0\end{bmatrix}\right)\\&=[1,1]^T,\end{aligned}$$

which agrees with the graphical result.

Capacity Method of Houthakker

An interesting method for solving quadratic programming problems, called the *capacity method*, is due to Houthakker. Consider the problem of minimizing $f(\mathbf{x})$ given by (10.4) subject to the constraint set M given by (10.1). Houthakker introduces a new constraint of the form

$$\sum_{i=1}^{n} x_i \leqslant u, \tag{10.11}$$

which is called a *capacity constraint*. If one of the existing constraints from $\mathbf{Bx}\leqslant\mathbf{c}$ is already in the form of a capacity constraint, then no new constraint is needed. The original problem is next modified to $\min_{\mathbf{x}\in M} f(\mathbf{x})$ such that $\sum_{i=1}^{n} x_i \leqslant u$, and the solution of this modified problem is denoted by $\mathbf{x}^*(u)$. As the parameter u is changed, we obtain different values for the solution $\mathbf{x}^*(u)$. The locus of $\mathbf{x}^*(u)$ with u is a polygonal line in M, and there exists a u^* such that $\mathbf{x}^*(u)=\mathbf{x}^*(u^*)$ for all $u \geqslant u^*$. Since the capacity constraint becomes inactive for all $\mathbf{x} \in M$ when $u \geqslant u^*$, we have $\mathbf{x}^* = \mathbf{x}^*(u^*)$. To start the algorithm we choose $u=0$ as the initial value of the parameter u.

EXAMPLES

10.5 Determine the minimum of the function $f(\mathbf{x})$ having circular level curves with $\mathbf{x}^0$ as the center, subject to the constraint set M shown in Fig. 10.12.

In Fig. 10.12, the straight lines $u_1, u_2, \ldots,$ indicate the level curves of $\sum_{i=1}^{n} x_i = u$ when $u=u_1$, $u=u_2$, etc. For all u, $0 \leqslant u \leqslant u_1$, $\mathbf{x}^*(u)$ lies on the x_2 axis. At $u=u_1$, the locus of $\mathbf{x}^*(u)$ changes direction and moves orthogonally to the capacity constraints until $u_1=u_2$. For all u, $u_2 \leqslant u \leqslant u^*$, $\mathbf{x}^*(u)$ lies on the line $x_2=a$; and for all $u \geqslant u^*$, $\mathbf{x}^*(u)=\mathbf{x}^*(u^*)$, i.e., the capacity constraint has become inactive and hence the minimum point of $f(\mathbf{x})$ is $\mathbf{x}^* = \mathbf{x}^*(u^*)$.

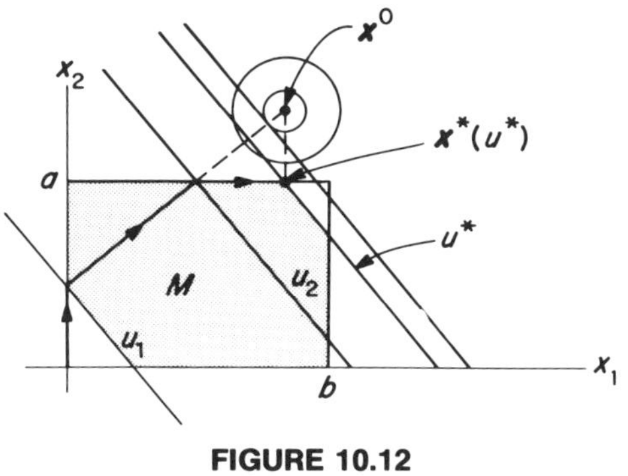

FIGURE 10.12

10.6 For $f(x_1,x_2)=x_1^2+x_2^2$ and $M=\{(x_1,x_2)|3x_1+2x_2 \leqslant 6,\ \ x_1,x_2 \geqslant 0\}$, solve $\max_{\mathbf{x}\in M} f(\mathbf{x})$ by Houthakker's capacity method.

The set M is shown in Fig. 10.13 along with the level curves of $f(\mathbf{x})$, which are circles with center at $[0,0]^T$. From geometric considerations we see that $f(\mathbf{x})$ attains a maximum at $\mathbf{x}^* = [0,3]^T$, which is an extreme point of M.

Now we consider the capacity constraint $x_1+x_2 \leqslant u$, with $0 \leqslant u \leqslant 2$. The level curve $x_1+x_2=1$ is shown in Fig. 10.13, where from geometric considerations we see that $\max_{\mathbf{x}\in M} f(x)$ such that $x_1+x_2 \leqslant 1$ has two solutions, $[1,0]^T$ and $[0,1]^T$. Thus we obtain two loci of $\mathbf{x}^*(u)$ as u is increased from 0 to 2.

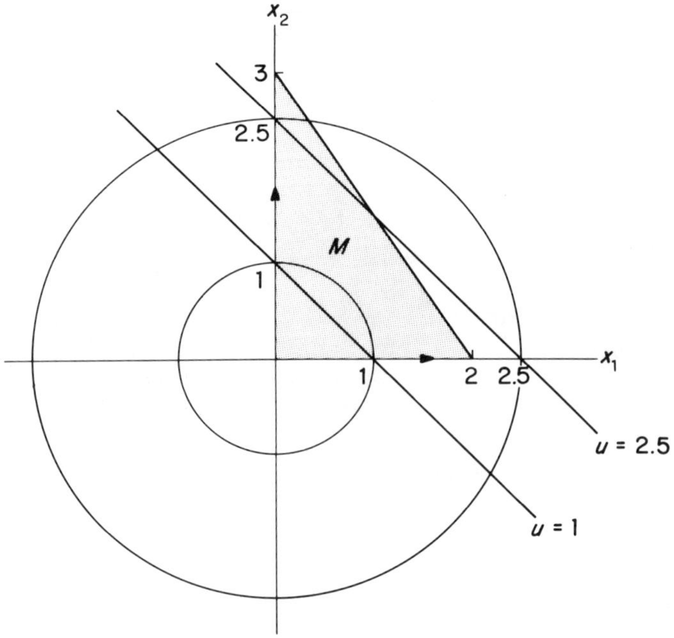

FIGURE 10.13

For $u>2$, part of the level curve of $x_1+x_2=u$ lies outside the set M, as shown in Fig. 10.13 for $u=2.5$. Thus for $u>2$ we see from geometric considerations that the problem $\max_{\mathbf{x}\in M} f(x)$ such that $x_1+x_2 \leqslant u$ has a unique solution given by $[0,u]^T$. The locus of $\mathbf{x}^*(u)$ along the x_1-axis stops at $[2,0]^T$, but the second locus along the x_2-axis continues until the point $[0,3]^T$ is reached. For $u>3$, the capacity constraint becomes inactive for all $x\in M$, and so $\mathbf{x}^*(3)=[0,3]^T$ is the optimal solution.

Modified Simplex Methods

The simplex method (see Appendix) is one very efficient algorithm for solving linear programming problems, and several modified versions have been developed for solving quadratic programming problems. The applicability of the simplex method becomes clear when we examine the Kuhn-Tucker conditions for quadratic programs.

To be specific, if M is defined by (10.2), then the Kuhn-Tucker conditions are (see p. 248)

$$\mathbf{B}\mathbf{x}^* = \mathbf{c}, \tag{10.6a}$$

$$\mathbf{A}\mathbf{x}^* + \mathbf{B}^T\boldsymbol{\lambda}^* - \mathbf{u}^* = -\mathbf{b}, \tag{10.6b}$$

$$(\mathbf{x}^*, \boldsymbol{\mu}^*) = 0, \tag{10.6c}$$

$$\mathbf{x}^*, \boldsymbol{\mu}^* \geqslant 0. \tag{10.6d}$$

According to the Kuhn-Tucker theorem, if we can find a solution to (10.6a) and (10.6b) which also satisfies (10.6c) and (10.6d), it will also be a solution to the quadratic programming problem. Every solution of (10.6a) and (10.6b) is an $(m+2n)$-dimensional vector $\mathbf{u} = [\mathbf{x}^T, \boldsymbol{\lambda}^T, \boldsymbol{\mu}^T]^T$, and we call such a vector a *basic solution* if it has at most $m+n$ nonzero components (see Example A.5, page 356).

Let $\mathbf{u} = [\mathbf{x}^T, \boldsymbol{\lambda}^T, \boldsymbol{\mu}^T]^T$ be a solution of (10.6a) and (10.6b) which also satisfies (10.6c) and (10.6d). From (10.6d), $x_i, \mu_i \geqslant 0$; and in order to satisfy (10.6c) either $x_i = 0$ or $\mu_i = 0$, $i = 1, \ldots, n$, which implies that $\mathbf{u}$ has at most $m+n$ nonzero components. Therefore, every solution of (10.6a) and (10.6b) that satisfies (10.6c) and (10.6d) is a basic solution of (10.6a) and (10.6b). However, since every basic solution of (10.6a) and (10.6b) does not necessarily satisfy (10.6c) and (10.6d), solving a quadratic programming problem is equivalent to finding a basic solution of (10.6a) and (10.6b) which also satisfies (10.6c) and (10.6d). The simplex method discussed in the Appendix is an efficient method for generating such basic solutions. However, the simplex method should be modified in such a way that we obtain a basic solution satisfying (10.6c). Such a modified simplex method was developed by Wolfe for solving quadratic programming problems.

WOLFE'S METHOD

First we introduce slack variables $\mathbf{w}$, $\mathbf{z}^1$, and $\mathbf{z}^2 \geqslant \mathbf{0}$ and rewrite (10.6a) and (10.6b) as follows:

$$\mathbf{B}\mathbf{x} + \mathbf{w} = \mathbf{c}, \tag{10.12a}$$

$$\mathbf{A}\mathbf{x} + \mathbf{B}^T\boldsymbol{\lambda} - \boldsymbol{\mu} + \mathbf{z}^1 - \mathbf{z}^2 = -\mathbf{b}. \tag{10.12b}$$

Clearly one basic solution of (10.12a) and (10.12b) is the one for which $\mathbf{x} = \boldsymbol{\mu} = \boldsymbol{\lambda} = \mathbf{0}$ and at least one of the components z_i^1 or z_i^2 equals zero for each i; this solution also satisfies (10.6c). The components of this basic solution are $w_i = c_i$, $i = 1, \ldots, m$, where we assume that $c_i \geqslant 0$, and that

$z_i^1 = -b_i$ if $b_i < 0$ or $z_i^2 = b_i$ if $b_i > 0$. If $b_i = 0$, then it is immaterial whether we consider z_i^1 or z_i^2 as part of the basic feasible solution.

We next minimize $\Sigma_{i=1}^m w_i$ by the simplex method (phase 1) subject to the constraints (10.12a) and (10.12b) with $\boldsymbol{\lambda} = \boldsymbol{\mu} = \mathbf{0}$. If this minimum is zero and the corresponding basic feasible solution is nondegenerate (i.e., has exactly $m+n$ nonzero components), then the constraints of the original problem are incompatible and no solution exists. However, if the basic feasible solution is degenerate (i.e., has less than $m+n$ nonzero components), then we continue with the simplex method until one of the w_i variables appears in the basic feasible solution even with a zero value. The final basic solution then contains at least m of the variables x_i and either z_i^1 or z_i^2 for each i. Now let $\mathbf{z}$ denote the vector whose components are either z_i^1 or z_i^2 for each i, where the vector $\mathbf{z}$ has at most n components, and let $\mathbf{D}$ denote a diagonal matrix whose ith diagonal element is 1 if $z_i = z_i^1$ and -1 otherwise. Then the final simplex tableau becomes

$$\mathbf{Bx} = \mathbf{c}, \tag{10.13a}$$

$$\mathbf{Ax} + \mathbf{Dz} = -\mathbf{b}, \tag{10.13b}$$

where $\mathbf{x}, \mathbf{z} \geqslant 0$; and we note that (10.6c) is satisfied, since $\boldsymbol{\mu} = \mathbf{0}$. We now use the simplex method to minimize $\Sigma_{i=1}^n z_i$ with $\mathbf{w}, \mathbf{z} \geqslant \mathbf{0}$ and (10.13a), (10.13b) as constraints (phase 2). However, we can no longer accept every possible basic solution, since (10.6c) must also be satisfied. Thus if x_i is already in the basic solution, μ_i is not included, and vice versa. In this way we guarantee that both x_i and μ_i are not basic variables simultaneously, and so $(\mathbf{x}, \boldsymbol{\mu}) = 0$ for all basic solutions considered. The simplex method is continued until $\Sigma_{i=1}^n z_i = 0$, i.e., $\mathbf{z} = \mathbf{0}$, at which time we have a solution of (10.6a)–(d) which also solves the quadratic programming problem.

EXAMPLE 10.7 Solve the quadratic programming problem discussed in Example 10.3 by Wolfe's modified simplex method.

It was shown in Example 10.3 that the positive definite quadratic function $f(\mathbf{x})$ attained its minimum at an interior point $[1,1]^T$ of the given set M (see Fig. 10.10). Now the simplex method discussed in the Appendix explores only the extreme points of the set M. Hence without Wolfe's modification, this method would be unable to generate an interior point such as $[1,1]^T$ of M. In this example we will see that the modification of the simplex method allows it to locate minimum points inside the set M.

We first introduce a slack variable $x_3 \geqslant 0$ so that the inequalities defining the set M can be written as $3x_1 + 2x_2 + x_3 = 6$, $x_1, x_2, x_3 \geqslant 0$, or equivalently as $\mathbf{Bx} = c$, $\mathbf{x} \geqslant \mathbf{0}$, where $\mathbf{B} = [3, 2, 1]$, $\mathbf{x} = [x_1, x_2, x_3]^T$, and $c = 6$. Now that M is given by (10.2), we can use (10.6a)–(d).

Introduction of the slack variable x_3 modifies the matrix **A** and vector **b** given in Example 10.3. In order that $f(\mathbf{x})$ remain a function of x_1 and x_2 as in Example 10.3, we define

$$\mathbf{b}=[-1,0,0]^T \quad \text{and} \quad \mathbf{A}=\begin{bmatrix} 2 & -1 & 0 \\ -1 & 1 & 0 \\ 0 & 0 & 0 \end{bmatrix}.$$

Equations (10.12a) and (10.12b) can be written as

$$\begin{aligned}
&3x_1+2x_2+x_3+w &&=6,\\
&2x_1 \;\; -x_2 &&+3\lambda-\mu_1+z_1^1-z_1^2=1,\\
&-x_1 \;\; +x_2 &&+2\lambda-\mu_2+z_2^1-z_2^2=0,\\
& &&\lambda-\mu_3+z_3^1-z_3^2=0.
\end{aligned}$$

Now following the first step of Wolfe's method, we set $\lambda=\mu_1=\mu_2=\mu_3=0$ and solve for x_1 and x_2 as functions of z_1^1, z_2^1, z_1^2, and z_2^2. Substituting in (10.12a), we have

$$w=1+5(z_1^1-z_1^2)+7(z_2^1-z_2^2)-x_3.$$

Now to minimize w as a function of z_1^1, z_1^2, z_2^1, z_2^2, and x_3 with the constraints $w\geqslant 0$ and $z_1^1,z_2^1,z_1^2,z_2^2,x_3\geqslant 0$, we could apply the simplex method. However, for clarity of presentation we will solve this problem by simple algebraic considerations.

Since $w\geqslant 0$, the minimum of w is attained by every choice of values of the variables that satisfies the nonnegativity constraints and gives $w=0$. One possible choice of the values is $z_1^1=z_2^1=1$, $z_1^2=z_2^2=0$, and $x_3=13$. However, there are many other possible choices for the values of these variables. In the second part of Wolfe's method the simplex method is used until $\Sigma_{i=1}^n z_i=0$, which means that we must determine a solution which minimizes w when all the z-variables are equal to zero. An obvious and unique choice is $z_1^1=z_2^1=z_1^2=z_2^2=0$ and $x_3=1$, in which case we have $x_1=1$ and $x_2=1$. Also since $\mu_1=\mu_2=\mu_3=0$, we have $x_1\mu_1+x_2\mu_2+x_3\mu_3=0$. By direct substitution we see that $\mathbf{x}=[1,1,1]^T$, $\boldsymbol{\mu}=[0,0,0]^T$, and $\lambda=0$ satisfy (10.6a)–(d). Thus by the Kuhn-Tucker theorem, $\mathbf{x}^*=[1,1,1]^T$, but since x_3 is a slack variable, the solution of the quadratic programming problem is given by $\mathbf{x}^*=[1,1]^T$.

Solved Problems

10.1 Let $f(\mathbf{x})$ be a positive definite quadratic function given by (10.4) and define a hyperplane $H=\{\mathbf{x}|\mathbf{x}\in E^n,\ \mathbf{Bx}=\mathbf{c}\}$. Find the solution of the problem $\min_{\mathbf{x}\in H} f(\mathbf{x})$.

We assume that $\mathbf{B}$ is an $m\times n$ matrtix of rank m, where $m<n$. Since H is a convex set and $f(\mathbf{x})$ is strictly convex, a unique constrained minimum exists. Now form the corresponding Lagrangian

$$L(\mathbf{x},\boldsymbol{\lambda})=\tfrac{1}{2}\mathbf{x}^T\mathbf{Ax}+(\mathbf{b}+\mathbf{B}^T\boldsymbol{\lambda})^T\mathbf{x}-\boldsymbol{\lambda}^T\mathbf{c}, \tag{10.14}$$

where $\boldsymbol{\lambda}\in E^m$ is a vector of Lagrange multipliers. For constant $\boldsymbol{\lambda}$, $L(\mathbf{x},\boldsymbol{\lambda})$ has a unique stationary point at

$$\mathbf{x}(\boldsymbol{\lambda})=-\mathbf{A}^{-1}(\mathbf{b}+\mathbf{B}^T\boldsymbol{\lambda}). \tag{10.15}$$

To ensure $\mathbf{x}(\boldsymbol{\lambda})\in H$, we choose $\boldsymbol{\lambda}$ in (10.15) to satisfy $\mathbf{Bx}(\boldsymbol{\lambda})=\mathbf{c}$. This results in

$$\boldsymbol{\lambda}^*=-(\mathbf{BA}^{-1}\mathbf{B}^T)^{-1}(\mathbf{c}+\mathbf{BA}^{-1}\mathbf{b}), \tag{10.16}$$

$$\mathbf{x}^*=-\mathbf{A}^{-1}(\mathbf{b}+\mathbf{B}^T\boldsymbol{\lambda}^*), \tag{10.17}$$

where $\mathbf{x}^*$ is the constrained minimum of $f(\mathbf{x})$ on H.

10.2 In Problem 10.1, show that there is an $n\times n$ symmetric matrix $\mathbf{W}$ such that for every $\mathbf{x}\in H$, $\mathbf{Wg}(\mathbf{x})=\mathbf{x}-\mathbf{x}^*$, where $g(\mathbf{x})=\nabla f(\mathbf{x})$.

Let $\mathbf{w}=\mathbf{B}^T(\mathbf{BB}^T)^{-1}\mathbf{c}$ and $\mathbf{z}=-\mathbf{g}(\mathbf{w})=-(\mathbf{Aw}+\mathbf{b})$. Then $\mathbf{BA}^{-1}\mathbf{z}=-(\mathbf{c}+\mathbf{BA}^{-1}\mathbf{b})$, and from (10.16)

$$\boldsymbol{\lambda}^*=-(\mathbf{BA}^{-1}\mathbf{B}^T)^{-1}\mathbf{BA}^{-1}\mathbf{g}(\mathbf{w}). \tag{10.18}$$

Since $f(\mathbf{x})$ is a quadratic function, we have $\mathbf{g}(\mathbf{w})=\mathbf{g}(\mathbf{x})+\mathbf{A}(\mathbf{w}-\mathbf{x})$. Substituting this expression for $\mathbf{g}(\mathbf{w})$ in (10.18) and noting that for $\mathbf{x}\in H$, $\mathbf{Bx}=\mathbf{c}$, we have

$$\boldsymbol{\lambda}^*=-(\mathbf{BA}^{-1}\mathbf{B}^T)^{-1}\mathbf{BA}^{-1}\mathbf{g}(\mathbf{x}). \tag{10.19}$$

Now substituting $\boldsymbol{\lambda}^*$ from (10.19) in (10.17) and rearranging terms, we obtain $\mathbf{Wg}(\mathbf{x})=\mathbf{x}-\mathbf{x}^*$, where

$$\mathbf{W}=\mathbf{A}^{-1}-\mathbf{A}^{-1}\mathbf{B}^T(\mathbf{BA}^{-1}\mathbf{B}^T)^{-1}\mathbf{BA}^{-1}. \tag{10.20}$$

Now since $\mathbf{A}$ is an $n\times n$ matrix, $\mathbf{A}^{-1}$ is also $n\times n$, and hence $\mathbf{W}$ is $n\times n$. Also $\mathbf{A}$ is symmetric, which implies that $\mathbf{A}^{-1}$ is symmetric and so $\mathbf{A}^{-1}\mathbf{B}^T(\mathbf{BA}^{-1}\mathbf{B}^T)^{-1}\mathbf{BA}^{-1}$ is symmetric by construction. Therefore $\mathbf{W}$ is symmetric.

10.3 Let M be defined by (10.3) and $f(\mathbf{x})$ by (10.4), and let $\mathbf{x}^*$ denote the unique minimum of $f(\mathbf{x})$ on M. As before, S^* is defined to be a subset of the integers $\{1,\ldots,m\}$ such that for $j\in S^*$, $(\mathbf{b}^j,\mathbf{x}^*)=c_j$. Show that $\mathbf{x}^*=\mathbf{x}[S^*]$.

We use the Kuhn-Tucker conditions to prove this result. Since $\mathbf{x}^* \in M$, we have $(\mathbf{b}^j, \mathbf{x}^*) < c_j$ whenever $j \notin S^*$. From (10.7b) we obtain

$$\mathbf{A}\mathbf{x}^* + \mathbf{b} = -\sum_{j=1}^{m} \lambda_j^* \mathbf{b}^j. \tag{10.21}$$

Furthermore, since $y_j^* > 0$ when $j \notin S^*$, from (10.7c) and (10.7d) we have $\lambda_j^* = 0$ when $j \notin S^*$. Hence (10.21) becomes

$$\mathbf{A}\mathbf{x}^* + \mathbf{b} = -\sum_{j \in S^*} \lambda_j^* \mathbf{b}^j. \tag{10.22}$$

It follows from (10.22) and the given hypothesis $(\mathbf{b}^j, \mathbf{x}^*) = c_j, j \in S^*$ that $\mathbf{x}^*$ satisfies the Kuhn-Tucker conditions for the minimum point of $f(\mathbf{x})$ with the constraints in S^* as equality constraints. This minimum point, denoted by $\mathbf{x}[S^*]$, is unique, since $f(\mathbf{x})$ is strictly convex and hence $\mathbf{x}^* = \mathbf{x}[S^*]$.

10.4 Let S be a subset of the integers $\{1, \ldots, m\}$ such that $\mathbf{x}[S] \in M$. If for every $k \in S$, $k \in V[\mathbf{x}[S-k]]$, then show that $\mathbf{x}[S] = \mathbf{x}^*$.

From Problem 10.1 we have

$$\mathbf{A}\mathbf{x}[S] + \mathbf{b} = -\sum_{j \in S} \lambda_j \mathbf{b}^j, \tag{10.23}$$

$$(\mathbf{b}^j, \mathbf{x}[S]) = c_j, \qquad j \in S. \tag{10.24}$$

Now we make an arbitrary choice of $k \in S$. Using the results of Problem 10.1 again, we obtain

$$\mathbf{A}\mathbf{x}[S-k] + \mathbf{b} = -\sum_{j \in S-k} \hat{\lambda}_j \mathbf{b}^j, \tag{10.25}$$

$$(\mathbf{b}^j, \mathbf{x}[S-k]) = c_j, \qquad j \in S-k. \tag{10.26}$$

The hypothesis states that

$$(\mathbf{b}^j, \mathbf{x}[S-k]) > c_k, \tag{10.27}$$

and using (10.23)–(10.26) we obtain

$$((\mathbf{x}[S] - \mathbf{x}[S-k]),\ \mathbf{A}(\mathbf{x}[S] - \mathbf{x}[S-k])) = \lambda_k\left[(\mathbf{b}^k, \mathbf{x}[S-k]) - c_k\right]. \tag{10.28}$$

Since $\mathbf{A}$ is positive definite, the left-hand side of (10.28) is nonnegative; and by (10.27) the factor within the outer brackets on the right-hand side of (10.28) is nonnegative. Therefore, $\lambda_k \geqslant 0$ for all $k \in S$. Combining this result with (10.23) and the hypothesis that $\mathbf{x}[S] \in M$, we conclude that $\mathbf{x}[S]$ satisfies the Kuhn-Tucker conditions for $\min_{\mathbf{x} \in M} f(\mathbf{x})$ and hence $\mathbf{x}[S] = \mathbf{x}^*$.

10.5 Suppose $\mathbf{x}^* = \mathbf{x}[S]$, where S is some nonempty subset of the integers $\{1,\dots,m\}$. Then for all $S_1 \subset S$, including $S_1 = \varnothing$, show that there is at least one $k \in S - S_1$ such that $k \in V\big[\mathbf{x}[S_1]\big]$.

We note that since $\mathbf{x}[S] = \mathbf{x}^*$, $\mathbf{x}[S]$ satisfies the Kuhn-Tucker conditions satisfied by $\mathbf{x}^*$. This means that $\mathbf{x}[S]$ satisfies (10.23) and (10.24) with $\lambda_j \geqslant 0$ and $(\mathbf{b}^j, \mathbf{x}[S]) < c_j$, $j \notin S$. Now for every $S_1 \subset S$, the Kuhn-Tucker conditions satisfied by $\mathbf{x}[S_1]$ are as follows:

$$\mathbf{A}\mathbf{x}[S_1] + \mathbf{b} = -\sum_{j \in S_1} \hat{\lambda}_j \mathbf{b}^j, \tag{10.29}$$

$$(\mathbf{b}^j, \mathbf{x}[S_1]) = c_j, \qquad j \in S_1. \tag{10.30}$$

From (10.23) and (10.29) we have

$$\mathbf{A}\big[\mathbf{x}[S_1] - \mathbf{x}[S]\big] = \sum_{j \in S_1} (\lambda_j - \hat{\lambda}_j)\mathbf{b}^j + \sum_{j \in S - S_1} \lambda_j \mathbf{b}^j.$$

Using (10.26) and (10.30) and algebraic manipulations, we obtain

$$\big((\mathbf{x}[S_1] - \mathbf{x}[S]), \mathbf{A}(\mathbf{x}[S_1] - \mathbf{x}[S])\big) = \sum_{j \in S - S_1} \lambda_j \big[(\mathbf{b}^j, \mathbf{x}[S_1]) - c_j\big]. \tag{10.31}$$

Since $\mathbf{A}$ is positive definite, the left-hand side of (10.31) is nonnegative; and for all $j \in S - S_1$, $\lambda_j \geqslant 0$. Hence there must exist at least one $k \in S - S_1$ such that $(\mathbf{b}^k, \mathbf{x}[S_1]) > c_k$.

10.6 In the method of Theil and Van de Panne, it is necessary to obtain $\mathbf{x}[S]$. Find $\mathbf{x}[S]$ as a function of $\mathbf{x}^0$, $\mathbf{A}^{-1}$, $\mathbf{B}$, and $\mathbf{c}$.

Since $\mathbf{x}[S]$ minimizes $f(\mathbf{x})$ with the constraints in S active, we can use the results of Problem 10.1. Assuming that the vectors $\mathbf{b}^j$, $j \in S$, forming the matrix $\mathbf{B}$ are linearly independent (nondegenerate), and using (10.15), we have

$$\mathbf{x}[S] = -\mathbf{A}^{-1}\mathbf{b} - \mathbf{A}^{-1}\mathbf{B}^T\boldsymbol{\lambda} = \mathbf{x}^0 - \mathbf{A}^{-1}\mathbf{B}^T\boldsymbol{\lambda}, \tag{10.32}$$

where $\mathbf{x}[S]$ must satisfy $\mathbf{B}\mathbf{x}[S] = \mathbf{c}$. Premultiplying (10.32) by $\mathbf{B}$ and solving for $\boldsymbol{\lambda}$, we get

$$\boldsymbol{\lambda} = (\mathbf{B}\mathbf{A}^{-1}\mathbf{B}^T)^{-1}(\mathbf{B}\mathbf{x}^0 - \mathbf{c}),$$

and so

$$\mathbf{x}[S] = \mathbf{x}^0 - \mathbf{A}^{-1}\mathbf{B}^T(\mathbf{B}\mathbf{A}^{-1}\mathbf{B}^T)^{-1}(\mathbf{B}\mathbf{x}^0 - \mathbf{c}).$$

10.7 Consider the problem of minimizing $f(\mathbf{x})$ defined by (10.4) on M defined by (10.3). Let $\boldsymbol{\lambda}^*$ be the solution of the problem

$$\min\big[\mathbf{d}^T\boldsymbol{\lambda} + \tfrac{1}{2}\boldsymbol{\lambda}^T\mathbf{D}\boldsymbol{\lambda}\big],$$

where $\mathbf{D}$ and $\mathbf{d}$ are defined after (10.9) on p. 253. Show that $\mathbf{x}^*$ given by (10.8) is a solution of the initial quadratic programming problem.

From the hypothesis, we have $\boldsymbol{\lambda}^* \geqslant 0$. Since $\mathbf{x}^*$ is given by (10.8), $\mathbf{x}^*$ and $\boldsymbol{\lambda}^*$ satisfy (10.7b). Since $\boldsymbol{\lambda}^*$ is the solution of the minimization problem given above, by the Kuhn-Tucker conditions, we know that there is a $\mathbf{y}^* \geqslant 0$ such that $\mathbf{D}\boldsymbol{\lambda}^* - \mathbf{y}^* = -\mathbf{d}$ and $(\boldsymbol{\lambda}^*, \mathbf{y}^*) = 0$. Using Eq. (10.8) and the definitions of $\mathbf{D}$ and $\mathbf{d}$, we can easily verify that $\mathbf{x}^*$ and $\mathbf{y}^*$ satisfy (10.7a). Hence by the Kuhn-Tucker theorem, $\mathbf{x}^*$ is the solution of the initial quadratic programming problem.

10.8 The method of Hildreth and D'Espo uses one-dimensional search techniques for solving the problem $\min_{\boldsymbol{\lambda} \geqslant 0}[\mathbf{d}^T\boldsymbol{\lambda} + \frac{1}{2}\boldsymbol{\lambda}^T\mathbf{D}\boldsymbol{\lambda}]$. Show that the sequence $\{\boldsymbol{\lambda}^k\}$ generated by this method converges to the desired minimum at $\boldsymbol{\lambda}^*$.

From Eq. (10.10) we observe that since $d_{ii} \neq 0$, $\hat{\lambda}_i$ is unique. Let $L[\boldsymbol{\lambda}] = \mathbf{d}^T\boldsymbol{\lambda} + \frac{1}{2}\boldsymbol{\lambda}^T\mathbf{D}\boldsymbol{\lambda}$. Then $L(\boldsymbol{\lambda})$ attains a minimum with respect to λ_i at either λ_i or zero when $\lambda_j = \lambda_j^k$, $j = 1, \dots, m$ $(j \neq i)$. It follows that $L[\boldsymbol{\lambda}^{k+1}] \leqslant L[\boldsymbol{\lambda}^k]$. Since $\boldsymbol{\lambda}^{k+1}$ is obtained from $\boldsymbol{\lambda}^k$, we can denote this process by $\boldsymbol{\lambda}^{k+1} = \mathbf{T}(\boldsymbol{\lambda}^k)$, where $\mathbf{T}$ denotes an operator representing the search technique. Then $L[\mathbf{T}(\boldsymbol{\lambda}^k)] \leqslant L[\boldsymbol{\lambda}^k]$, and furthermore $L[\mathbf{T}(\boldsymbol{\lambda})] = L[\boldsymbol{\lambda}]$ if and only if $\boldsymbol{\lambda} = \boldsymbol{\lambda}^*$. If we define a bounded region $\Lambda = \{\boldsymbol{\lambda} | \boldsymbol{\lambda} \geqslant 0, L[\boldsymbol{\lambda}] \leqslant L[\boldsymbol{\lambda}^0]\}$, then $\boldsymbol{\lambda}^k \in \Lambda$ for all $k > 0$ and hence the set $\{\boldsymbol{\lambda}^k\}$ has a limit point, denoted by $\bar{\boldsymbol{\lambda}}$. Since $L[\boldsymbol{\lambda}]$ is continuous in $\boldsymbol{\lambda}$, $L[\bar{\boldsymbol{\lambda}}]$ is a limit point of $L[\boldsymbol{\lambda}^k]$; and since $\mathbf{T}(\boldsymbol{\lambda})$ is continuous in $\boldsymbol{\lambda}$, $\mathbf{T}(\bar{\boldsymbol{\lambda}})$ is a limit point of $\mathbf{T}(\boldsymbol{\lambda}^k)$. Consequently $L[\mathbf{T}(\bar{\boldsymbol{\lambda}})]$ is a limit point of $L[\mathbf{T}(\boldsymbol{\lambda}^k)]$. However, $L[\mathbf{T}(\boldsymbol{\lambda}^k)]$ is the same sequence as $L[\boldsymbol{\lambda}^k]$. Since $L[\boldsymbol{\lambda}^k]$ is monotone, we have $L[\mathbf{T}(\bar{\boldsymbol{\lambda}})] = L[\boldsymbol{\lambda}]$ and hence $\bar{\boldsymbol{\lambda}} = \boldsymbol{\lambda}^*$.

10.9 (Dual problem). Let $F(\mathbf{x}, \boldsymbol{\lambda}) = -[\frac{1}{2}\mathbf{x}^T\mathbf{A}\mathbf{x} + \boldsymbol{\lambda}^T\mathbf{c}]$, and define $N = \{(\mathbf{x}, \lambda) | \mathbf{A}\mathbf{x} + \mathbf{B}^T\boldsymbol{\lambda} = -\mathbf{b}, \boldsymbol{\lambda} \geqslant \mathbf{0}\}$. Then $\max_{(\mathbf{x},\boldsymbol{\lambda}) \in N} F(\mathbf{x}, \boldsymbol{\lambda})$ is called the dual of $\min_{\mathbf{x} \in M} f(\mathbf{x})$, where $f(\mathbf{x})$ is given by (10.4) and M is defined by (10.3). Suppose $\mathbf{x} \in M$ is such that there is some $\boldsymbol{\lambda} \geqslant 0$ satisfying $(\mathbf{x}, \boldsymbol{\lambda}) \in N$. Show that $f(\mathbf{x}) \geqslant F(\mathbf{x}, \boldsymbol{\lambda})$.

Since $(\mathbf{x}, \boldsymbol{\lambda}) \in N$, we have $\mathbf{A}\mathbf{x} + \mathbf{B}^T\boldsymbol{\lambda} = -\mathbf{b}$. Premultiplying this equation by $\mathbf{x}^T$, transposing and rearranging the result, we obtain $\boldsymbol{\lambda}^T\mathbf{B}\mathbf{x} = -[\mathbf{b}^T\mathbf{x} + \mathbf{x}^T\mathbf{A}\mathbf{x}]$. Since $\mathbf{x} \in M$, we have $\mathbf{B}\mathbf{x} \leqslant \mathbf{c}$. Then noting that $\boldsymbol{\lambda} \geqslant \mathbf{0}$, we can write $\boldsymbol{\lambda}^T\mathbf{B}\mathbf{x} \leqslant \boldsymbol{\lambda}^T\mathbf{c}$. Hence $-[\mathbf{b}^T\mathbf{x} + \mathbf{x}^T\mathbf{A}\mathbf{x}] \leqslant \boldsymbol{\lambda}^T\mathbf{c}$, or $[\frac{1}{2}\mathbf{x}^T\mathbf{A}\mathbf{x} + \mathbf{b}^T\mathbf{x}] \geqslant -[\frac{1}{2}\mathbf{x}^T\mathbf{A}\mathbf{x} + \boldsymbol{\lambda}^T\mathbf{c}]$, i.e., $f(\mathbf{x}) \geqslant F(\mathbf{x}, \boldsymbol{\lambda})$.

10.10 (Duality theorem). Show that if the primal problem has a solution $\mathbf{x}^* \in M$, then its dual has a solution $(\mathbf{x}^*, \boldsymbol{\lambda}^*) \in N$ and $f(\mathbf{x}^*) = F(\mathbf{x}^*, \boldsymbol{\lambda}^*)$.

Since $\mathbf{x}^* \in M$ is a solution of the primal minimization problem, we know from the Kuhn-Tucker conditions that there are some $\boldsymbol{\lambda}^*, \mathbf{y}^* \geqslant 0$ satisfying (10.7a–c); and since $\mathbf{x}^*, \boldsymbol{\lambda}^*$ satisfy (10.7b), we have $(\mathbf{x}^*, \boldsymbol{\lambda}^*) \in N$. Now using (10.7a), (10.7c), and arguments similar to those in Problem 10.9, we can eaily obtain $f(\mathbf{x}^*) = F(\mathbf{x}^*, \boldsymbol{\lambda}^*)$. Assuming there is at least one $(\mathbf{x}, \boldsymbol{\lambda}) \in N$ satisfying $F(\mathbf{x}, \boldsymbol{\lambda}) > F(\mathbf{x}^*, \boldsymbol{\lambda}^*)$, and using the expression for $F(\mathbf{x}, \boldsymbol{\lambda})$ given in Problem 10.9 and the positive definiteness of $\mathbf{A}$, we obtain $\mathbf{c}^T(\boldsymbol{\lambda}^* - \boldsymbol{\lambda}) > 12[\mathbf{x}^T\mathbf{A}\mathbf{x} - (\mathbf{x}^*)^T\mathbf{A}\mathbf{x}^*] > (\mathbf{x}^*)^T\mathbf{A}(\mathbf{x} - \mathbf{x}^*)$.

Since $(\mathbf{x}, \boldsymbol{\lambda}) \in N$, we have $\mathbf{A}\mathbf{x} + \mathbf{B}^T\boldsymbol{\lambda} = -\mathbf{b}$. Combining this equation with (10.7b), we obtain $\mathbf{A}(\mathbf{x} - \mathbf{x}^*) = \mathbf{B}^T(\boldsymbol{\lambda}^* - \boldsymbol{\lambda})$ and hence $\mathbf{c}^T(\boldsymbol{\lambda}^* - \boldsymbol{\lambda}) > (\mathbf{x}^*)^T\mathbf{B}^T(\boldsymbol{\lambda}^* - \boldsymbol{\lambda})$. Applying

(10.7a) and (10.7c) to this inequality, we obtain $(\mathbf{y}^*)^T\boldsymbol{\lambda}<0$. Since $\mathbf{y}^*\geqslant 0$, at least one component of $\boldsymbol{\lambda}$ must be negative. However, this contradicts our assumption that $(\mathbf{x},\boldsymbol{\lambda})\in N$; hence $F(\mathbf{x}^*,\boldsymbol{\lambda}^*)=\max_{(\mathbf{x},\boldsymbol{\lambda})\in N}F(\mathbf{x},\boldsymbol{\lambda})$.

10.11 Show that if $(\mathbf{x}^*,\boldsymbol{\lambda}^*)\in N$ is a solution of the dual problem and $\mathbf{A}$ is positive definite, then $\mathbf{x}^*$ is a solution of the primal problem.

We form the Lagrangian of the dual problem, which is $-(\frac{1}{2}\mathbf{x}^T\mathbf{A}\mathbf{x}+\mathbf{c}^T\boldsymbol{\lambda})+\mathbf{p}^T(\mathbf{A}\mathbf{x}+\mathbf{B}^T\boldsymbol{\lambda}+\mathbf{b})+\mathbf{q}^T\boldsymbol{\lambda}$, where $\mathbf{p}$ and $\mathbf{q}$ are vectors of Lagrange multipliers. Then the Kuhn-Tucker conditions for the dual problem become

$$-\mathbf{A}\mathbf{x}^*+\mathbf{A}\mathbf{p}^*=0, \tag{10.33a}$$

$$\mathbf{B}\mathbf{p}^*+\mathbf{q}^*=\mathbf{c}, \tag{10.33b}$$

$$\mathbf{A}\mathbf{x}^*+\mathbf{B}^T\boldsymbol{\lambda}^*=-\mathbf{b}, \tag{10.33c}$$

$$(\boldsymbol{\lambda}^*,\mathbf{q}^*)=0, \tag{10.33d}$$

$$\boldsymbol{\lambda}^*,\mathbf{q}^*\geqslant \mathbf{0}. \tag{10.33e}$$

Since $\mathbf{A}$ is positive definite, from (10.33a) we have $\mathbf{x}^*=\mathbf{p}^*$. Substituting $\mathbf{x}^*$ for $\mathbf{p}^*$ in (10.33b) and renaming $\mathbf{q}^*=\mathbf{y}^*$, we observe that (10.33b)–(10.33e) are identical to (10.7a)–(d). Hence by the Kuhn-Tucker theorem, $\mathbf{x}^*$ is a solution to the primal problem.

Supplementary Problems

10.12 Find the constrained minimum of $f(\mathbf{x})$ on M by the method of Theil and Van de Panne when the set M is defined by the inequalities $x_1,x_2\geqslant 0$, $x_1+x_2\leqslant 3$, $2x_1-x_2\leqslant 4$, and $x_2\leqslant 2$, and when the objective function is $f(\mathbf{x})=\frac{1}{2}\mathbf{x}^T\mathbf{A}\mathbf{x}+\mathbf{b}^T\mathbf{x}$, where

$$\mathbf{b}=[-15,-7]^T \quad\text{and}\quad \mathbf{A}=\begin{bmatrix}3&1\\1&1\end{bmatrix}.$$

10.13 Solve Problem 10.12 by the method of Hildreth and D'Espo.

10.14 Solve Problem 10.12 by the capacity method of Houthakker.

10.15 Solve Problem 10.12 by Wolfe's modified simplex method.

Chapter 11

Integer Programming

In the previous chapters we have discussed methods of optimization for both unconstrained and constrained problems where all the variables were assumed to be real-valued. In this chapter we consider a class of optimization problems where at least some of the variables are constrained to be integer-valued. When a variable is integer-valued it can no longer be changed by infinitesimal increments, and hence such concepts as continuity and differentiability can no longer be used.

The integer constraint can be applied to any optimization problem, and so we can have *linear integer programs* and *nonlinear integer programs*. If some of the variables are allowed to have noninteger values, then we have a *mixed integer program*. Since there are relatively few methods for obtaining integer solutions to nonlinear programming problems, we confine our attention in this chapter to integer and mixed integer linear programs. Several well-known and practical problems of this type are formulated in the following examples. Methods for their solution are developed later in the chapter.

EXAMPLES

11.1 (Knapsack problem) Let $S=\{1,\ldots,M\}$ denote a set of objects, where the ith object weighs w_i units and has a value of v_i units. The problem is to load a knapsack with objects from the set S so as to maximize the total value of its contents without exceeding a given upper bound on the total weight. We wish to formulate this problem as an integer programming problem.

Let W denote the given upper bound on the total weight, and define a variable x_i such that $x_i=1$ if the ith item is selected and $x_i=0$ otherwise. We can now formulate the problem as follows:

$$\max_x \sum_{i=1}^{M} x_i v_i$$

such that

$$\sum_{i=1}^{M} x_i w_i \leqslant W,$$

where $0 \leqslant x_i \leqslant 1$ and x_i is an integer for all i, $1 \leqslant i \leqslant M$. Since x_i is integer-valued for all i, $1 \leqslant i \leqslant M$, we have obtained an integer programming problem. Note that if x_i were real-valued for all i, $1 \leqslant i \leqslant M$, then we would have a standard linear programming problem.

11.2 (Traveling-salesman problem) Consider a traveling salesman who is expected to travel to $N+1$ cities starting from city 0. The problem is to find a round-trip tour of all $N+1$ cities that includes every city once and only once and also has the lowest cost. Formulate this problem as an integer programming problem.

Let c_{ij} denote the cost of traveling from city i to city j, and introduce a variable x_{ij} such that $x_{ij}=1$ if the salesman travels from city i to city j and $x_{ij}=0$ otherwise. Without loss of generality we assume that the salesman starts from city 0. Now we formulate the constraints to be satisfied by the integer-valued variables x_{ij}. Since each city on the round-trip tour must be visited once, we see that for every j there must be some $i \neq j$ such that $x_{ij}=1$ and $x_{pj}=0$, $p \neq i$. This condition is equivalent to the following constraint

$$\sum_{i=1}^{N-1} x_{ij} = 1 \qquad \text{for all} \quad j \neq 0, \quad i \neq j.$$

In a similar manner, noting that the traveling salesman must leave every city once, we have

$$\sum_{j=1}^{N} x_{ij} = 1 \qquad \text{for all} \quad i \neq N-1, \quad i \neq j.$$

These constraints on the variables x_{ij} ensure that the salesman starts from city 0 and ends his journey in city N. Thus instead of a round-trip tour we have an open tour. However, every round-trip tour problem can be converted to an open-tour problem. We introduce an artificial city, say city N, such that $c_{0N}=0$ and $c_{0j}=c_{Nj}$ for all j, which implies that the last city in the open tour is the same as the first city and we have a round-trip tour. The constraints we have introduced thus far ensure that the salesman will visit each city once and only once and will start and finish his journey in city 0. However, these constraints do not exclude the possibility of obtaining more than one nonoverlapping round-trip tour rather than a single round-trip tour of all $N+1$ cities. To prevent the occurrence of nonoverlapping round-trip tours, we associate with each city i a real number y_i, $0 \leqslant y_i \leqslant N$. Then we introduce the following constraints:

$$y_i - y_j + Nx_{ij} \leqslant N-1; \qquad 0 \leqslant i \leqslant N-1, \quad 1 \leqslant j \leqslant N, \quad i \neq j.$$

To see that this last set of constraints rules out any nonoverlapping round-trip tours, we consider such a subtour consisting of k cities for which $x_{ij}=1$ in the tour. Each arc in the tour with $x_{ij}=1$ satisfies the above inequality; and when we add these inequalities corresponding to the k arcs in the subtour, the differences $y_i - y_j$ cancel each other and we obtain $Nk \leqslant (N-1)k$, a contradiction. Thus the last set of constraints will rule out any subtours.

The traveling-salesman problem now becomes that of minimizing the cost of the tour given by

$$\sum_{i=0}^{N-1} \sum_{j=1}^{N} c_{ij} x_{ij},$$

where the variables x_{ij} are integer-valued and are required to satisfy the given constraints. There are approximately N^2 constraints, which makes the numerical solution of the traveling-salesman problem extremely difficult for large values of N.

11.3 (Assignment problem) Consider an electronics manufacturer whose company has an order for N different electronic devices. The company employs M persons, each of whom is capable of manufacturing every one of these devices. The problem is to minimize the time required to supply the customer. We wish to formulate this problem as an integer programming problem.

Let d_j denote the total number of devices of the jth type being ordered, let x_{ij} denote the number of devices of the jth type manufactured by the ith employee, and let t_{ij} denote the amount of time required by the ith employee to produce the jth device. Then the standard assignment problem is to minimize $\sum_{i=1}^{M} \sum_{j=1}^{N} x_{ij} t_{ij}$ subject to $\sum_{i=1}^{M} x_{ij} \geqslant d_j$, $j=1,\ldots,N$. Note that by definition the variables x_{ij} must be nonnegative integers for all i and j.

Truncation and Rounding

On the surface it seems possible to treat a problem of integer linear programming as one of ordinary linear programming and to solve it by the standard simplex method. However, to obtain an integer solution, we must

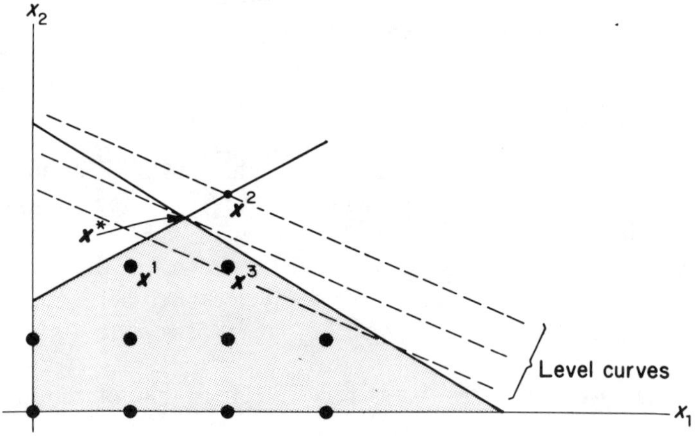

FIGURE 11.1

either truncate or round off the solution obtained by the simplex method. Unfortunately, this simple procedure may not provide an optimal integer solution, as is shown in Fig. 11.1.

In Fig. 11.1, the bold dots represent feasible solutions to an integer linear programming problem. The optimal solution of the standard linear program is denoted by $\mathbf{x}^*$. If $\mathbf{x}^*$ is truncated, we obtain $\mathbf{x}^1$. However, $\mathbf{x}^3$ is an integer feasible solution having a higher function value, and is in fact the optimal solution to the integer linear program. If $\mathbf{x}^*$ is rounded off, we obtain $\mathbf{x}^2$, which is not even a feasible solution. Hence these simple procedures of truncation and rounding are not recommended unless there is some simple way of testing the accuracy of the final solution. However, they are often applicable when the optimal solution will be insensitive to a change of plus or minus one, e.g., when the solution is a large number.

Branch-and-Bound Method

The *branch-and-bound algorithm* is particularly well suited to solving integer programming problems. This method utilizes the concept of a *tree of solutions,* defined as a *graph* containing all the nodes and branches of an integer program, where a *node* represents a subset of the set of all feasible solutions and a *branch* is a directed line segment joining any two nodes. In the next example we develop the tree of solutions for an assignment problem.

EXAMPLE 11.4 Consider three bank robbers planning a robbery where there are three jobs involved: (1) driving the getaway car, (2) collecting the money, and (3) providing cover. Each robber can be assigned to any one of the three jobs. Our problem is to construct a tree of solutions showing all possible assignments.

We introduce variables x_{ij}, where $x_{ij}=1$ if the ith robber is assigned the jth job, $1 \leqslant (i,j) \leqslant 3$, and $x_{ij}=0$ otherwise. The tree of solutions is shown in Fig. 11.2, where each node has an identifying number on it. The construction of the tree is described below.

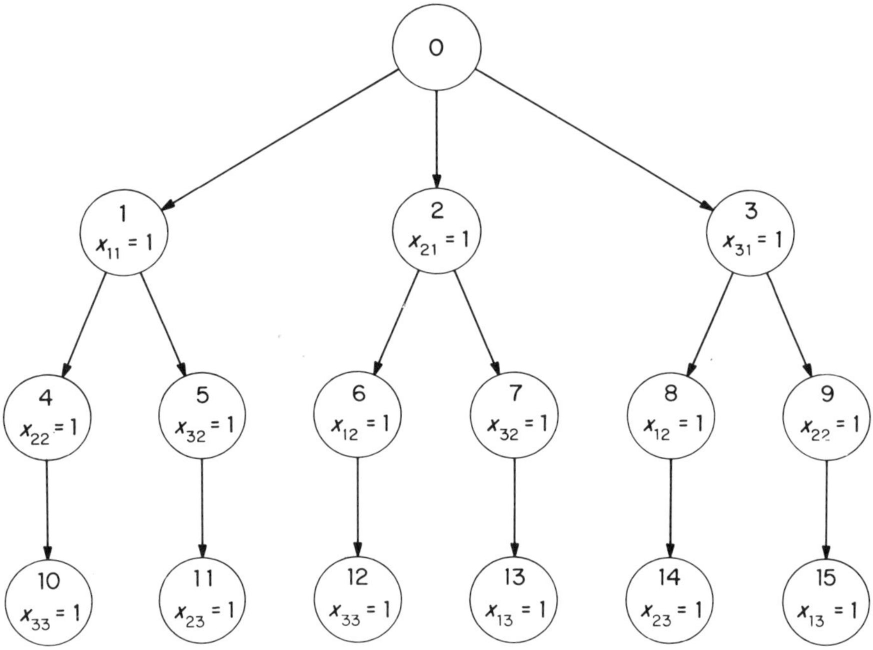

FIGURE 11.2

Node 0 is the set of all possible assignments. Node 1 is the set of all assignments where robber 1 is assigned to job 1. Similar interpretations are applicable to nodes 2 and 3. Node 4 is the set of all assignments where robber 1 is assigned to job 1 and robber 2 is assigned to job 2. Similar interpretations are applicable to nodes 5 through 9. Finally, the terminal node 10 is the unique assignment where robbers 1, 2, and 3 are assigned to jobs 1, 2, and 3 respectively. Similar interpretations are applicable to nodes 11 through 15.

Consider now a problem of function minimization where we can find the minimizing solution by using a tree of solutions as discussed in Example 11.4. One approach is to calculate the function value for each terminal node and then obtain the minimizing solution by direct inspection. However, this procedure requires us to find every possible sequence of branches ending in a terminal node. This can be a formidable task for problems involving a large number of variables. The branch-and-bound method simplifies this task by using a criterion for rejecting certain sequences of branches (paths) and thereby reducing the effort needed to locate the minimizing solution.

Consider an arbitrary node in the tree, say the kth node. Then find a feasible solution in the kth node and evaluate the objective function $f(\mathbf{x})$ at this point, say f_k. Now suppose node k is further subdivided into other nodes, including among them node p. Compute a lower bound of $f(\mathbf{x})$ in the set of all feasible solutions contained in node p, and denote this lower bound by f_p. The method of obtaining this lower bound depends on the problem under consideration. However, if $f_k < f_p$, then none of the paths in the tree of solutions originating from the pth node can lead to a lower function value than f_k. Hence all such paths in the tree can be removed from further considerations until we arrive at a stage where $f_q > f_p$ for all nodes q currently under consideration. This elimination results in an economy of search effort for the minimizing solution.

EXAMPLES

11.5 (Integer linear program) We assume that the set of all feasible solutions is defined by the constraints

$$-2x_1 + 3x_2 \leqslant 3,$$

$$x_1 + 2x_2 \leqslant 8,$$

where $x_1, x_2 \geqslant 0$. The problem is to maximize $x_1 + 4x_2$, where the optimal values of x_1 and x_2 are required to be integers.

The set of all feasible solutions is shown by the lightly shaded region in Fig. 11.3, where the bold dots indicate integer feasible solutions. Two level curves of the function $f(\mathbf{x}) = x_1 + 4x_2$ are also shown in the figure, and evidently $f(\mathbf{x})$ attains a maximum value of 12 at $x_1 = 4$, $x_2 = 2$. We denote the set of all feasible solutions (disregarding the integer constraint) as node 0 in the branch-and-bound method; and by using a simple graphical analysis we verify that $f(\mathbf{x})$ attains a maximum value of 13.42 at $x_1^* = 2.5$ and $x_2^* = 2.7$ (see Fig. 11.3). Since x_1^* and x_2^* are not integers, we proceed to subdivide node 0 into nodes 1 and 2 as shown in Fig. 11.4.

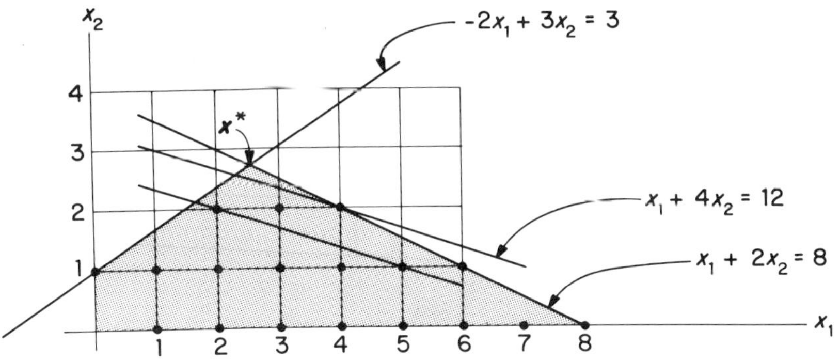

FIGURE 11.3

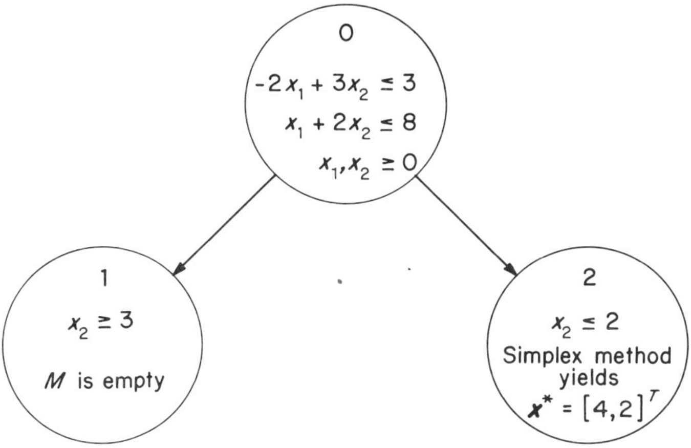

FIGURE 11.4

Since $x_2^* = 2.7$, we conclude that the optimal integer solution will satisfy either $x_2 \geqslant 3$ or $x_2 \leqslant 2$. So we generate node 1 by adding the constraint $x_2 \geqslant 3$ to those defining node 0, and from Fig. 11.3 we see that node 1 is an empty set, M, where M is the set of all feasible solutions.

We obtain node 2 by adding the constraint $x_2 \leqslant 2$ to those defining node 0, and so node 2 contains all feasible solutions in the darkly shaded area shown in Fig. 11.3. Graphically we verify that $f(\mathbf{x})$ attains a maximum value of 12 at $x_1 = 4$ and $x_2 = 2$.

The optimal solution obtained in node 2 is integer-valued, and further subdivision of node 2 can only result in lower function values. Hence we conclude that node 2 provides the maximizing solution to our problem. Notice that if we round off x_1^* and x_2^*, we obtain $x_1 = 3$, $x_2 = 3$, which is not a feasible solution. Truncation results in $x_1=2$, $x_2=2$ with a function value of 10, which is also not the optimal solution.

11.6 (Mixed integer linear program) Consider the problem stated in Example 11.5, where we now require that only x_1 be integer-valued. Solve this mixed integer linear programming problem by the branch-and-bound method.

As in the case of Example 11.5, we start with node 0 and obtain $x_1^*=2.58$ and $x_2^*=2.71$. Since only x_1 is required to have an integer value, we obtain nodes 1 and 2 in the following manner.

Node 1 is obtained by adding the constraint $x_1 \leqslant 2$ to those defining node 0. Graphically we verify that $f(\mathbf{x})$ attains a maximum value of 11.36 at $x_1=2$, $x_2=2.34$.

Node 2 is obtained by adding the constraint $x_1=3$ to those defining node 0. Graphically we verify that $f(\mathbf{x})$ attains a maximum value of 13 at $x_1=3$ and $x_2=2.5$.

Node 2 provides a higher function value than node 1, so the maximizing solution is given by $x_1=3$, $x_2=2.5$.

Cutting-Plane Methods

As shown in Fig. 11.1, the optimal solution to an integer linear program can very well be an interior point of the set of all feasible solutions. Hence algorithms like the simplex method, which explore only the extreme points of this set, will be unable to locate such a solution. One way to eliminate this difficulty is by cutting off sections of the initial set of feasible solutions with hyperplanes such that the optimal integer solution eventually becomes an extreme point of the new set of feasible solutions. The hyperplanes used for this purpose are called *cutting planes*. There are many ways of obtaining such cutting planes; we discuss only the method due to Gomory.

Consider now the general problem of linear programming, where in addition we require that the variables be integer-valued. To solve this integer linear programming problem, we first ignore the integer constraints and solve the linear programming problem by the simplex method. If the optimal solution is integer-valued, then we are finished; otherwise we introduce a new constraint hyperplane which shrinks the original set of feasible solutions.

In order to be effective in solving the integer linear program, the new constraint hyperplane should have two properties. First of all, it should not eliminate any integer-valued feasible solution of the original problem to obtain the new set of feasible solutions. Secondly, it should make the optimal solution of the original problem infeasible. We next solve the linear programming problem over this new set of feasible solutions by the simplex method, and if the new optimal solution is integer-valued, then we are finished; otherwise we introduce another constraint hyperplane and continue. We deal now with the question of how to choose these additional hyperplanes.

Consider the linear programming problem in the following form:

$$\max_{\mathbf{y}\in Y}(\mathbf{c},\mathbf{y}),$$

where

$$Y=\{\mathbf{y}|\mathbf{y}\in E^{m+n},\ \mathbf{B}\mathbf{y}=\mathbf{b},\ \mathbf{y}\geqslant\mathbf{0}\},$$

in which $\mathbf{b}\in E^m$, $\mathbf{b}\geqslant\mathbf{0}$, and $\mathbf{B}$ is an $m\times(n+m)$ matrix. The equations $\mathbf{B}\mathbf{y}=\mathbf{b}$ can also be written as

$$\begin{aligned} b_{11}y_1+\cdots+b_{m,m+n}y_{m+n}&=b_1,\\ \vdots\qquad\qquad\vdots\qquad&\quad\vdots\\ b_{m1}y_1+\cdots=b_{m,m+n}y_{m+n}&=b_m, \end{aligned}\tag{11.1}$$

a set of m linear algebraic equations in $m+n$ unknown $y_1,\ldots,y_{m+n}$. We can arbitrarily set any n of the $m+n$ variables equal to zero and solve (11.1) for the remaining m variables. Without loss of generality we assume that $y_{m+1}=\cdots=y_{m+n}=0$, and then, solving (11.1), we obtain unique values of $y_1,\ldots,y_m$ provided the vectors $\mathbf{b}^j=[b_{1j},\ldots,b_{mj}]^T$, $j=1,\ldots,m$, form a basis in E^m. If $y_j\geqslant 0$, $j=1,\ldots,m$, we call the corresponding solution of (11.1) a *basic feasible solution*, and y_j the *basic variables.* The problem is now to find a basic feasible solution of (11.1) which maximizes $(\mathbf{c},\mathbf{y})$.

Let $\mathbf{y}^*=[(\mathbf{y}^1)^T,(\mathbf{y}^2)^T]^T$ denote the optimal basic feasible solution obtained by the simplex method, where $\mathbf{y}^1$ contains the basic variables and $\mathbf{y}^2=\mathbf{0}$. We partition the matrix $\mathbf{B}$ as $\mathbf{B}=[\mathbf{B}_1|\mathbf{B}_2]$, where the column vectors of $\mathbf{B}_1$ correspond to the basic variables in $\mathbf{y}^1$. Since the column vectors of $\mathbf{B}_1$ form a basis in E^m, $\mathbf{B}_1$ has an inverse; and from $\mathbf{B}\mathbf{y}=\mathbf{b}$ we have $\mathbf{y}^1=\mathbf{B}_1^{-1}\mathbf{b}-\mathbf{B}_1^{-1}\mathbf{B}_2\mathbf{y}^2$. Now letting $\mathbf{v}^0=\mathbf{B}_1^{-1}\mathbf{b}$ and $\mathbf{v}^j=\mathbf{B}_1^{-1}\mathbf{b}^j$, where $\mathbf{b}^j$ denotes the jth column vector of $\mathbf{B}_2$, we get

$$\mathbf{y}^1=\mathbf{v}^0-\sum_{j=1}^{m}y_j^2v^j.\tag{11.2}$$

If every component of $\mathbf{y}^1$ is an integer, then we have an optimal solution. So we assume y_k^1 is not an integer; then from (11.2),

$$y_k^1 = v_k^0 - \sum_{j=1}^{m} y_j^2 v_k^j. \tag{11.3}$$

Letting w_k^j and θ_k^j denote the integer and fractional parts respectively of v_k^j so that $w_k^j \geqslant 0$ and $0 < \theta_k^j < 1$, Eq. (11.3) can be rewritten as

$$y_k^1 = \left(w_k^0 - \sum_{j=1}^{m} y_j^2 w_k^j \right) + \left(\theta_k^0 - \sum_{j=1}^{m} y_j^2 \theta_k^j \right). \tag{11.4}$$

Now if we want y_k^1 to be an integer, then from (11.4) we see that $y_j^2 > 0$ for at least one j. Since $\sum_{j=1}^{m} y_j^2 \theta_k^j \geqslant 0$ and $0 < \theta_k^0 < 1$, we introduce the constraint

$$\theta_k^0 - \sum_{j=1}^{m} y_j^2 \theta_k^j \leqslant 0 \tag{11.5}$$

in order to obtain an integer value for y_k^1. When Eq. (11.5) is active, it defines the hyperplane which we introduce as a new constraint in the original linear programming problem. This constraint is often referred to as a *Gomory cut* after the originator of this method. Since the original optimal solution was $y_j^2 = 0$ for all j, this solution does not satisfy the constraint (11.5) and so becomes infeasible when the new constraint is added. It is clear from our derivation that all integer-valued feasible solutions to the original problem will also satisfy (11.5).

EXAMPLE 11.7 Consider the problem of integer linear programming where the set of all feasible solutions is defined by the constraints $x_1, x_2 \geqslant 0$, $x_2 \leqslant 3$, $2x_1 + x_2 \leqslant 6$, and x_1, x_2 are integer-valued. We want to find the optimal solution by the cutting-plane method, where the objective function is $f(x_1, x_2) = x_1 + 2x_2$.

The feasible solutions are shown by the bold dots in Fig. 11.5, which also shows the level curves of the given objective function. Graphically we see that the optimal solution is $\mathbf{x}^* = [1, 3]^T$. Applying the cutting-plane technique to this problem, and using u_1 and u_2 as slack variables, the inequality constraints can be written as

$$0x_1 + x_2 + u_1 = 3,$$

$$2x_1 + x_2 + u_2 = 6,$$

where $x_1, x_2, u_1, u_2 \geqslant 0$. Letting $\mathbf{b}=[3,6]^T$, $\mathbf{y}=[x_1,x_2,u_1,u_2]^T$, and

$$\mathbf{B}=\begin{bmatrix} 0 & 1 & 1 & 0 \\ 2 & 1 & 0 & 1 \end{bmatrix},$$

and using the simplex method (ignoring integer constraints), we obtain the optimal basic feasible solution $\mathbf{y}^*=[1.5,3,0,0]^T$. In this case the basic variables are x_1 and x_2, $\mathbf{y}^1=[1.5,3]^T$, and the basis vectors are $\mathbf{b}^1=[0,2]^T$ and $\mathbf{b}^2=[1,1]^T$. Hence

$$\mathbf{B}_1=\begin{bmatrix} 0 & 1 \\ 2 & 1 \end{bmatrix} \quad \text{and} \quad \mathbf{B}_1^{-1}=\begin{bmatrix} -\frac{1}{2} & \frac{1}{2} \\ 1 & 0 \end{bmatrix}.$$

By direct computation we obtain $\mathbf{v}^0=\mathbf{B}_1^{-1}\mathbf{b}=[1.5,3]^T$, $\mathbf{v}^3=\mathbf{B}_1^{-1}\mathbf{b}^3=[-\frac{1}{2},1]^T$, and $\mathbf{v}^4=\mathbf{B}_1^{-1}\mathbf{b}^4=[\frac{1}{2},0]^T$. Thus (11.2) becomes

$$\begin{bmatrix} y_1^1 \\ y_2^1 \end{bmatrix}=\begin{bmatrix} 1.5 \\ 3 \end{bmatrix}-y_3^2\begin{bmatrix} -\frac{1}{2} \\ 1 \end{bmatrix}-y_4^2\begin{bmatrix} \frac{1}{2} \\ 0 \end{bmatrix}.$$

Since y_1^1 is not an integer,, we set $k=1$, and (11.4)–(11.5) become

$$\begin{aligned} &y_1^1=1+\tfrac{1}{2}(1+y_3^2-y_4^2), \\ &1+y_3^2-y_4^2 \leqslant 0, \end{aligned} \tag{11.6}$$

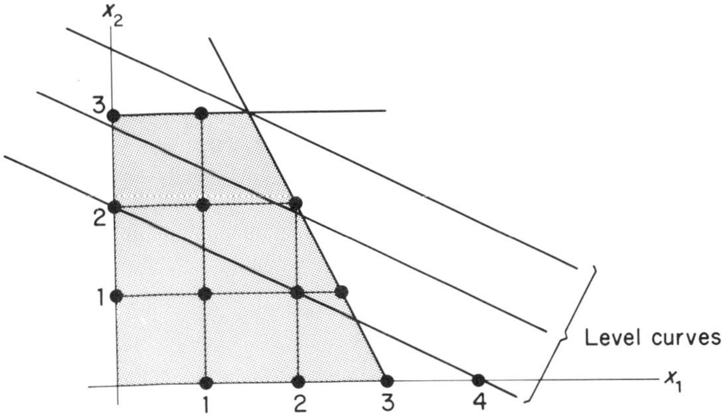

FIGURE 11.5

where y_3^2 and y_4^2 correspond to the slack variables in the vector $\mathbf{y}$; hence $y_3^2 = 3 - x_2$ and $y_4^2 = 6 - 2x_1 - x_2$. Using these expressions for y_3^2 and y_4^2 in (11.6), we obtain the new constraint $x_1 \leqslant 1$. The hyperplane $x_1 = 1$ cuts off the portion of the feasible region shown by the shaded area in Fig. 11.6. Clearly when $x_1 \leqslant 1$ is introduced as a new constraint, the old constraint $2x_1 + x_2 \leqslant 6$ is redundant.

The new linear programming problem is now $\max f(x_1, x_2) = x_1 + 2x_2$ on the set given by $0 \leqslant x_1 \leqslant 1$ and $0 \leqslant x_2 \leqslant 3$. Using the simplex method, we obtain the optimum solution of this problem as $\mathbf{x}^* = [1, 3]^T$, which is also the optimal solution of the integer linear programming problem.

Solved Problems

11.1 (Assignment problem) Consider the assignment problem introduced in Example 11.1 and suppose the efficiency of the jth robber in performing the ith job is given by e_{ij}. The problem is now to find a job assignment that maximizes the total efficiency.

Suppose the efficiency of a single job is measured on a scale of zero to one. For convenience, we arrange the numbers e_{ij} in the form of a matrix

$$\mathbf{E} = \begin{bmatrix} 0.7 & 0.3 & 0.4 \\ 0.5 & 0.3 & 0.6 \\ 0.2 & 0.6 & 0.4 \end{bmatrix}.$$

When robber j is assigned to job i, the ith row and the jth column of the matrix E are deleted, and we can now define the sets

$$I_k = \{ i | \text{row } i \text{ of } E \text{ is not deleted at node } k \},$$

$$J_k = \{ j | \text{column } j \text{ of } E \text{ is not deleted at node } k \},$$

$$B_k = \{ (i,j) | \text{row } i \text{ and column } j \text{ of } E \text{ are deleted at node } k \}.$$

Then at node k, the set B_k denotes the partial assignment already made, and the efficiency of this assignment is given by $\sum_{i,j \in B_k} e_{ij}$. The efficiency of the remaining assignments can be bounced by the sum over jobs of the maximum efficiency with which each job can be done, and this is given by $\sum_{i \in I_k} [\max_{j \in J_k} e_{ij}]$. Hence an upper bound for the total efficiency is given by

$$e_k = \sum_{(i,j) \in B_k} e_{ij} + \sum_{i \in I_k} \left[\max_{j \in J_k} e_{ij} \right]. \tag{11.7}$$

Using Fig. 11.2 as a guide to the solution of this problem by the branch-and-bound method, we can first obtain a lower bound on the maximum value of the total efficiency. This bound is computed from the assignment $x_{11} = x_{22} = x_{33} = 1$, and we have $e = 0.7 + 0.3 + 0.4 = 1.4$.

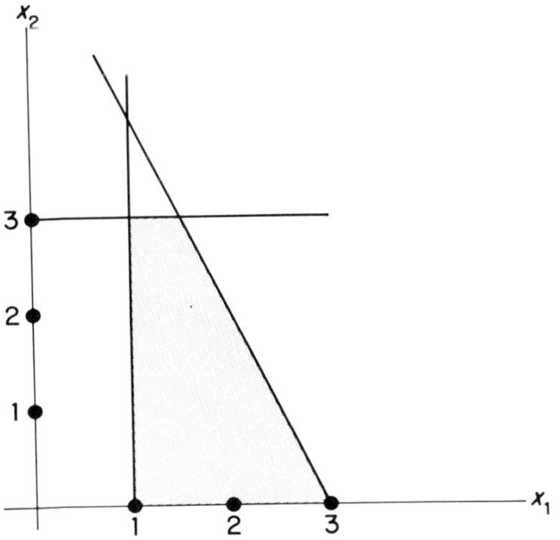

FIGURE 11.6

At node 1 in Fig. 11.2, the upper bound on the total efficiency is obtained from (11.7) as $e_1 = 0.7 + \max(0.3, 0.6) + \max(0.6, 0.4) = 1.9$. Since e_1 is greater than e, we subdivide node 1 into nodes 4 and 5; and by (11.7), $e_4 = 0.7 + 0.3 + \max(0.4) = 1.4$. Since e_4 is not greater than e, we return to node 1 and proceed to node 5. By (11.7), $e_5 = 0.7 + 0.6 + \max(0.6) = 1.9$; but since e_5 is greater than e, we subdivide node 5 to obtain node 11. By (11.7), $e_{11} = 1.9$; and since node 11 is a terminal node, the total efficiency will be at least 1.9 with the assignment $x_{11} = x_{32} = x_{23} = 1$, and so we set $e = 1.9$.

Now at node 2 we have $e_2 = 0.5 + \max(0.3, 0.4) + \max(0.6, 0.4) = 0.5 + 0.4 + 0.6 = 1.5$. Since e_2 is less than $e = 1.9$, we need not subdivide node 2 any further.

At node 3, using (11.7), we have $e_3 = 0.2 + \max(0.3, 0.4) + \max(0.3, 0.6) = 0.2 + 0.4 + 0.6 = 1.2$; and since e_3 is less than $e = 1.9$, we need not subdivide node 3 any further. Since we have now examined all possible subdivisions of node 0, we conclude that the optimal assignment is $x_{11} = x_{32} = x_{23} = 1$ with a maximum total efficiency of 1.9 units. This tree of solutions is shown in Fig. 11.7.

11.2 Consider the problem of transporting 750 automobiles from Japan to the United States, and assume that two types of ships can be used having the following characteristics:

	TYPE 1	TYPE 2
No. of autos/ship	200	100
Fuel (gallons/trip)	1200	700
Size of crew	25	10

The total amount of fuel available is 5500 gallons, and the maximum number of sailors available is 90. The profit from each ship of type 1 is \$2000 and from each ship of type 2 is \$1000. Assuming the supply of ships is unlimited, find the number of ships of each type which should be used in order to maximize profit.

Let x_1 and x_2 denote the number of ships of types 1 and 2 respectively, where clearly x_1 and x_2 must be integer-valued. The above constraints can be written as

$$200x_1 + 100x_2 \geqslant 750,$$

$$1200x_1 + 700x_2 \leqslant 5500,$$

$$25x_1 + 10x_2 \leqslant 90,$$

$$x_1, x_2 = 0, 1, 2, \ldots,$$

and the objective function to be maximized is $f(x_1, x_2) = 2000x_1 + 1000x_2$. Allowing x_1 and x_2 to be real-valued and solving the corresponding linear program by the simplex method, we have $x_1 = 1.45$, $x_2 = 5.4$ with an optimal profit of \$8300. Figure 11.8 depicts the tree of solutions, where this case is denoted by node 0. We now subdivide node 0 into nodes 1 and 2 by considering the cases $x_1 \leqslant 1$ and $x_1 \geqslant 2$ respectively. For node 1 the optimal solution by the simplex method is $x_1 = 1$, $x_2 = 6.15$ with a profit of \$8150. For node 2 the optimal solution is given by $x_1 = 2$,

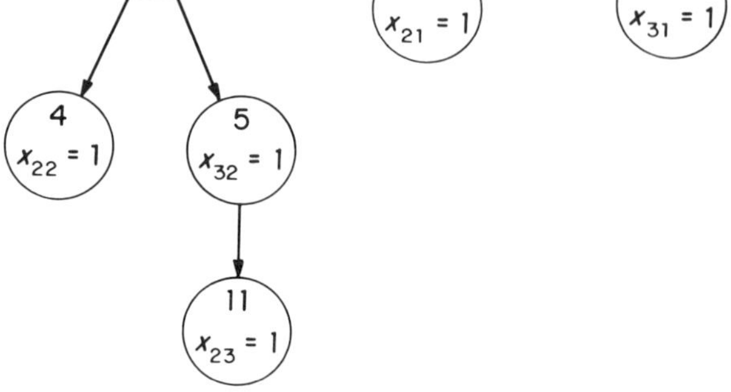

FIGURE 11.7

$x_2=4$ with a profit of \$8000. Since x_1 and x_2 are integer-valued, we need not subdivide node 2 any further.

The profit obtained at node 1 is greater than that at node 2, and hence we subdivide node 1 into nodes 3 and 4 as shown in Fig. 11.8. At node 4 the optimum solution is $x_1=1$, $x_2=6$ with a profit of \$8000, and at node 5 the optimum solution is $x_1=0.5$, $x_2=7$ with a profit of \$8000. Further subdivisions of nodes 4 and 5 can yield only lower profits, and hence we obtain two optimum solutions: either $x_1^*=2$, $x_2^*=4$ or $x_1^*=1$, $x_2^*=6$; in either case the profit is \$8000.

11.3 (Knapsack problem) Consider a set of five indivisible items, each having the weight and value units given below:

Item No.	Weight	Value
1	18	27
2	12	9
3	15	30
4	16	16
5	13	6.5

The problem is to fill a knapsack with these items so as to maximize its total value without exceeding a total weight of 45 units.

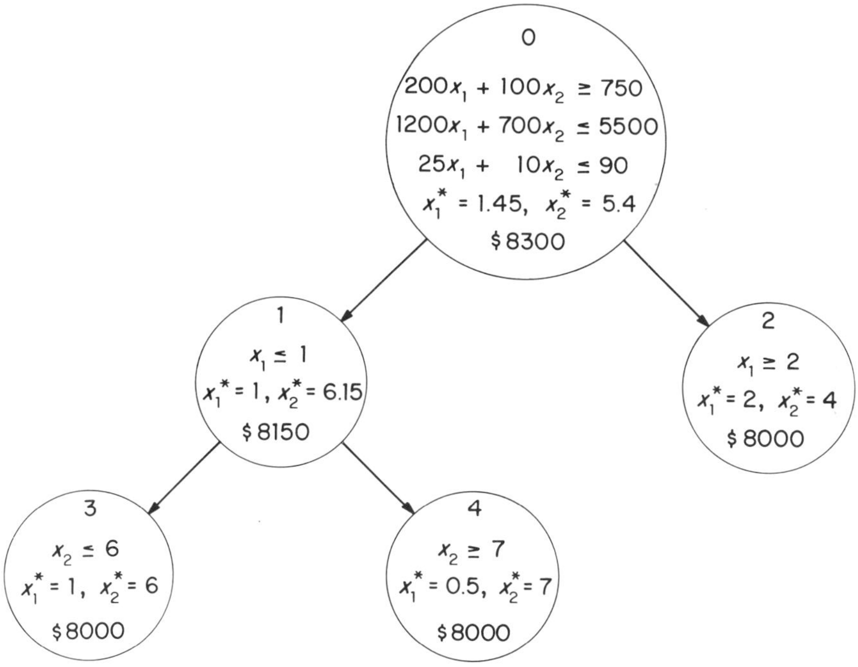

FIGURE 11.8

We first arrange these items in terms of their value per unit weight as shown below:

INDEX NO.	ITEM NO.	WEIGHT	VALUE	VALUE/WEIGHT
1	3	15	30	2
2	1	18	27	1.5
3	4	16	16	1
4	2	12	9	0.75
5	5	13	6.5	0.5

If the items were divisible, then we could fill the knapsack with item number 3, which has the largest value per unit weight, until the weight of the knapsack reached its allowable limit. If this weight limit were not reached, then we should start putting item number 1 into the knapsack, since it has the second largest value per unit weight. If we continued with this procedure, the knapsack contents having the maximum value would comprise all of items 3 and 1 plus $\frac{3}{4}$ of item 4. The total weight and value of this knapsack would be $15+18+\frac{3}{4}(16)=45$ units and $30+27+\frac{3}{4}(16)=69$ units respectively.

Since the items are indivisible, the above solution is not valid. However, we can proceed to take the items from the top of the above list until no more undivided items can be taken without exceeding the weight limit. This procedure results in taking all of items 3 and 1, with a total weight and value of 33 and 57 units respectively. Clearly this solution is not optimal, since a combination of all of items 3, 1, and 2 results in a total weight and value of 45 and 66 units respectively. Note also that the value of any combination of indivisible items with a total weight limit can always be increased if division of items is allowed.

Figure 11.9 shows the tree of solutions of this problem as created by the branch-and-bound method. In this figure, a node marked $(i,j,\bar{k})$ represents all possible combinations of items which include indices i and j and exclude index k. The upper bound V of the total value is computed by allowing divisibility of items.

Node 1 contains all possible combinations of items that exclude index 1, and the upper bound V for node 1 is computed as shown below:

INDEX NO.	ITEM NO.	WEIGHT	VALUE
2	1	18	27
3	4	16	16
4	2	12	9

Now if we take all of items numbered 1 and 4 and $\frac{11}{12}$ of item number 2, then the total weight of the knapsack becomes $18+16+\frac{11}{12}(12)=45$ units and the corresponding value $V=27+16+\frac{11}{12}(9)=51.25$ units.

On the other hand, node 2 represents all possible combinations of items that include index 1, and the upper bound V for node 2 is computed as shown below:

INDEX NO.	ITEM NO.	WEIGHT	VALUE
1	3	15	30
2	1	18	27
3	4	16	16

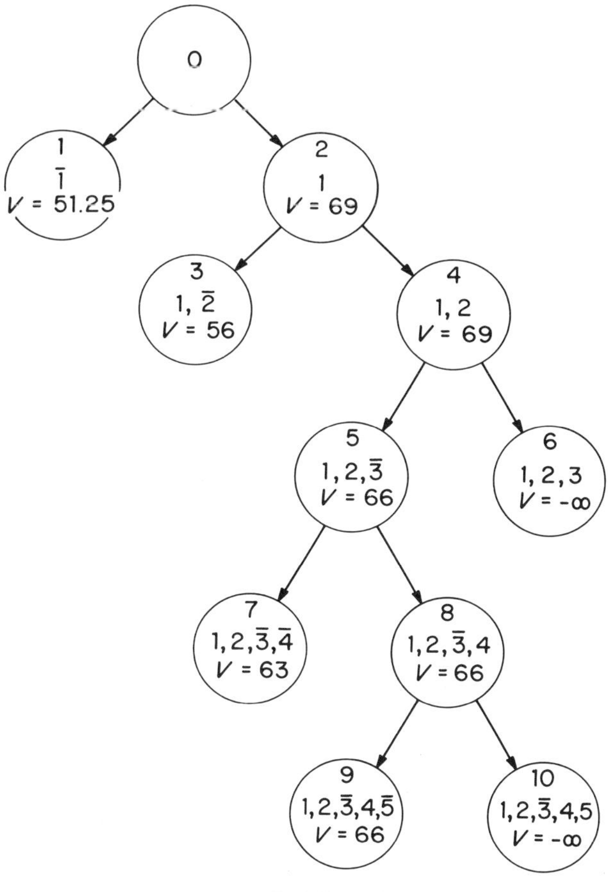

FIGURE 11.9

If we take all of items 3 and 1 and $\frac{3}{4}$ of item 4, then the total weight of the knapsack is $15+18+\frac{3}{4}(16)=45$ units and the corresponding value $V=30+27+\frac{3}{4}(16)=69$ units.

Since the upper bound on the total value at node 2 is larger than that at node 1, we subdivide node 2 into nodes 3 and 4. Similarly, since node 4 has a higher upper bound, we subdivide node 4 into nodes 5 and 6. Node 6 represents an inadmissible combination of items, since it violates the total-weight constraint. Thus we set $V=-\infty$ for node 6 and proceed in this manner until terminal nodes 9 and 10 are reached, where node 9 provides the optimum combination of the given items.

11.4 Consider the traveling-salesman problem shown in Figure. 11.10. In this figure, A, B, C, and D represent the four cities and the number next to the line joining each city pair represents the time required to travel between these two cities. Find the minimum-time round-trip tour.

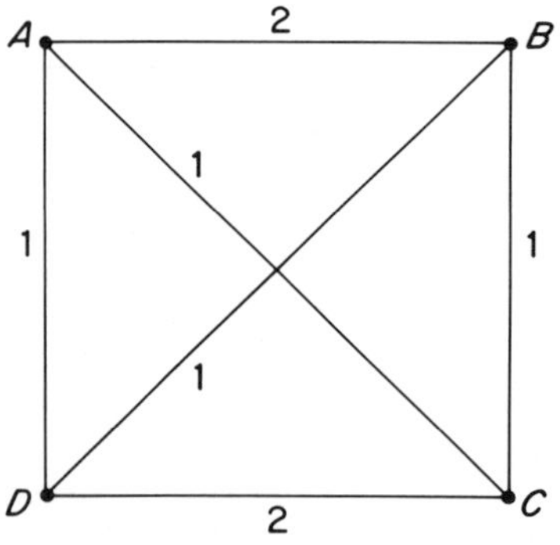

FIGURE 11.10

We represent the time data given in Fig. 11.10 in the form of a matrix $\mathbf{T}_1$ given below:

$$\mathbf{T}_1 = \begin{array}{c} \\ A \\ B \\ C \\ D \end{array} \begin{array}{c} \begin{array}{cccc} A & B & C & D \end{array} \\ \begin{bmatrix} 10^7 & 2 & 1 & 1 \\ 2 & 10^7 & 1 & 1 \\ 1 & 1 & 10^7 & 2 \\ 1 & 1 & 2 & 10^7 \end{bmatrix} \end{array}.$$

In the matrix $\mathbf{T}_1$, the element 2 in the Ath row and Bth column gives the time required to travel from city A to city B. The other elements in $\mathbf{T}_1$ are interpreted in a similar way. All diagonal elements in $\mathbf{T}_1$ are taken to be a large number, 10^7, in order to rule out obviously unnecessary travels such as from city B to city B.

Now if we subtract a constant from each element of the ith row of the matrix $\mathbf{T}_1$, then in effect we reduce the travel time from city i to each of the remaining cities by the value of that constant. We shall call this operation a row reduction. In a similar way we can define the operation of a column reduction. Evidently a row or column reduction does not alter the optimal round-trip tour, although it reduces the travel time by the amount used for the purpose of reducing the matrix $\mathbf{T}_1$.

We now reduce each row of $\mathbf{T}_1$ by unity, so that the total amount of reduction achieved is 4. The reduced matrix $\mathbf{T}_2$ is

$$\mathbf{T}_2 = \begin{array}{c} \\ A \\ B \\ C \\ D \end{array} \begin{array}{c} \begin{array}{cccc} A & B & C & D \end{array} \\ \begin{bmatrix} 10^7 & 1 & 0 & 0 \\ 1 & 10^7 & 0 & 0 \\ 0 & 0 & 10^7 & 1 \\ 0 & 0 & 1 & 10^7 \end{bmatrix} \end{array}.$$

Now interpreting the elements of $\mathbf{T}_2$ like those of $\mathbf{T}_1$, we note from the Ath row that the travel from city A to either city C or city D takes less time than that from city A to city B. Similar conclusions can be drawn from the other rows of $\mathbf{T}_2$. Since we have at least one zero in each row and each column of $\mathbf{T}_2$, we can form a round-trip tour taking zero time, and one such tour is A, C, B, D, A. Since the total time for a round-trip tour is a nonnegative quantity, the tour A, C, B, D, A is optimal. However, the real value of the minimum time is not zero but equals the total amount of reduction used, in this case 4.

In some cases, the pattern of zeros produced in the reduced matrix by the reduction procedure does not allow the formation of a zero-time round-trip tour. In such cases, the simple reduction procedure does not yield the optimal solution.

11.5 Consider the traveling-salesman problem where the matrix $\mathbf{C}_1$ represents the costs of traveling between cities:

$$\mathbf{C}_1 = \begin{array}{c} \\ A \\ B \\ C \\ D \end{array} \begin{array}{c} \begin{array}{cccc} A & B & C & D \end{array} \\ \begin{bmatrix} 10^7 & 5 & 3 & 3 \\ 4 & 10^7 & 6 & 1 \\ 2 & 5 & 10^7 & 2 \\ 1 & 3 & 4 & 10^7 \end{bmatrix} \end{array}.$$

Find the minimum-cost round-trip tour.

First we reduce the rows A, B, C, D by amounts 2, 1, 2, 1 and reduce the columns B and C by amounts 2 and 1 respectively. After a total reduction of $2+1+2+1+2+1=9$ units, the reduced matrix is:

$$\mathbf{C}_2 = \begin{array}{c} \\ A \\ B \\ C \\ D \end{array} \begin{array}{c} \begin{array}{cccc} A & B & C & D \end{array} \\ \begin{bmatrix} 10^7 & 1 & 0 & 1 \\ 3 & 10^7 & 4 & 0 \\ 0 & 1 & 10^7 & 0 \\ 0 & 0 & 2 & 10^7 \end{bmatrix} \end{array}.$$

Although in $\mathbf{C}_2$ each row and each column has at least one zero, still we are unable to form a zero-cost round-trip tour. So the cost of the minimum-cost round-trip tour must be greater than 9. In Fig. 11.11 we form node 0 consisting of all tours with a lower bound (L.B.) on the cost given by 9 units.

We now form node 1 by excluding the path AC from consideration, and this is denoted in the figure by $\overline{AC}$. Since the path AC is excluded, we must go from city A to some city other than C; and from $\mathbf{C}_2$, the minimum cost of such a travel is 1 unit from city A to city B. Also we must arrive at city C from some city other than A; and from $\mathbf{C}_2$, the minimum cost of such an arrival is 2 units. Thus we must pay a total of 3 units for excluding AC, and the lower bound in node 1 becomes L.B. $=9+3=12$.

We now form node 2 by considering all tours in node 0 which include the path AC. Since AC is included in our tour, we can eliminate row A and column C from $\mathbf{C}_2$, obtaining $\mathbf{C}_3$:

$$\mathbf{C}_3 = \begin{array}{c} \\ A \\ B \\ C \\ D \end{array} \begin{array}{c} \begin{array}{cccc} A & B & C & D \end{array} \\ \begin{bmatrix} - & - & - & - \\ 3 & 10^7 & - & 0 \\ 0 & 1 & - & 0 \\ 0 & 0 & - & 10^7 \end{bmatrix} \end{array}.$$

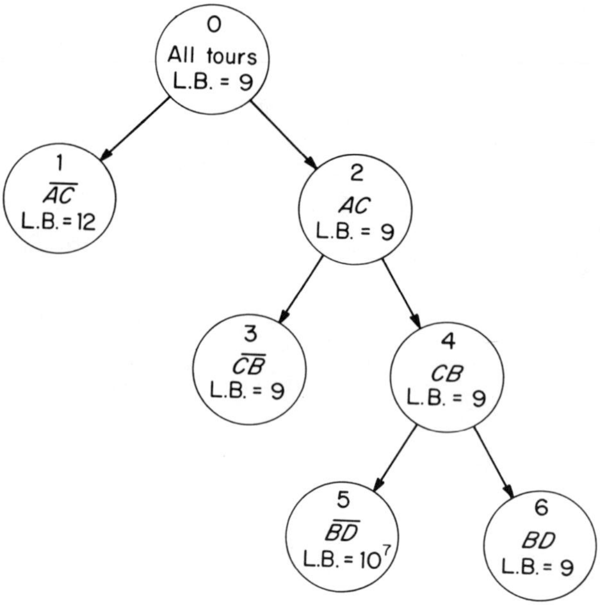

FIGURE 11.11

In $\mathbf{C}_3$ each row and each column have at least one zero, and hence row and column reductions are not necessary. So we can say that $\mathbf{C}_3$ is reduced by a total amount of 0 units, and consequently the lower bound on costs of all tours in node 2 becomes L.B. $=9+0=9$.

We repeat this procedure by dividing node 2 into nodes 3 and 4 as shown in Fig. 11.11, and the corresponding lower bounds are computed to be 9. Nodes 3 and 4 have the same lower bounds, and we can branch off from either one of these nodes. Branching off node 4 as shown in the figure, we obtain nodes 5 and 6 with the indicated lower bounds. Since node 5 has a higher lower bound than node 6, we terminate at node 6. The optimum round-trip tour is A, C, B, D, A, with a cost of 10 units, obtained by direct computation from $\mathbf{C}_1$.

Supplementary Problems

11.6 Solve Problem 11.2 by Gomory's cutting-plane method.

11.7 Solve Example 11.4 by the branch-and-bound method.

11.8 Solve Example 11.2 by Gomory's cutting-plane method.

11.9 Consider the traveling-salesman problem where the salesman has to travel to four cities. Letting c_{ij} denote the cost of traveling from city i to city j, the cost matrix $\mathbf{C}=[c_{ij}]$ is:

$$\mathbf{C}=\begin{bmatrix} 0 & 6 & 4 & 5 \\ 2 & 0 & 5 & 4 \\ 5 & 8 & 0 & 3 \\ 7 & 3 & 6 & 0 \end{bmatrix}.$$

Find the minimum-cost round-trip tour by the branch-and-bound method.

The minimum-cost round-trip tour consists of the sequence of cities 1, 3, 4, 2, 1.

11.10 Solve the knapsack problem by the branch-and-bound method with the data given below and with an upper bound on weight of 65 units.

ITEM NO.	WEIGHT	VALUE
1	40	400
2	10	150
3	23	280
4	31	340

11.11 Using the cutting-plane method, solve:

$$\max\left[f(x_1,x_2)=15x_1+32x_2\right]$$

such that

$$7x_1+16x_2\leqslant 52,$$

$$3x_1+2x_2\leqslant 9,$$

$$x_1,x_2\geqslant 0,$$

and x_1,x_2 are integer-valued.

11.12 Maximize $f(x_1,x_2)=x_1+3x_2$ subject to $0\leqslant x_2\leqslant 1.87$, $60x_1+28x_2\leqslant 165$, $x_1\geqslant 0$, and x_1,x_2 integer-valued, by the cutting-plane method.

11.13 Solve Problem 11.12 by the branch-and-bound method.

Chapter 12

Dynamic Programming

Multistage Decision Processes

Dynamic programming is an optimization method which is particularly applicable to multistage decision processes. In this section we introduce the concept of a general multistage decision process. Such a process contains three variables, called the *stage variable*, the *state variable*, and the *decision variable*. The process is described by a mathematical equation involving these three variables. The stage variable is scalar-valued and can be defined either on an interval or on a discrete set of points. For simplicity we assume that the stage variable is a real-valued variable defined on a discrete set of points. The values of the stage variable are considered to be discrete instants of time so we shall refer to the stage variable as the time variable in the rest of this chapter.

The state variable completely describes the decision process at any instant of time. We assume that at every instant of time the values of the state variable are finite-dimensional vectors called *state vectors*. A decision variable represents the independent decision inputs inherent in a multistage decision process. Its values are also assumed to be finite-dimensional vectors, called *decision vectors*. Letting $\mathbf{x}^i$ and $\mathbf{u}^i$ denote the state and decision vectors respectively at the ith instant of time, a multistage decision process is represented by an equation of the form

$$\mathbf{x}^{i+1} = \mathbf{g}[\mathbf{x}^i, \mathbf{u}^i, i], \tag{12.1}$$

where $i = 0, 1, 2, \ldots,$ and the initial state vector $\mathbf{x}^0$ is given.

The state vector $\mathbf{x}^i$ has the following fundamental characteristics: (1) Given $\mathbf{x}^i$ and $\mathbf{u}^i$, $\mathbf{x}^{i+1}$ is unique; (2) $\mathbf{x}^i$ contains all relevant information about the decision process at the ith instant of time. This condition means that for a given $\mathbf{x}^i$ and $\mathbf{u}^{i+k}$, $k = 0, 1, 2, \ldots,$ any future state $\mathbf{x}^j$, $j > i$, can be computed without any knowledge of the past states $\mathbf{x}^p$, $p < i$.

Suppose the decision-maker chooses $\mathbf{u}^i$ at the ith instant of time. Then at the $(i+1)$st instant, the state of the process changes from $\mathbf{x}^i$ to $\mathbf{x}^{i+1}$ according to Eq. (12.1). Due to this change of state, the performance of this process is either improved or degraded according to some objective function specified by the decision-maker. The multistage decision process consists of finding a sequence of decision vectors over a given period of time that results in the best performance of the process according to the objective-function measure used by the decision maker.

We assume that the decision-maker uses an objective function, which is a function of $\mathbf{x}^i$, $\mathbf{u}^i$, and i and is denoted by $J(\mathbf{x}^i,\mathbf{u}^i,i)$, to measure the performance of the system over the time period $[i,i+1]$. The cumulative measure of the performance of the process over N stages is given by

$$\sum_{i=0}^{N-1} J(\mathbf{x}^i,\mathbf{u}^i,i). \tag{12.2}$$

According to (12.1), for each $i>0$, $\mathbf{x}^i$ is a function of $\mathbf{x}^0$ and $\mathbf{u}^j, j=0,\ldots,i-1$, and hence $J(\mathbf{x}^i,\mathbf{u}^i,i)$ is also a function of $\mathbf{x}^0$ and $\mathbf{u}^j$, $j=0,\ldots,i-1$. The cumulative measure of the performance of the process given by (12.2) then becomes a function of $\mathbf{x}^0$, N, and $\mathbf{u}^j, j=0,\ldots,N-1$. Denoting this function by $J(\mathbf{x}^0,N,\mathbf{u}^0,\ldots,\mathbf{u}^{N-1})$, we have

$$J(\mathbf{x}^0,N,\mathbf{u}^0,\ldots,\mathbf{u}^{N-1})=\sum_{i=0}^{N-1} J(\mathbf{x}^i,\mathbf{u}^i,i). \tag{12.3}$$

Frequently the value of $J(\mathbf{x}^i,\mathbf{u}^i,i)$ is referred to as the *incremental cost* of the process at the ith stage, and the value of $J(\mathbf{x}^0,N,\mathbf{u}^0,\ldots,\mathbf{u}^{N-1})$ is called the *total cost* of the process over N stages. The multistage decision process consists of finding an extreme value of $J(\mathbf{x}^0,N,\mathbf{u}^0,\ldots,\mathbf{u}^{N-1})$ with respect to the decision vectors $\mathbf{u}^0,\ldots,\mathbf{u}^{N-1}$. To be specific we consider the problem of minimizing $J(\mathbf{x}^0,N,\mathbf{u}^0,\ldots,\mathbf{u}^{N-1})$ with respect to $\mathbf{u}^0,\ldots,\mathbf{u}^{N-1}$.

This problem of minimization can be further complicated by the presence of constraints on the allowable values of $\mathbf{x}^i$ and $\mathbf{u}^i$. For example, $\mathbf{x}^i$ and $\mathbf{u}^i$ may be required by physical considerations to be in some given subsets:

$$\mathbf{x}^i \in X(i), \tag{12.4}$$

$$\mathbf{u}^i \in U(\mathbf{x}^i,i). \tag{12.5}$$

The simplest such constraints are magnitude constraints of the form $\boldsymbol{\alpha}^i \leqslant \mathbf{u}^i \leqslant \boldsymbol{\beta}^i$ and $\boldsymbol{\mu}^i \leqslant \mathbf{x}^i \leqslant \boldsymbol{\lambda}^i$. As will be seen later, the presence of such constraints actually simplifies the numerical work involved in obtaining a solution.

Method of Direct Enumeration

One obvious method for solving the minimization problem is to evaluate the cost function $J(\mathbf{x}^0,N,\mathbf{u}^0,\ldots,\mathbf{u}^{N-1})$ for all allowable input sequences

$(\mathbf{u}^0,\ldots,\mathbf{u}^{N-1})$ and obtain the optimal solution by direct inspection. This method is called *direct enumeration* and is feasible only if $U(\mathbf{x}^i,i)$ contains a limited number of elements for every i. Letting $\mathbf{x}^i \in X(i)$ denote an admissible state at the ith instant of time, we obtain a finite set of states $\mathbf{x}^{i+1}$ from $\mathbf{x}^i$ by applying $\mathbf{u}^i \in U(\mathbf{x}^i,i)$ in Eq. (12.1). We then compute the incremental cost of the ith stage given by $J(\mathbf{x}^i,\mathbf{u}^i,i)$ for all $\mathbf{u}^i$ such that $\mathbf{x}^{i+1} \in X(i+1)$, and add this incremental cost to the previously computed cost of arriving at $\mathbf{x}^i$ from $\mathbf{x}^0$ to obtain the total cost of arriving at $\mathbf{x}^{i+1}$ from $\mathbf{x}^0$. This procedure is repeated until $i=N-1$. After termination of this procedure we have available the values of $J(\mathbf{x}^0,N,\mathbf{u}^0,\ldots,\mathbf{u}^{N-1})$ for all allowable choices of $\mathbf{u}^0,\ldots,\mathbf{u}^{N-1}$, and the optimal sequences of decision vectors are obtained by direct inspection of the values of $J(\mathbf{x}^0,N,\mathbf{u}^0,\ldots,\mathbf{u}^{N-1})$.

EXAMPLE 12.1. Consider a multistage decision process whose state is given by a scalar variable x^i at each time instant i, and where there are only two possible choices for the scalar decision variable u^i. Show graphically the application of the direct-enumeration method to this process when $N=5$ and x^i is constrained to be in $[0,\beta]$ for all i.

Figure 12.1 shows the change of the state variable x^i with time. In this figure, an admissible value of x^i at $i=2$ and 4 results in a value of x^{i+1} which violates the given constraint $0 \leqslant x^i \leqslant \beta$. Hence the corresponding values of u^i are not admissible. Each path in Fig. 12.1, shown by a sequence of arrows, has associated with it a total cost of arriving at an allowable final state x^5 from an allowable initial state x^0. We can find the

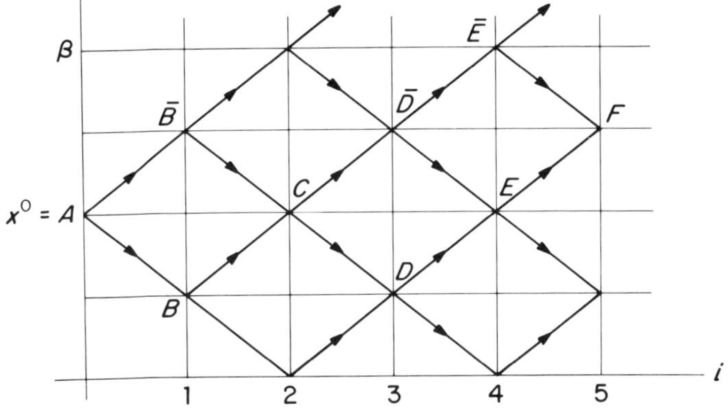

FIGURE 12.1

minimum cost and the corresponding paths in Fig. 12.1 by direct comparison of these total costs. Note that the constraint $0 \leqslant x^i \leqslant \beta$ for all i actually helps in solving the problem by limiting the range of possible values of x^i.

Principle of Optimality

The principal disadvantage of directly enumerating all allowable sequences of decision vectors lies in the huge storage required for reasonably large values of N and/or high dimensionality of $\mathbf{x}^i$. The method of dynamic programming was developed by Bellman for sequentially solving minimization problems which can be formulated as multistage decision processes. This method is based on the principle of optimality, which we will now derive from Fig. 12.1.

Consider the path shown in Fig. 12.1 by the sequence of states $ABCDEF$, and suppose that this is an optimal path for arriving at $x^5 = F$ from $x^0 = A$. Then the section of the path given by $CDEF$ is also an optimal path for arriving at F from $x^2 = C$. To prove this fact by contradiction, suppose $C\bar{D}\bar{E}F$ is the optimal path from C to F, which implies that the total cost associated with $C\bar{D}\bar{E}F$ is less than the cost associated with $CDEF$. Then the total cost associated with $ABC\bar{D}\bar{E}F$ is less than that associated with $ABCDEF$, which contradicts our assumption that $ABCDEF$ is an optimal path from A to F. Note that the optimal path $CDEF$ starting from C is not altered whether $x^2 = C$ is arrived at from $x^1 = B$ or from $x^1 = \bar{B}$. Summarizing these observations, we arrive at the following principle.

> *Principle of Optimality*: An optimal decision sequence has the property that whatever the past states and decisions, the remaining decisions form an optimal decision sequence with respect to the state resulting from the past decisions.

Functional Equation

Let $I(\mathbf{x}^j, N-j)$ denote the minimum cost of the $(N-j)$-stage decision process with $\mathbf{x}^j \in X^j$, i.e.,

$$I(\mathbf{x}^j, N-j) = \min_{\substack{\mathbf{u}^i \in U \\ i=j,\ldots,N-1}} \left\{ \sum_{i=j}^{N-1} J(\mathbf{x}^i, \mathbf{u}^i, i) \right\}. \tag{12.6}$$

We can rewrite the above expression for $I(\mathbf{x}^j, N-j)$ as

$$I(\mathbf{x}^j, N-j) = \min_{\substack{\mathbf{u}^i \in U \\ i=j,\ldots,N-1}} \left\{ J(\mathbf{x}^j, \mathbf{u}^j, j) + \sum_{i=j+1}^{N-1} J(\mathbf{x}^i, \mathbf{u}^i, i) \right\},$$

and by the principle of optimality

$$I(\mathbf{x}^j, N-j) = \min_{\mathbf{u}^j \in U} \min_{\substack{\mathbf{u}^i \in U \\ i=j+1,\ldots,N-1}} \left\{ J(\mathbf{x}^j, \mathbf{u}^j, j) + \sum_{i=j+1}^{N-1} J(\mathbf{x}^i, \mathbf{u}^i, i) \right\}. \tag{12.7}$$

Now in Eq. (12.7) the first factor $J(\mathbf{x}^j, \mathbf{u}^j, j)$ is independent of the choice of $\mathbf{u}^i$, $i = j+1, \ldots, N-1$. Hence we can rewrite this equation as

$$I(\mathbf{x}^j, N-j) = \min_{\mathbf{u}^j \in U} \left\{ J(\mathbf{x}^j, \mathbf{u}^j, j) + \min_{\substack{\mathbf{u}^i \in U \\ i=j+1,\ldots,N-1}} \sum_{i=j+1}^{N-1} J(\mathbf{x}^i, \mathbf{u}^i, i) \right\}. \tag{12.8}$$

By the definition in (12.6),

$$\min_{\substack{\mathbf{u}^i \in U \\ i=j+1,\ldots,N-1}} \left\{ \sum_{i=j+1}^{N-1} J(\mathbf{x}^i, \mathbf{u}^i, i) \right\} = I(\mathbf{x}^{j+1}, N-j-1),$$

and hence Eq. (12.8) becomes

$$I(\mathbf{x}^j, N-j) = \min_{\mathbf{u}^j \in U} \left\{ J(\mathbf{x}^j, \mathbf{u}^j, j) + I(\mathbf{x}^{j+1}, N-j-1) \right\}, \tag{12.9}$$

where $\mathbf{x}^{j+1}$ is given by (12.1). The equation (12.9) is called the *functional equation of dynamic programming* and provides a recursive scheme for obtaining $I(\mathbf{x}^j, N-j)$ for all $\mathbf{x}^j \in X^j$ given a knowledge of $I(\mathbf{x}^{j+1}, N-j-1)$ for all $\mathbf{x}^{j+1} \in X^{j+1}$. We obtain the optimal $\mathbf{u}^j$ by minimizing the right-hand side of Eq. (12.9). In order to use (12.9) recursively to generate $I(x, \cdot)$, it is necessary to start at stage $N-1$ and compute

$$I(\mathbf{x}^{N-1}, 1) = \min_{\mathbf{u}^{N-1} \in U} J(\mathbf{x}^{N-1}, \mathbf{u}^{N-1}, N-1). \tag{12.10}$$

The functional equation (12.9) can also be derived intuitively in the following manner. Suppose $\mathbf{x}^i$ is the state of the process at the ith instant of time and we want to find the optimal decision vector $\mathbf{u}^i$. For each choice of $\mathbf{u}^i$ we arrive at a specific value for $\mathbf{x}^{i+1}$ given by (12.1), and each of these states at the $(i+1)$st instant of time has associated with it at least one optimal path ending in an admissible value for $\mathbf{x}^N$. Therefore, the total cost starting from $\mathbf{x}^i$ should be no greater than $J(\mathbf{x}^i,\mathbf{u}^i,i)+I(\mathbf{x}^{i+1},N-i-1)$, and the optimal $\mathbf{u}^i$ is obtained by minimizing this expression with respect to $\mathbf{u}^i$ as shown on the right-hand side of (12.9). Equations (12.9) and (12.10) define a sequential minimization process where each iteration provides the optimal value for one decision vector. The optimal values of $\mathbf{u}^{j+k}$, $k=1,\ldots,N-j-1$, must be computed sequentially before the optimal value of $\mathbf{u}^j$ can be determined. This method is clearly different from direct enumeration, which provides the optimal values of all the decision vectors simultaneously. The amount of numerical computation in the minimization problem is also reduced, since $\min_{\mathbf{u}^0,\ldots,\mathbf{u}^{N-1}} J(\mathbf{x}^0,N,\mathbf{u}^0,\ldots,\mathbf{u}^{N-1})$ is now replaced by a sequence of minimization problems each involving only one decision vector. Note that in order to obtain the optimal value of $\mathbf{u}^j$ from (12.9) it is necessary to know $I(\mathbf{x}^{j+1},N-j-1)$. Therefore, the values of $I(\mathbf{x}^{j+1},N-j-1)$ must be stored for every feasible value of $\mathbf{x}^{j+1}$. In general, the number of feasible values of $\mathbf{x}^{j+1}$ increases with increasing number of components of $\mathbf{x}^{j+1}$, and hence the storage problem becomes critical with large state vectors.

EXAMPLES

12.2 Consider an N-stage decision process where the state variable and the decision variable can have P values at each stage. Find the total number of additions and comparisons needed to obtain the optimal values of the decision variable by (a) direct enumeration and (b) dynamic programming.

(a) Since the variables $\mathbf{x}^0$ and $\mathbf{u}^0,\ldots,\mathbf{u}^{N-1}$ can each have P values, there are P^{N+1} feasible solutions to the optimization problem. For each feasible solution, the total cost is obtained by adding the incremental costs at each stage. If these incremental costs are added two at a time, then we have $(N-1)P^{N+1}$ additions to be performed. To obtain the optimal value of the total cost by direct enumeration we require $P^{N+1}-1$ comparisons of the P^{N+1} values of the total cost. Hence for direct enumeration we require $NP^{N+1}-1$ additions and comparisons.

(b) In the case of dynamic programming we note form (12.9) that for each value of $\mathbf{x}^j$ and $\mathbf{u}^j$, $0 \leqslant j \leqslant N-1$, we require one addition, so that we require a total of $(N-1)P^2$ additions. We observe also from (12.9) that at each stage, for each value of $\mathbf{x}^j$, we require $P-1$ comparisons, and in (12.10) we require an additional $P-1$ comparisons. Therefore, the total number of comparisons equals $NP(P-1)+(P-1)$, and the total number of additions and comparisons is given by $(2N-1)P^2-(N-1)P-1$. We observe that as N increases, the total number of additions and comparisons increases exponentially for direct enumeration and linearly for dynamic programming.

12.3 (Resource allocation) Consider an investment corporation that invests money in real estate and oil wells. There are ceilings, denoted by b_1 and b_2 respectively, on the amount of capital that can be invested in the real-estate and oil-well ventures. The total amount of capital available for investment is limited by b_3. The problem is to maximize the total profit of the investment corporation under the assumption that the profit is linearly related to the investment.

The problem as stated is not necessarily a multistage decision process and can be solved by other means (see Example A.1 in the Appendix). However, we shall formulate this problem as a multistage decision process and solve it by dynamic programming. Let u_1 and u_2 denote respectively the investments in the real estate and oil-well ventures. Then from physical considerations $u_1, u_2, b_1, b_2, b_3 \geqslant 0$ and the constraints on the investments are $0 \leqslant u_1 \leqslant b_1$, $0 \leqslant u_2 \leqslant b_2$, and $0 \leqslant u_1 + u_2 \leqslant b_3$. Letting $2u_1 + 4u_2$ denote the total profit from the investments u_1 and u_2, the problem is to maximize $2u_1 + 4u_2$ with respect to u_1 and u_2 subject to the given constraints.

We interpret each investment decision as the outcome of a decision rule at a single stage of a two-stage decision process. The state variable is given by the capital available for investment at the beginning of each stage. Letting x_j denote this state variable, we have

$$x_0 = b_3,$$
$$x_1 = x_0 - u_1,$$
$$x_2 = x_1 - u_2,$$

and Fig. 12.2 shows the two-stage decision process we have formulated.

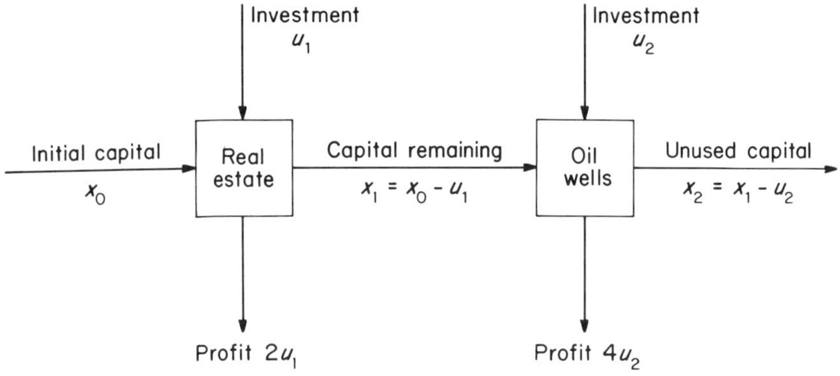

FIGURE 12.2

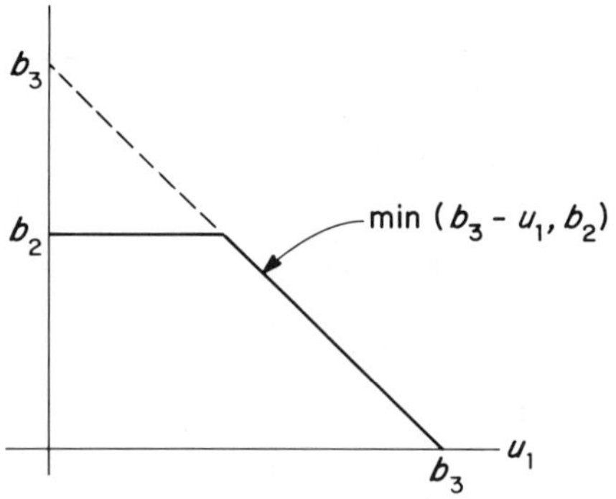

FIGURE 12.3

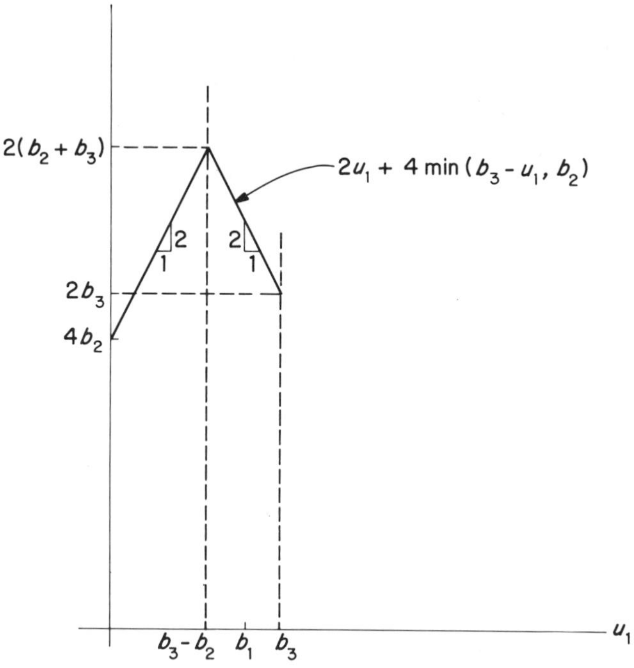

FIGURE 12.4

Since there is no profit to be made by saving capital, at the last stage of the two-stage process we should invest all the capital we have subject to the given constraints, and therefore $u^*_2 = \min(x_1, b_2)$. The maximum value of profit from this investment is $I(x_1, 1) = 4u^*_2 = 4\min(x_1, b_2)$.

Next the optimal value of u_1 is obtained by using the recursive scheme (12.9), and we have

$$\begin{aligned} I(x^0, 2) &= \max_{0 \leqslant u_1 \leqslant b_1} \left[2u_1 + 4\min(x_1, b_2)\right] \\ &= \max_{0 \leqslant u_1 \leqslant b_1} \left[2u_1 + 4\min(x_0 - u_1, b_2)\right] \\ &= \max_{0 \leqslant u_1 \leqslant b_1} \left[2u_1 + 4\min(b_3 - u_1, b_2)\right], \end{aligned}$$

where $I(x^0, 2)$ denotes the maximum value of the total profit. Figure 12.3 shows the function $\min(b_3 - u_1, b_2)$ and Fig. 12.4 shows the function $2u_1 + 4\min(b_3 - u_1, b_2)$. Assuming $b_1 \geqslant b_3 - b_2$, from Fig. 12.4 we have $u^*_1 = b_3 - b_2$, $u^*_2 = \min(b_3 - u^*_1, b_2) = b_2$, and the optimal profit $I(x^0, 2) = 2u^*_1 + 4u^*_2 = 2b_3 + 2b_2$. Note that since the oil wells produce twice as much profit as the real estate for the same amount of investment, it is obvious that one should invest as much as allowed in the oil wells and the remaining capital in real estate. The optimal values of u_1 and u_2 obtained by dynamic programming satisfy this distribution of capital between the two investments.

12.4 (Traveling salesman) Consider a salesman who is expected to travel to $N+1$ cities starting from city 0. Letting $d_{ij} \geqslant 0$ denote the distance between cities i and j, $0 \leqslant (i,j) \leqslant N$, find a round-trip tour of minimum length that includes every city once and only once.

The traveling salesman problem as stated here can also be solved by integer programming and the branch-and-bound technique (see Chapter 11). However, we will formulate this problem as a multistage decision process and obtain a recursive scheme similar to (12.9) for its solution.

We assume that following an optimal tour, the salesman has reached city i and that there remain k cities, $j_1, \ldots, j_k$, to be visited before he returns to city 0. Since the tour is optimal, the route from city i through cities $j_1, \ldots, j_k$ in some order to be determined and then to city 0 must be of minimum length. We define $D(i|j_1, \ldots, j_k)$ to be the length of the shortest route from city i to city 0 which passes through each of the remaining k unvisited cities once and only once. Thus the problem is solved when we compute $D(0|j_1, \ldots, j_k)$. Now using the above remark on the length of the optimal route from city i to city 0 via cities $j_1, \ldots, j_k$ and the definition

$D(i|j_1,\ldots,j_k)$, we obtain

$$D(i|j_1,\ldots,j_k)=\min_{1\leqslant m\leqslant k}\left[d_{ij_m}+D(j_m|j_1,\ldots,j_{m-1},j_{m+1},\ldots,j_k)\right], \quad (12.11)$$

which is a sequential or recursive scheme for computing $D(i|j_1,\ldots,j_k)$ for all i. To begin the calculations we use

$$D(i|j)=d_{ij}+d_{j0}$$

and terminate iteration when $D(0|j_1,\ldots,j_N)$ is obtained.

12.5 Estimate the amount of storage required to solve the traveling-salesman problem by dynamic programming.

To apply the sequential scheme of (12.11) we need to store the values of the function $D(i|j_1,\ldots,j_k)$, where it is only necessary to take into account the totality of values $j_1,\ldots,j_k$ and not the order in which these k cities are visited from city i before returning to city 0. Therefore storing the values of $D(i|j_1,\ldots,j_k)$ is equivalent to storing all possible ways of choosing k quantities from $N-1$ quantities. This required storage is at a maximum when k is the integer nearest to $(N-1)/2$. Letting $2m+1$ denote the number of cities and using Stirling's formula to approximate the value of $\binom{2m}{m}$, we obtain the following table giving upper bounds on the amount of storage required.

No. of cities $2m+1$	Upper bound on storage
11	300
17	12,000
21	200,000

This table shows one of the major drawbacks of dynamic programming. As the number of components of the state vector of a multistage decision process increases, the total number of stored values of the objective function also increases and can soon exceed the rapid-access memory of many modern computers.

Computational Requirements with One State Variable

In this section we consider in more detail the computational requirements imposed by the recursive scheme of (12.9). Equation (12.9) reduces the computational effort of minimization, since in it we minimize only with respect to $\mathbf{u}^j$ and not with respect to $\mathbf{u}^0,\ldots,\mathbf{u}^{N-1}$. However, as a price for this reduction in computational effort, we must store the values of

$I(\mathbf{x}^{j+1}, N-j-1)$ for all feasible values of $\mathbf{x}^{j+1}$. We noticed in Example 12.5 that this storage requirement can become critical even for a modern computer.

To gain more insight into the computational scheme, let us consider the following problem: maximize $\Sigma_{j=1}^{N} g_j(u_j)$ with respect to $u_j \geqslant 0$, $1 \leqslant j \leqslant N$, subject to $\Sigma_{j=1}^{N} u_j = x$. To transform this problem into a multistage decision process, we define $J(u_1, \ldots, u_k) = \Sigma_{j=1}^{k} g_i(u_j)$ and $I(x,k) = \max J(u_1, \ldots, u_k)$ subject to $u_j \geqslant 0$ and $\Sigma_{j=1}^{k} u_j = x$. The recursive scheme given by (12.9) becomes

$$I(x,k) = \max_{0 \leqslant u_k \leqslant x} \left[g_k(u_k) + I(x-u_k, k-1) \right], \tag{12.12}$$

where we set $I(x,1) = g_1(x)$.

In order to apply the recursive scheme given by (12.12), we must compute and store the values of $I(x,k)$ for all feasible values of x. Assuming that the feasible values of x are in the interval $[0, x_0]$, we choose a set of grid points given by $0, \Delta, 2\Delta, \ldots, R\Delta = x_0$ in $[0, x_0]$. Next we compute and store the values of $I(x,k)$ at these grid points in $[0, x_0]$. To obtain the value of $I(x,k)$ for values of x in $[0, x_0]$ different from the grid points, we can interpolate using the formula

$$I(x,k) = I(m\Delta, k) + (x - m\Delta)\frac{I((m+1)\Delta, k) - I(m\Delta, k)}{\Delta}.$$

We now turn our attention to the maximization operation in (12.12). The maximum value of the objective function can be obtained by direct enumeration if the decision variable ranges over a finite set of values. Therefore we discretize the decision variable u_k over the range $[0, x]$ by the same set of grid points common to $[0, x_0]$, and then obtain the maximum value of the objective function with respect to u_k by direct enumeration over these grid points. Due to the discretization of continuous variables, this maximization scheme provides only an approximation to the true maximum value of the objective function. Note that if the interval Δ is too small, we have a large number of grid points which yield a good approximation but require a large number of function evaluations and much storage. On the other hand, if Δ is too large, we have a relatively poor approximation with fewer function evaluations and less storage. We now consider the computational scheme in more detail, beginning with the calculation of $g_j(m)$ for $0 \leqslant m \leqslant R$ and $1 \leqslant j \leqslant N$. Then $I(x,1) = g_1(x)$ is given by a discrete set of values at the grid points in $[0, x_0]$. Using (12.12), we have

$$I(x,2) = \max_{0 \leqslant u_2 \leqslant x} \left[g_2(u_2) + I(x-u_2, 1) \right].$$

Since we have discretized the range of values of u_2, the above equation is replaced by the approximate equation

$$I(x,2)=\max_{m=0,\ldots,R}\left[g_2(m\Delta)+I(x-m\Delta,1)\right], \tag{12.13}$$

where x ranges over the set of grid points in $[0,x_0]$ and the maximization in (12.13) can be done by direct enumeration. This maximization process is easily carried out by pairwise comparison of function values. In this process, the values of $g_2(0)+I(x,1)$ and $g_2(\Delta)+I(x-\Delta,1)$ are compared and only the larger one is retained. This value is compared with $g_2(2\Delta)+I(x-2\Delta,1)$ and again the larger one is retained. This process yields the maximum value in (12.13) after all comparisons are made. In this manner the values of $I(x,2)$ are obtained for all the grid points in $[0,x_0]$, and in the process we also obtain the optimal values of u_2, which evidently depend on the value of x.

At the end of two stages, the computational results can be stored in the format shown in Table 12.1. Evidently Table 12.1 will include more columns if results for more than two stages are computed. This table of values can now be used to obtain the optimal solutions of a two-stage decision process. Given the value of x, we obtain the optimal value of $u_2(x)$ from the table and since $u_1(x)+u_2(x)=x$, the optimal value of $u_1(x)$ is given by $u_1(x)=x-u_2(x)$.

Table 12.1

x	$I(x,1)$	$u_1(x)$	$I(x,2)$	$u_2(x)$
0	×	×	×	×
Δ	×	×	×	×
2Δ	×	×	×	×
⋮				
$R\Delta$	×	×	×	×

EXAMPLE 12.6 Letting $g_j(x)=xe^{-jx}$, $j=1,2$, solve

$$\max \sum_{j=1}^{2} g_j(u_j)$$

subject to

$$\sum_{j=1}^{2} u_j=x, \qquad \text{where} \quad 0\leqslant x\leqslant 2.$$

The function $g_j(x)$ is shown in Fig. 12.5. Letting $I(x,1)=g_1(x)$, we have $I(x,1)=xe^{-x}$. We choose $\Delta=\frac{1}{2}$, resulting in the grid points 0, $\frac{1}{2}$, 1, $\frac{3}{2}$, and 2 in the interval $[0,2]$. Table 12.2 lists the function values of $I(x,1)$ and

$u_1(x)$ at these grid points. Next, using (12.13), we compute $I(x,2)$; a sample calculation is given below:

$$I\left(\tfrac{3}{2},2\right)=\max\left[m\Delta e^{-2m}+I\left(\tfrac{3}{2}-m\Delta,1\right)\right]$$

$$=\max\left[I\left(\tfrac{3}{2},1\right),\tfrac{1}{2}e^{-1}+I(1,1),e^{-2}+I\left(\tfrac{1}{2},1\right),\tfrac{3}{2}e^{-3}+I(0,1)\right]$$

$$=\max\left[\tfrac{3}{2}e^{-3/2},\tfrac{3}{2}e^{-1},e^{-2}+\tfrac{1}{2}e^{-1/2},\tfrac{3}{2}e^{-3}\right]=\tfrac{3}{2}e^{-1},$$

which corresponds to $u_2(\tfrac{3}{2})=\tfrac{1}{2}$. The values in Table 12.2 can now be used to obtain the value of $I(x,2)$ by interpolation for any value of x in $[0,2]$. For example, if $x=\tfrac{5}{4}$, then by the linear interpolation formula we have

$$I\left(\tfrac{5}{4},2\right)=I(1,2)+\left(\tfrac{5}{4}-1\right)\frac{I\left(\tfrac{3}{2},2\right)-I(1,2)}{\tfrac{1}{2}}$$

$$=I(1,2)+\tfrac{1}{2}\left[I\left(\tfrac{3}{2},2\right)-I(1,2)\right]$$

$$=\tfrac{1}{2}\left[I(1,2)+I\left(\tfrac{3}{2},2\right)\right]=e^{-1}+\tfrac{1}{4}e^{-1/2}.$$

Table 12.2

x	$I(x,1)$	$u_1(x)$	$I(x,2)$	$u_2(x)$
0	0	0	0	0
$\tfrac{1}{2}$	$\tfrac{1}{2}e^{-1/2}$	$\tfrac{1}{2}$	$\tfrac{1}{2}e^{-1/2}$	0
1	e^{-1}	1	$\tfrac{1}{2}e^{-1}+\tfrac{1}{2}e^{-1/2}$	$\tfrac{1}{2}$
$\tfrac{3}{2}$	$\tfrac{3}{2}e^{-3/2}$	$\tfrac{3}{2}$	$\tfrac{3}{2}e^{-1}$	$\tfrac{1}{2}$
2	$2e^{-2}$	2	$\tfrac{1}{2}e^{-1}+\tfrac{3}{2}e^{-3/2}$	$\tfrac{1}{2}$

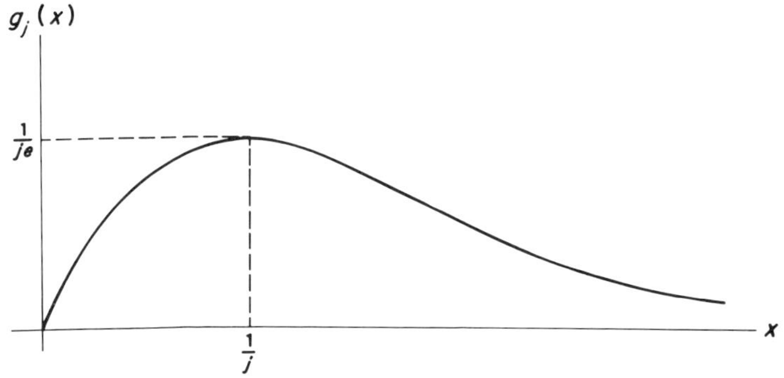

FIGURE 12.5

Computational Requirements with Two State Variables

In this section we discuss the difficulties that arise when the dimensionality of the state vector of a multistage decision process increases. To be specific let us consider the problem:

$$\max \sum_{j=1}^{N} g_j(u_j)$$

subject to

$$u_j \geqslant 0, \qquad 1 \leqslant j \leqslant N,$$

$$\sum_{j=1}^{N} u_j = x \quad \text{and} \quad \sum_{j=1}^{N} u_j = y.$$

To transform this problem into a multistage decision process, we define $J(u_1,\ldots,u_k) = \sum_{j=1}^{k} g_j(u_j)$ and $I(x,y,k) = \max J(u_1,\ldots,u_k)$ subject to $u_j \geqslant 0$, $1 \leqslant j \leqslant k$, $\sum_{j=1}^{k} u_j = x$, and $\sum_{j=1}^{k} u_j = y$. Note that $I(x,y,k)$ is a function of two state variables x and y, and the values of $I(x,y,k)$ for every feasible pair of values of x and y must be stored. In other words, for every feasible value of y we must store a table of values similar to Table 12.1. As the number of states increases further, this storage requirement becomes even more severe.

One way to alleviate this difficulty is by the use of Lagrange multipliers. Using a Lagrange multiplier λ, we form a new objective function

$$F(u_1,\ldots,u_N) = \sum_{j=1}^{N} g_j(u_j) + \lambda\left(\sum_{j=1}^{N} u_j - y\right) = \sum_{j=1}^{N} \left[\, g_j(u_j) + \lambda u_j \right] - \lambda y.$$

We now choose a value of λ and maximize this new objective function $F(u_1,\ldots,u_N)$ by dynamic programming subject to $u_j \geqslant 0$, $1 \leqslant j \leqslant N$, and $\sum_{j=1}^{N} u_j = x$. For a given value of λ, this new maximization problem is similar to the one discussed in the previous section and requires less storage than the original problem. Thus we have replaced a problem involving two state variables by a sequence of problems in only one state variable. It is shown in the following example that a proper choice of the value of λ will lead to the solution of the original problem.

EXAMPLE 12.7. Solve the following problem:

$$\min \frac{1}{2} \sum_{j=1}^{2} u_j^2$$

subject to

$$u_j \geqslant 0, \qquad 1 \leqslant j \leqslant 2,$$

$$u_1 + u_2 = 8 \quad \text{and} \quad 2u_1 + u_2 \leqslant 10.$$

Using the Lagrange multiplier λ, we obtain

$$F(u_1, u_2) = \tfrac{1}{2}u_1^2 + 2\lambda u_1 + \tfrac{1}{2}u_2^2 + \lambda u_2 - \lambda\, 10$$

and this new objective function must be minimized subject to $u_1, u_2 \geqslant 0$ and $u_1 + u_2 = 8$. This new problem is similar to the resource allocation problem discussed in Example 12.3. Letting x_k denote the state variable x at the kth stage, we have

$$x_0 = 8,$$

$$x_1 = x_0 - u_1,$$

$$x_2 = x_1 - u_2,$$

and Fig. 12.6 shows the two-stage decision process that we have formulated.

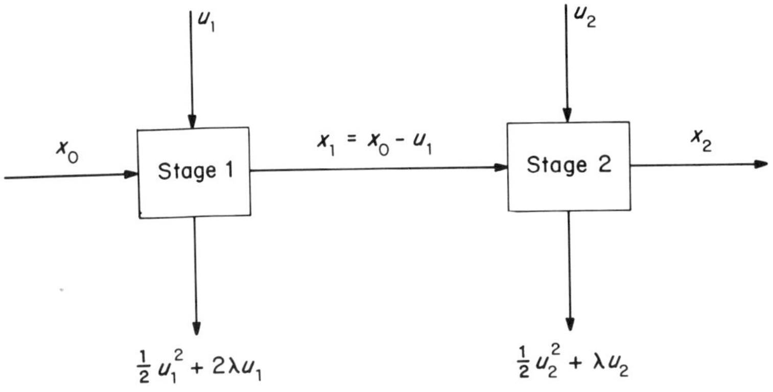

FIGURE 12.6

Setting $\lambda = 0$, we obtain $F(u_1, u_2) = \tfrac{1}{2}(u_1^2 + u_2^2)$, and the computations are similar to Example 12.2, but now the optimal solutions are $u^*_1 = u^*_2 = 4$. Clearly $2u^*_1 + u^*_2 = 12$, and the optimal solutions do not satisfy the given constraint $2u_1 + u_2 \leqslant 10$. In order to improve these solutions, we set $\lambda = 5$ and obtain $F(u_1, u_2) = \tfrac{1}{2}u_1^2 + 10u_1 + \tfrac{1}{2}u_2^2 + 5u_2 - 50$. Now minimizing this

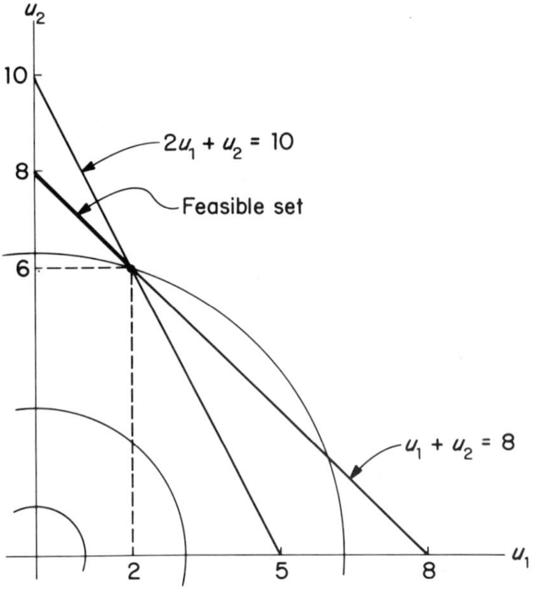

FIGURE 12.7

objective function by methods similar to Example 12.2, we have $u^*_1 = 1.5$ and $u^*_2 = 6.5$. Clearly $2u^*_1 + u^*_2 = 9.5 < 10$, and the constraint $2u_1 + u_2 \leqslant 10$ is satisfied. However, the objective function $\frac{1}{2}(u_1^2 + u_2^2)$ may not have attained the lowest possible value subject to the given constraints. To test this possibility we must use another value of λ. From the previous calculations we know that the proper value of λ is in $[0,5]$, so we set $\lambda = 4$ and obtain $F(u_1,u_2) = \frac{1}{2}u_1^2 + 8u_1 + \frac{1}{2}u_2^2 + 4u_2 - 40$. Minimizing this function by the method of Example 12.2, we have $u^*_1 = 2$, $u^*_2 = 6$. Clearly $2u^*_1 + u^*_2 = 10$ and the constraint $2u_1 + u_2 \leqslant 10$ is satisfied. The values of the original objective function $\frac{1}{2}(u_1^2 + u_2^2)$ for the different values of λ and u_1, u_2 are given below.

λ	u^*_1	u^*_2	$\frac{1}{2}(u_1^2 + u_2^2)$	Constraint violated?
0	4	4	16	Yes
4	2	6	20	No
5	1.5	6.5	20.25	No

It can be seen graphically in Fig. 12.7 that the proper value of λ is 4 and the minimum value of the objective function subject to the given constraints is 20.

Infinite-Stage Processes and Successive Approximations

A multistage decision process becomes an infinite-stage process if $N=\infty$. Such processes may not be of direct practical interest, but they are often useful as approximations. We assume that $g(\mathbf{x}^j,\mathbf{u}^j,j)$ and $J(\mathbf{x}^j,\mathbf{u}^j,j)$ do not depend explicitly on j and that as $N\to\infty$, $I(\mathbf{x},N)\to I(\mathbf{x})$. Then as $N\to\infty$, (12.9) becomes

$$I(\mathbf{x})=\min_{\mathbf{u}\in U}\left[J(\mathbf{x},\mathbf{u})+I(g(\mathbf{x},\mathbf{u}))\right]. \tag{12.14}$$

Equation (12.14) represents a functional equation in terms of the unknown function $I(\mathbf{x})$. For many multistage decision processes, (12.14) can be solved analytically. In other cases (12.14) can be solved approximately by numerical methods. The solution of (12.14) given by $I(\mathbf{x})$ and the minimizing decision vector $u^*(\mathbf{x})$ are used as approximations to $I(\mathbf{x},i)$ and $\mathbf{u}^i(\mathbf{x})$ respectively. These approximations are good for large values of i and save the expense of computing $I(\mathbf{x},i)$ and $\mathbf{u}^i(\mathbf{x})$ by (12.9) and (12.10).

We now discuss two successive-approximation techniques for solving Eq. (12.14). The first method is to define the ith successive approximation to $I(\mathbf{x})$ as

$$I_i(\mathbf{x})=\min_{\mathbf{u}\in U}\left[J(\mathbf{x},\mathbf{u})+I_{i-1}(g(\mathbf{x},\mathbf{u}))\right], \tag{12.15}$$

where the iteration starts with an initial guess $I_0(\mathbf{x})$. Taking $I_0(\mathbf{x})=0$, we obtain $I_i(\mathbf{x})=I(\mathbf{x},i)$ as given by Eq. (12.9). Thus (12.15) leads to economy in the computation only when we start with a good initial guess to the solution of (12.14).

The second method starts with an initial guess $\mathbf{u}^0(\mathbf{x})$ to the optimal input $\mathbf{u}^*(\mathbf{x})$. We then obtain $I_0(\mathbf{x})$ by solving the functional equation

$$I_0(\mathbf{x})=J\left(\mathbf{x},\mathbf{u}^0(\mathbf{x})\right)+I_0\left(g\left(\mathbf{x},\mathbf{u}^0(\mathbf{x})\right)\right) \tag{12.16}$$

and obtain the next approximation $\mathbf{u}^1(\mathbf{x})$ by solving

$$\min_{\mathbf{u}\in U}\left[J(\mathbf{x},\mathbf{u})+I_0(g(\mathbf{x},\mathbf{u}))\right]. \tag{12.17}$$

Then we use $\mathbf{u}^1(\mathbf{x})$ to compute $I_1(\mathbf{x})$, and the iterations are continued in this manner. These approximate methods for solving (12.14) become very useful when in particular it can be shown that the approximate solution sequences $I_i(\mathbf{x})$ and $\mathbf{u}_i(\mathbf{x})$ are monotonically convergent as illustrated in Figs. 12.8 and 12.9. In such cases every iteration of the approximation scheme produces a better solution of (12.14), and the iterations can be

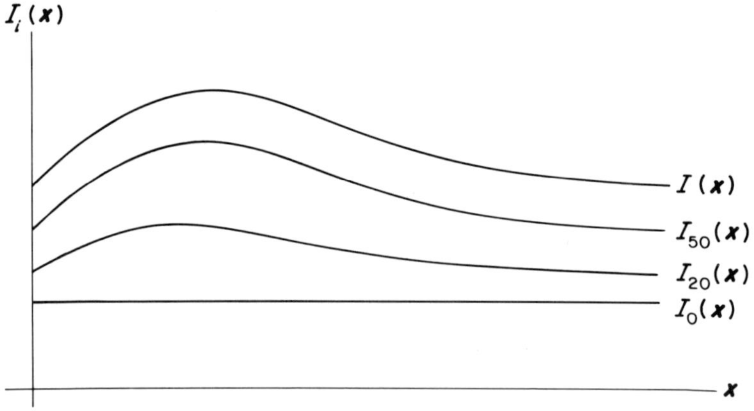

FIGURE 12.8

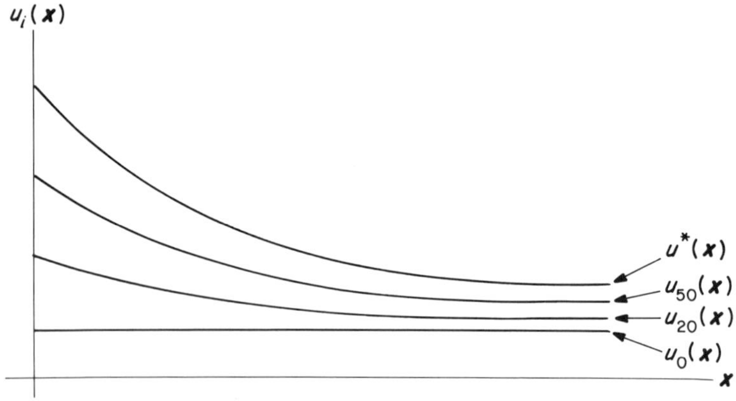

FIGURE 12.9

stopped after a sufficiently accurate solution is obtained. We next introduce a resource allocation problem and discuss a successive-approximation technique for solving (12.14).

Consider an investor with a total resource of x units. He invests an amount u, $0 \leqslant u \leqslant x$, in oil wells with a resulting profit given by $g(u)$. With the remaining resource of $x-u$ units he buys real estate which produces a profit of $h(x-u)$. Hence the total profit from his one-stage investment program is $g(u)+h(x-u)$. We assume that after these investments his initial capital is depreciated to $au+b(x-u)$, where $0 \leqslant (a,b) < 1$, and it can now be reinvested. The problem of resource allocation is to maximize the investor's profit over N stages.

EXAMPLE 12.8 Show that the successive-approximation technique for solving (12.14) for the resource allocation problem generates a monotonically convergent sequence of $I_i(\mathbf{x})$.

For the resource allocation problem, (12.4) becomes

$$I(x) = \max_{0 \leqslant u \leqslant x} \left[g(u) + h(x-u) + I(au + b(x-u)) \right] \qquad (12.18)$$

Since we have no advance knowledge of the form of the solution $I(x)$ of (12.18), we make an initial guess of $I_0(x) = 0$, and from (12.18) we have

$$I_1(x) = \max_{0 \leqslant u \leqslant x} \left[g(u) + h(x-u) \right].$$

In any practical resource allocation problem, we can assume that the profit functions $g(x)$ and $h(x)$ satisfy the conditions $g(x), h(x) \geqslant 0$ for all $x \geqslant 0$ and $g(0) = h(0) = 0$. Then from the above equation for $I_1(x)$, we conclude $I_1(x) \geqslant I_0(x) = 0$. Furthermore, if $g(x)$ and $h(x)$ are convex, then since $g(u) + h(x-u)$ will also be convex in x, we conclude that $I_1(x)$ is convex. Next, using mathematical induction, we conclude that $I_i(x)$ is convex in x and furthermore $I_i(x) \geqslant I_{i-1}(x)$. Thus the successive-approximation technique produces a monotonically nondecreasing sequence of convex solutions $I_i(x)$. We can also show that $I_i(x)$ converges to the solution of (12.18), but the details are complex and beyond the scope of this text.

Solved Problems

12.1 (Shortest route) Consider a network of roads, where the points $1, 2, \ldots, N$ represent cities. Let $d_{ij} \geqslant 0$ denote the distance from city i to city j, and f_i denote the length of the shortest route from city i to city N. Assuming that every pair of cities is joined by a road, find f_i, $i = 1, \ldots, N-1$.

The length of the route from city i to city j followed by the shortest route to city N is $d_{ij} + f_j$, and since f_i is the shortest route from city i to city N, we have $f_i \leqslant d_{ij} + f_j$ for all $j \neq i$. On the other hand, the shortest route from city i to city N must pass through some city m and then follow some route to city N whose length is denoted by l_{mN}, and hence $f_i = d_{im} + l_{mN}$. However, since $l_{mN} \geqslant f_m$, we have $f_i \geqslant d_{im} + f_m$ for some $m = 1, \ldots, N$, $m \neq i$. Combining these two inequalities, we have $f_i = d_{im} + f_m$, and since f_i cannot be greater than $d_{ij} + f_j$, $j \neq i$, we have

$$f_i = \min_{j \neq i} (d_{ij} + f_j).$$

We observe that the above are nonlinear algebraic equations and there are no general methods for solving them. So we develop a successive-approximation technique for solving these equations which provides us with a recursive equation similar to (12.9).

Letting $f_i^{(k)}$ denote the kth approximation to f_i and using the above set of equations as a guide, we define a recursive scheme

$$f_i^{(k)} = \min_{j \neq i} \left(d_{ij} + f_j^{(k-1)} \right), \tag{12.19}$$

where initially we set $f_j^{(1)} = \infty$, $i \neq N$, and $f_N^{(1)} = 0$. Thus we have recast the original problem as a multistage decision process where the kth approximation corresponds to the kth stage, the decision variable ranges over the set of cities $1, \ldots, N$, and the objective function is the length of a route from city i to city N.

12.2 Show that the approximate solution $f_i^{(k)}$ given by the recursion equation (12.19) converges to f_i.

Although an analytical proof can be given, we provide a physical interpretation of the kth approximation $f_i^{(k)}$ which shows intuitively that $f_i^{(k)}$ converges to f_i in a finite number of stages. We prove by induction that $f_i^{(k)}$ is the shortest route from city i to city N with at most $k-1$ cities in between.

The last statement is clearly valid for $k=2$, since from (12.19) we have $f_i^{(2)} = d_{iN}$. Next we assume that $f_i^{(k)}$ is the shortest route from city i to city N with at most $k-1$ cities in between. Consider a route from city i to city N with at most k cities in between. This route goes from city i to some city j, and the shortest route from city j to city N with at most $k-1$ cities in between is of length $f_j^{(k)}$. Then the length of the shortest route from city i to city N with at most k cities in between is given by $\min_{j \neq i}(d_{ij} + f_j^{(k)})$, and by (12.19) $f_i^{(k+1)} = \min_{j \neq i}(d_{ij} + f_j^{(k)})$. Hence $f_i^{(k+1)}$ is the shortest route from city i to city N with at most k cities in between.

Next we note that since there is a total of N cities, the shortest route from any city i to city N can contain at most $N-2$ cities in between. Therefore $f_i = f_i^{(k)}$ for some k, $1 \leqslant k \leqslant N$, and f_i can be obtained from (12.19) in at most N iterations.

12.3 Consider the network of roads shown in Fig. 12.10, where each line represents a road joining two cities and the number beside each line represents the length of the road. Find the lengths of the shortest routes from cities 1, 2, and 3 to city 4.

Following the recursion scheme of (12.19), we set $f_1^{(1)} = f_2^{(1)} = f_3^{(1)} = \infty$ and $f_4^{(1)} = 0$ to obtain

$$\begin{aligned} f_1^{(2)} &= \min_{j \neq 1} \left(d_{ij} + f_j^{(1)} \right) \\ &= \min\left(d_{12} + f_2^{(1)}, d_{13} + f_3^{(1)}, d_{14} + f_4^{(1)}\right) \\ &= \min(4 + \infty, 2 + \infty, 4 + 0) \\ &= 4, \end{aligned}$$

and similarly $f_2^{(2)} = 1$, $f_3^{(2)} = 4$, and $f_4^{(2)} = 0$. These values of $f_j^{(2)}$, $j = 1, \ldots, 4$, are tabulated along the second row of Table 12.3. Using these values of $f_j^{(2)}$ in the

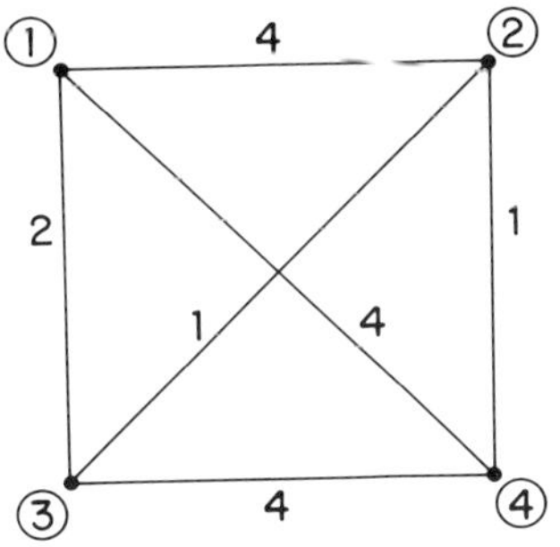

FIGURE 12.10

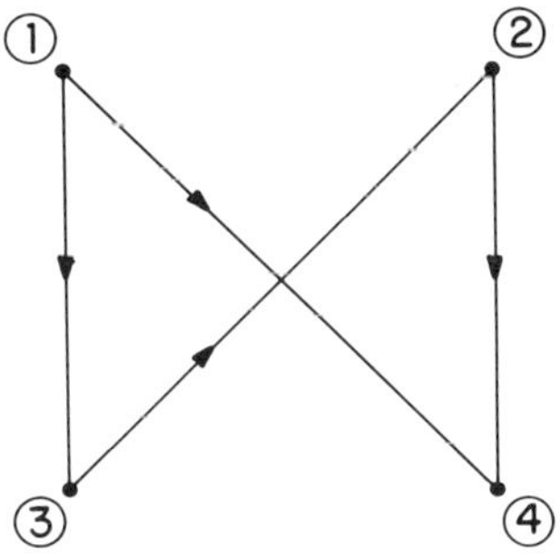

FIGURE 12.11

recursion scheme, we obtain

$$\begin{aligned} f_1^{(3)} &= \min_{j \neq i} \left(d_{ij} + f_j^{(2)} \right) \\ &= \min\left(d_{12} + f_2^{(2)}, d_{13} + f_3^{(2)}, d_{14} + f_4^{(2)} \right) \\ &= \min(4+1, 2+4, 4+0) \\ &= 4, \end{aligned}$$

and similarly $f_2^{(3)} = 1$, $f_3^{(3)} = 2$, and $f_4^{(3)} = 0$. These values of $f_j^{(3)}$, $j = 1, \ldots, 4$, are tabulated along the third row of Table 12.3. Using the recursion scheme once again in a similar manner, we obtain $f_1^{(4)} = 4$, $f_2^{(4)} = 1$, $f_3^{(4)} = 2$, $f_4^{(4)} = 0$. Since f_j must be the least value of $f_j^{(k)}$, $1 \leqslant k \leqslant N$, we have $f_1 = f_1^{(4)} = f_1^{(2)} = 4$, $f_2 = f_2^{(2)} = 1$, $f_3 = f_3^{(3)} = 2$, and $f_4 = 0$. The shortest routes from cities 1, 2, and 3 to city 4 are shown in Fig. 12.11, where the arrows indicate the direction of travel between city pairs.

Table 12.3

$f_i^{(k)}$ (k \ i)	1	2	3	4
1	∞	∞	∞	0
2	4	1	4	0
3	4	1	2	0
4	4	1	2	0

12.4 Find the length of the shortest round-trip tour for a salesman traveling among the cities shown in Fig. 12.10.

For consistency in notation, we replace city 4 by city 0. To start the recursive scheme, we compute $D(i|j) = d_{ij} + d_{j0}$ for $0 \leqslant (i,j) \leqslant 3$, and the values of $D(i|j)$ are given in Table 12.4. Then using the values of $D(i|j)$, we compute recursively

$$D(i|j_1 j_2) = \min_{1 \leqslant m \leqslant 2} \left[d_{ij_m} + D(j_m | j_{m-1}, j_{m+1}) \right].$$

Table 12.4

$D(i\|j)$ i \ j	0	1	2	3
0	0	8	2	8
1	4	4	5	6
2	1	8	1	5
3	4	6	2	4

For example,

$$\begin{aligned} D(2|1,3) &= \min\left[d_{21} + D(1|3), d_{23} + D(3|1)\right] \\ &= \min[4+6, 1+6] \\ &= 7 \end{aligned}$$

The values of $D(i|j_1,j_2)$ are given in Table 12.5. The length of the shortest round-trip tour is given by $D(0|1,2,3)$. Using the recursive scheme and the values of $D(i|j_1,j_2)$ from Table 12.5, we obtain

$$\begin{aligned} D(0|1,2,3) &= \min\left[d_{01} + D(1|2,3), d_{02} + D(2|1,3), d_{03} + D(3|1,2)\right] \\ &= \min[4+4, 1+7, 4+7] \\ &= 8 \end{aligned}$$

The optimum round-trip tour is the same as that shown in Fig. 12.12 but without the arrows, since a round-trip tour can be traversed in either direction.

Table 12.5

$D(i\|j_1,j_2)$ i \ (j_1,j_2)	(1,2)	(1,3)	(2,3)
0	6	7	6
1	5	6	4
2	8	7	3
3	7	6	2

12.5 (Reliability problem) Consider the problem of designing a minimum-weight structure consisting of N elements. The ith element has cross-sectional area A_i, and the probability that the element will *not* fail is given by $p_i(A_i)$. The weight of the ith element is denoted by $w_i(A_i)$. If one element fails, the whole structure is assumed to fail, and the events of failure are considered to be statistically independent. The problem is to minimize the total weight of the structure with a lower bound on the probability that it will not fail. Derive the functional equation (12.9) for this problem.

The total weight of the structure is $\Sigma_{i=0}^{N-1} w_i(A_i)$, and the probability that the structure will not fail is given by $\Pi_{i=0}^{N-1} p_i(A_i)$. Assuming that the lower bound on the probability of not failing is p, we define

$$P_{N-j}=\prod_{i=j}^{N-1} p_i(A_i) \quad \text{and} \quad I(p,N-j)=\min\sum_{i=j}^{N-1} w_i(A_i)$$

with $P_{N-j}\geqslant p$. For $j=N-1$, it is evident that $I(p,1)=w_{N-1}(A_{N-1})$, where A_{N-1} satisfies $p_{n-1}(A_{N-1})=p$.

Now suppose $j=N-2$. Given that the $(N-2)$nd element has a cross-sectional area A_{N-2}, the probability that it will not fail is $p_{N-2}(A_{N-2})$. Then the probability that the $(N-1)$st element does not fail is given by $p_{N-1}=p/p_{N-2}(A_{N-2})$. The minimum weight for the $(N-1)$st element is then $I(p_{N-1},1)$. Thus

$$I(p,2)=\min_{A_{N-2}}\left[w_{N-2}(A_{N-2})+I(p_{N-1},1)\right],$$

and in general,

$$I(p,N-j)=\min\left[w_j(A_j)+I(p_{j+1},N-j-1)\right].$$

12.6 Consider the structure shown in Fig. 12.12, consisting of two elements. We assume that the probability of failure of each element as a function of the

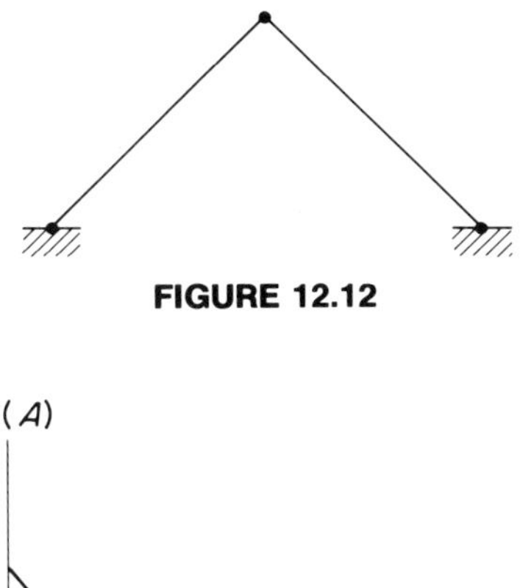

FIGURE 12.12

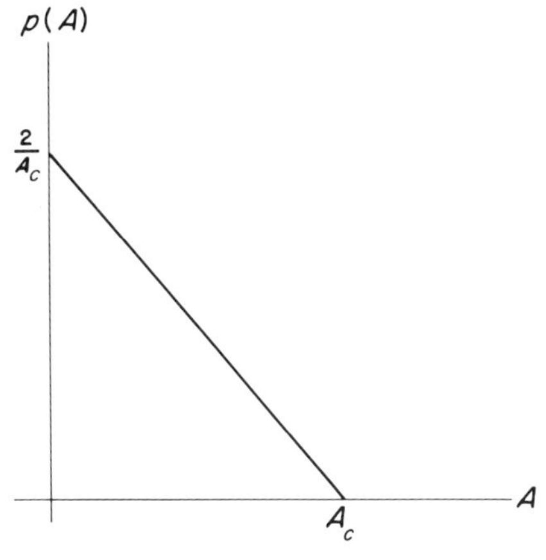

FIGURE 12.13

cross-sectional area is given by $q(A)=1-p(A)$, where

$$p(A)=\begin{cases}\dfrac{2}{A_c}\left(1-\dfrac{A}{A_c}\right) & \text{for} \quad 0\leqslant A\leqslant A_c,\\ 0 & \text{for} \quad A\geqslant A_c,\end{cases}$$

as shown in Fig. 12.13. The weight of each element is given by $w(A)=wA$, where w is a constant. Find the minimum-weight structure with a probability of failure less than $1-p$.

From Problem 12.5 we have $I(p,1)=w_1(A_1)=wA^*$, where $p_1(A^*{}_1)=p(A^*{}_1)=p$. Now solving the equation $p(A_1)=p$, we have $A^*{}_1=A_c-pA_c^2/2$ and $I(p,1)=w[A_c-pA_c^2/2]$. Next we use

$$I(p,2)=\min_{A_0}\left[wA_0+I(p_1,1)\right],$$

where

$$p_1=\frac{p}{p(A_0)}=\frac{pA_c}{2(1-A_0/A_c)},$$

and by direct computation we have

$$I(p_1,1)=w\left[A_c-\frac{p_1A_c^2}{2}\right]=w\left[A_c-\frac{pA_c^3}{4(1-A_0/A_c)}\right].$$

Substituting this expression for $I(p_1,1)$ in the functional equation above, we get $A^*{}_0=A_c-\sqrt{p}\,A_c^2/2$ and $I(p,2)=w[2A_c-\sqrt{p}\,A_c^2]$.

12.7 Now reconsider Problem 12.5 to maximize $P_N=\prod_0^{N-1}p_i(A_i)$ with constraints on the weight and cost of the structure. Formulate this problem as a multistage decision process and obtain the functional equation (12.9).

We assume that the total weight of the structure is W, and the total capital available for the strucutre is C. Then using $c_i(A_i)$ to denote the cost of the ith element, we obtain the following weight and cost constraints:

$$W_{i+1}=W_i-w_i(A_i),$$

$$C_{i+1}=C_i-c_i(A_i),$$

where $A_i\geqslant 0$, $W_0=W$, $C_0=C$, $0\leqslant w_i(A_i)\leqslant W_i\leqslant W$, and $0\leqslant c_i(A_i)\leqslant C_i\leqslant C$. If $I(W,C,N)=\max\prod_{i=0}^{N-1}p_i(A_i)$, then the functional equation (12.9) becomes

$$I(W_j,C_j,N-j)=\max_{A_j}\left[p_j(A_j)I(W_{j+1},C_{j+1},N-j-1)\right].$$

12.8 (Bottleneck problem) Consider an N-stage manufacturing process where at the ith stage the amount of raw material available is x^i. A fraction of x^i, denoted by y^i, is used to continue production of the raw material itself, while another fraction of x^i, denoted by z^i, is used to increase the capacity of production of the raw material denoted by c^i. The problem is to maximize the total amount of raw material available for the manufacturing process as given by $\Sigma_{i=0}^{N-1}(x^i - y^i - z^i)$.

We assume that an increase in x^i and c^i can be represented by the equations

$$x^{i+1} = (1+\alpha)y^i,$$

$$c^{i+1} = c^i + \beta z^i,$$

where $\alpha, \beta > 0$. Considering first the last stage of the process, we see that there is no point in increasing x^i and c^i, since this is the last stage, and so we use all the available raw material in the manufacturing process. It follows that

$$I(x^{N-1}, c^{N-1}, 1) = x^{N-1},$$

where $I(x^j, c^j, N-j)$ denotes the maximum amount of raw material that can be used for $N-j$ stages of the manufacturing process. Next, assuming that the manufacturing process is entering the second to last stage of its duration, we obtain from (12.9)

$$I(x^{N-2}, c^{N-2}, 2) = \max(x^{N-2} - y^{N-2} - z^{N-2} + x^{N-1}),$$

where $x^{N-1} = (1+\alpha)y^{N-2}$, and so

$$I(x^{N-2}, c^{N-2}, 2) = \max(x^{N-2} + \alpha y^{N-2} - z^{N-2}).$$

Since $y^{N-2}, z^{N-2} \geqslant 0$, the optimal policy is $z^{N-2} = 0$, and y^{N-2} should take on its maximum possible value. Now $y^{N-2} \leqslant x^{N-2}$ and $y^{N-2} \leqslant c^{N-2}$, so if $x^{N-2} \leqslant c^{N-2}$, then $y^{N-2} = x^{N-2}$ and

$$I(x^{N-2}, c^{N-2}, 2) = x^{N-2} + \alpha x^{N-2}.$$

Otherwise, when $c^{N-2} \leqslant x^{N-2}$,

$$y^{N-2} = c^{N-2} \quad \text{and} \quad I(x^{N-2}, c^{N-2}, 2) = x^{N-2} + \alpha c^{N-2}.$$

The functional equation can now be written as

$$I(x^{N-2}, c^{N-2}, 2) = x^{N-2} + \alpha \min(x^{N-2}, c^{N-2}).$$

We continue these iterations in a similar manner until all optimal policies and the optimal return $I(x^0, c^0, N)$ are obtained.

12.9 (Pest control) We consider now a problem of pest control where X and Y denote two kinds of insects, and we assume that insect X preys on insect Y. A pesticide is available for use against X, but it is not effective against Y. Let x^i and

y^i denote respectively the populations of X and Y at the ith instant of time. The problem is to minimize $ax^N + by^N$ with $a,b>0$ and N, an an arbitrary constant, by using the optimal dosage of the pesticide over a time period $[0, N-1]$.

Letting u^i denote the dosage of the pesticide at the ith instant of time, we can write the equations describing the population dynamics of X and Y as follows:

$$x^{i+1} = x^i + \alpha(u^i)x^i,$$

$$y^{i+1} = (1+\gamma)y^i - \beta(u^i)x^i,$$

where $\gamma \geqslant 0$ and the functions $\alpha(u^i)$ and $\beta(u^i)$ show the effects of the pesticide on the population and preying characteristics of X respectively. From physical considerations it is reasonable to assume that $\beta(u^i)$ is a decreasing function of u^i and that for large values of u^i, $\alpha(u^i)<0$. To be specific, we assume that $\alpha(u^i) = -\alpha(u^i)^2$ and $\beta(u^i) = e^{-\beta u^i}$. Now for $i = N-1$ and using the equations of population dynamics, we obtain

$$ax^N + by^N = ax^{N-1} - \left[be^{-\beta u^{N-1}} + a\alpha(u^{N-1})^2\right]x^{N-1} + b(1+\gamma)y^{N-1}.$$

since

$$\frac{d^2(ax^N + by^N)}{d(u^{N-1})^2} = b\beta^2 e^{\beta u^{N-1}} + 2a\alpha > 0,$$

we obtain the optimal value of u^{N-1} by setting $d(ax^N + by^N)/du^{N-1} = 0$. This leads to the equation

$$2a\alpha u^{N-1} - b\beta e^{-\beta u^{N-1}} = 0,$$

which can be solved graphically, as shown in Fig. 12.14, to obtain $(u^{N-1})^*$. Knowing $(u^{N-1})^*$, we can then compute the minimum value of $ax^N + by^N$ and use

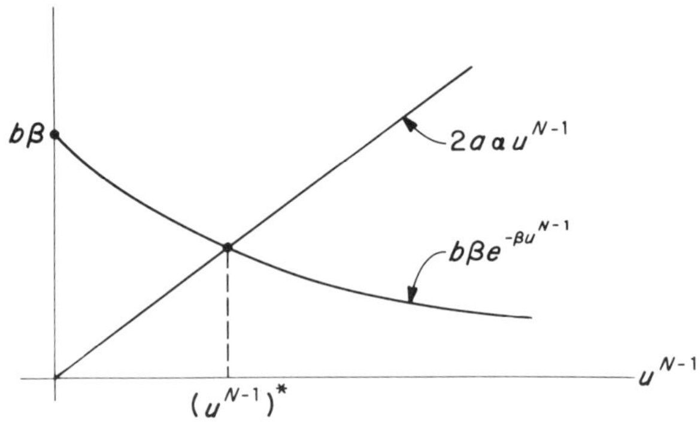

FIGURE 12.14

the functional equation

$$I(x^j,y^j,N-j)=\min_{u^j\geqslant 0} I(x^{j+1},y^{j+1},N-j-1)$$

to compute recursively the optimal values of u^j, $j=0,\ldots,N-1$.

12.10 (Economic growth) Consider a simplified version of the problem of stabilizing the economic growth of a nation by controlling the government's expenditure. Letting N denote the number of years between successive national elections, the problem of stabilizing the national economy can be formulated as that of maximizing the minimum national income over a period of N years. Determine the functional equation (12.9) for this problem.

In order to formulate this problem as a multistage decision process, we let x^i, c^i, and y^i denote the national income, consumer expenditure, and private investment respectively at the end of the ith year, and let u^i denote the government expenditure during the $(i+1)$st year. Then

$$x^i=c^i+y^i+u^{i+1} \quad \text{and} \quad c^i=\alpha(u^{i-1})x^{i-1},$$

where $\alpha(u^{i-1})$ is called the marginal propensity to consume, and we assume that it can be influenced by government actions. The private investment is assumed to depend on the net difference of the consumer expenditure over two successive years. Thus

$$y^i=\beta(u^{i-1})(c^i-c^{i-1}).$$

Combining the above equations, we obtain

$$x^i=\left[1+\beta(u^{i-1})\right]\alpha(u^{i-1})x^{i-1}-\alpha(u^{i-2})x^{i-2}+u^{i-1},$$

which can be simplified to

$$x^i=(1+\beta)\alpha x^{i-1}-\alpha\beta x^{i-2}+u^{i-1}$$

if we assume that $\alpha(u^{i-1})$ and $\beta(u^{i-1})$ are independent of government expenditure. The state variables of the economic system at the beginning of the ith year are given by x^{i-1} and x^{i-2}. Introducing a new variable $z^{i-1}=x^{i-1}-x^{i-2}$, we can rewrite the equation of the national income as

$$x^i=\alpha x^{i-1}+\alpha\beta z^{i-1}+u^{i-1}$$

and

$$z^i=(\alpha-1)x^{i-1}+\alpha\beta z^{i-1}+u^{i-1},$$

where the new state variables are x^{i-1} and z^{i-1}.

From practical considerations we assume that there is a ceiling, denoted by U, on the total government expenditure, which introduces the constraint $\sum_{i=0}^{N-1}u^i\leqslant U$. If we let $U_i=U-\sum_{j=0}^{i-1}u^j$, then $U_{i+1}=U_i-u^i$, and the yearly constraints on the government expenditure become $0 \leqslant u^i \leqslant U_i \leqslant U$. Defining

$$J(x^j, N-j, u^j, \ldots, u^{N-1}) = \min(x^{j+1}, \ldots, x^N)$$

and

$$I(x^j, N-j) = \max_{\substack{0 \leqslant u^i \leqslant U \\ i=j,\ldots,N-1}} J(x^j, N-j, u^j, \ldots, u^{N-1}),$$

we have from (12.9)

$$I(x^j, N-j) = \max_{0 \leqslant u^j \leqslant U_j} \min\left[x^{j+1}, I(x^{j+1}, N-j-1)\right].$$

We can obtain the optimal yearly government expenditures by starting with $j = N-1$ and iterating with the above functional equation.

12.11 In Problem 12.10, find the optimal yearly government expenditures over a two-year period. Take $\alpha = \beta = \frac{1}{2}$ and $U = 1$.

Consider first the problem of finding the optimal government expenditure over the 2nd year. Then

$$x^2 = \tfrac{1}{2}x^1 + \tfrac{1}{4}z^1 + u^1,$$

$$J(x^1, 1, u^1) = \min x^2 = \tfrac{1}{2}x^1 + \tfrac{1}{4}z^1 + u^1,$$

and

$$I(x^1, 1) = \max_{0 \leqslant u^1 \leqslant U_1} \left(\tfrac{1}{2}x^1 + \tfrac{1}{4}z^1 + u^1\right).$$

Now $J(x^1, 1, u^1)$ is linear in u^1, and hence it is maximized by $(u^1)^* = U_1 = 1 - u^0$, where U_1 is the amount of capital left for government expenditure at the beginning of the second year. By direct substitution of $(u^1)^*$ in $J(x^1, 1, u^1)$, we obtain $I(x^1, 1) = \frac{1}{2}x^1 + \frac{1}{4}z^1 + (1 - u^0)$.

Next consider the problem of finding the optimal government expenditure over a two-year period. In this case we have

$$x^1 = \tfrac{1}{2}x^0 + \tfrac{1}{4}z^0 + u^0,$$

and using the functional equation for $I(x^j, N-j)$ from Problem 12.10, we obtain

$$\begin{aligned} I(x^0, 2) &= \max_{0 \leqslant u^0 \leqslant 1} \left\{\min\left[x^1, I(x^1, 1)\right]\right\} \\ &= \max_{0 \leqslant u^0 \leqslant 1} \left\{\min\left[\tfrac{1}{2}x^0 + \tfrac{1}{4}z^0 + u^0, \tfrac{1}{2}x^1 + \tfrac{1}{4}z^1 + (1 - u^0)\right]\right\} \\ &= \max_{0 \leqslant u^0 \leqslant 1} \left\{\min\left[\tfrac{1}{2}x^0 + \tfrac{1}{4}z^0 + u^0, \tfrac{3}{8}x^0 + \tfrac{3}{16}z^0 + 1 - \tfrac{1}{4}u^0\right]\right\}. \end{aligned}$$

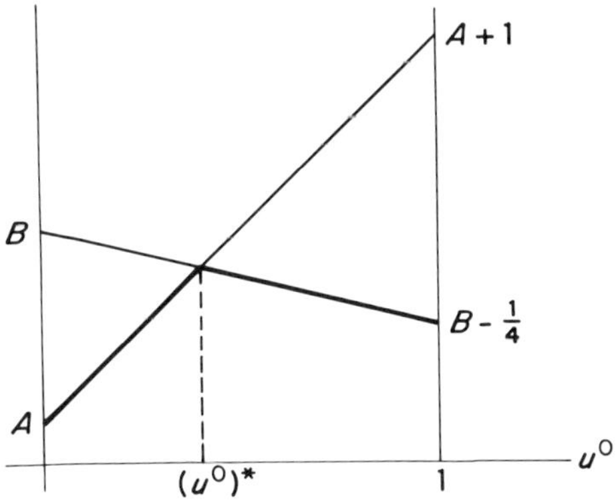

FIGURE 12.15

Letting $A=\frac{1}{2}x^0+\frac{1}{4}z^0$ and $B=\frac{3}{8}x^0+\frac{3}{16}z^0+1$, we have

$$I(x^0,2)=\max_{0\leqslant u^0\leqslant 1}\left[\min\left(A+u^0,B-\tfrac{1}{4}u^0\right)\right].$$

Figure 12.15 shows a plot of $A+u^0$ and $B-\frac{1}{4}u^0$ as functions of u^0 for the case $B>A$. The function $\min(A+u^0,B-\frac{1}{4}u^0)$ is also shown in Fig. 12.15 by the heavy line. From the figure we see that this function attains a maximum value inside the interval [0, 1], and the corresponding optimal value of u^0 is obtained from the equation $A+u^0=B-\frac{1}{4}u^0$ and is $(u^0)^*=\frac{4}{5}(B-A)$. For the case when $B\leqslant A$, the reader should verify by similar graphical means that $(u^0)^*=0$. The optimal value of u^1 in both cases is computed from the constraint equation $(u^1)^*=1-(u^0)^*$.

12.12 Consider the multistage decision process where the state equation is given by

$$X^{i+1}=x^i-u^i$$

and the objective function is

$$J(x^0,N,u^0,\ldots,u^{N-1})=\sum_{i=0}^{N-1}\left[(u^i)^2+(x^i-u^i)^2\right].$$

Find the optimal decision sequence.

Considering the decision process with only one stage remaining, we have

$$J(x^{N-1},1,u^{N-1})=(u^{N-1})^2+(x^{N-1}-u^{N-1})^2,$$

and since

$$\frac{d^2J(x^{N-1},1,u^{N-1})}{d(u^{N-1})^2}=4>0,$$

$J(x^{N-1},1,u^{N-1})$ is a convex function of u^{N-1}. Setting $dJ(x^{N-1},1,u^{N-1})/du^{N-1}=0$, we obtain the optimal value $(u^{N-1})^*=\frac{1}{2}x^{N-1}$ and $I(x^{N-1},1)=J(x^{N-1},1,(u^{N-1})^*)=\frac{1}{2}(x^{N-1})^2$.

Next, considering the decision process with two stages left, and using (12.9), we get

$$I(x^{N-2},2)=\min_{u^{N-2}}\left[(u^{N-2})^2+(x^{N-2}-u^{N-2})^2+I(x^{N-1},1)\right]$$

$$=\min_{u^{N-2}}\left[(u^{N-2})^2+(x^{N-2}-u^{N-2})^2+\tfrac{1}{2}(x^{N-1})^2\right],$$

where according to the state equation, $x^{N-1}=x^{N-2}-u^{N-2}$. Carrying out the minimization procedure by calculus as shown above, we obtain $(u^{N-2})^*=\frac{3}{5}x^{N-2}$ and $I(x^{N-2},2)=\frac{3}{5}(x^{N-2})^2$. Now we show by induction that $I(x,N)$ is a quadratic form in x and that the minimizing u^* is linear in x. Assuming that $I(x^{j+1},N-j-1)=K(x^{j+1})^2$ and using (12.9), we obtain

$$I(x^j,N-j)=\min_{u^j}\left[(u^j)^2+(x^j-u^j)^2+I(x^{j+1},N-j-1)\right]$$

$$=\min_{u^j}\left[(u^j)^2+(x^j-u^j)^2+K(x^{j+1})^2\right].$$

According to the state equation, $x^{j+1}=x^j-u^j$; and completing the minimization procedure, we obtain

$$(u^j)^*=\frac{(1+K)x^j}{2+K}\quad \text{and}\quad I(x^j,N-j)=\frac{(1+K)(x^j)^2}{2+K}.$$

The desired result follows by induction. If we denote by K_{N-j-1} the value of the constant K at the $(N-j-1)$st stage, then according to the above results

$$K_{N-j}=\frac{1+K_{N-j-1}}{2+K_{N-j-1}},$$

with $K_1=\frac{1}{2}$, $(u^j)^*=K_{N-j}x^j$, and $I(x^j,N-j)=K_{N-j}(x^j)^2$. Thus using the recursive equation for K_{N-j}, we can obtain $(u^j)^*$ and $I(x^j,N-j)$ without performing a minimization operation at each stage.

12.13 Find a solution to the infinite-stage approximation to the multistage decision process discussed in Problem 12.12.

Using (12.14) we get

$$I(x)=\min_u\left[u^2+(x-u)^2+I(x-u)\right].$$

To solve the above functional equation, we take $I^0(x-u)=K(x-u)^2$ and define

$$I^1(x)=\min_u\left[u^2+(x-u)^2+I^0(x-u)\right]$$

$$=\min_u\left[u^2+(K+1)(x-u)^2\right].$$

Now carrying out the minimization operation, we obtain $u^*=(1+K)x/(2+K)$ and $I^1(x)=(1+K)x^2/(2+K)$. We note that this approximation procedure will terminate at once if $I^0(x)=I^1(x)$, which requires that

$$\frac{1+K}{2+K}=K.$$

Solving this equation we get $K=0.62$, and so $u^*=0.62x$ and $I(x)=0.62x^2$.

12.14 (Curve fitting) Let $g(x)$ denote a real-valued function of the real variable x over the interval $[a,b]$. Then the problem of curve fitting consists of approximating $g(x)$ by a series of straight lines over $[a,b]$. Assuming we are allowed to use only two straight lines, formulate and solve this problem as a multistage decision process.

First consider the problem of approximating $g(x)$ by a single straight line on $[a,b]$, and let the equation of this straight line be $mx+c$. The best approximation is obtained by minimizing

$$J_0(a,b,m,c)=\int_a^b[g(x)-mx-c]^2\,dx$$

with respect to m and c. Since the integral is continuous in m and c, this minimization procedure can be carried out by means of calculus. Letting

$$M_0=\int_a^b g(x)\,dx,\qquad M_1=\int_a^b xg(x)\,dx$$

$$I_1=(b-a),\qquad I_2=\tfrac{1}{2}(b^2-a^2),\qquad I_3=\tfrac{1}{3}(b^3-a^3),$$

the optimal values of m and c must satisfy the equations

$$mI_3+cI_2=M_1,$$

$$mI_2+cI_1=M_0,$$

which yield

$$m^*=\frac{M_1I_1-M_0I_2}{I_3I_1-I_2^2}=\chi_1(a,b)\tag{12.22}$$

and

$$c^*=\frac{M_0I_3-M_1I_2}{I_3I_1-I_2^2}=\chi_2(a,b).\tag{12.23}$$

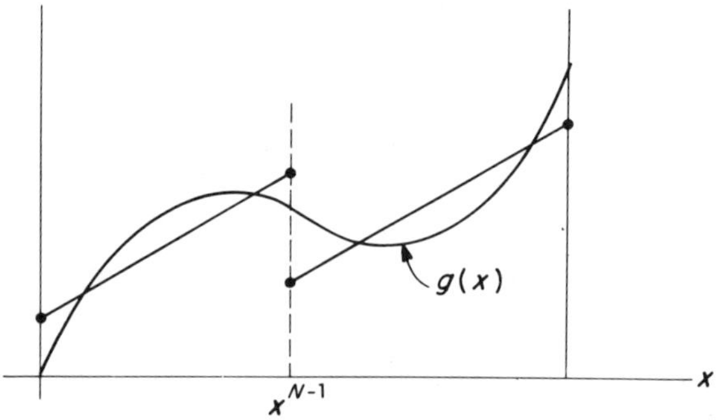

FIGURE 12.16

We introduce the functions $\chi_1(a,b)$ and $\chi_2(a,b)$ to remind the reader that m^* and c^* are functions of a and b.

Next we consider the problem of approximating $g(x)$ on $[a,b]$ by two straight lines. We divide the interval $[a,b]$ into two subintervals denoted by $[a,x_{N-1}]$ and $[x_{N-1},b]$, as shown in Fig. 12.16. The equation of the straight line on $[a,x_{N-1}]$ is given by $m_{N-1}x+c_{N-1}$, and that of the straight line on $[x_{N-1},b]$ is given by $m_N x+c_N$. The best approximation is obtained by minimizing

$$J_1(a,b,x_{N-1},m_{N-1},c_{N-1})=\int_a^{x_{N-1}}\left[g(x)-m_{N-1}x-c_{N-1}\right]^2dx$$
$$+\int_{x_{N-1}}^{b}\left[g(x)-m_N x-c_N\right]^2dx.$$

Defining $I_0(a,b)=J_0(a,b,m^*,c^*)$ and $I_1(a,b)=\min J_1(a,b,x_{N-1},c_{N-1})$, and using (12.9), we obtain

$$I_1(a,b)=\min_{m_{N-1},c_{N-1},x_{N-1}}\left[\int_a^{x_{N-1}}\left[g(x)-m_{N-1}x-c_{N-1}\right]^2dx+I_0(x_{N-1},b)\right].$$

Notice that for every value of x_{N-1} the optimal values of m_{N-1} and c_{N-1} are given by $m^*_{N-1}=\chi_1(a,x_{N-1})$ and $c^*_{N-1}=\chi_2(a,x_{N-1})$. Substituting these expressions for m_{N-1} and c_{N-1} into $J_1(a,b,x_{N-1},m_{N-1},c_{N-1})$, we obtain a function $\hat{J}_1(a,b,x_{N-1})$ which depends on x_{N-1} alone. Minimizing $\hat{J}_1(a,b,x_{N-1})$ with respect to x_{N-1}, we obtain the optimal value x^*_{N-1} and the corresponding optimal values of m_{N-1}, c_{N-1}, m_N, and c_N.

12.15 Find the best approximation to $g(x)=x$ over $[a,b]$ by two constant functions.

Defining $J_0(a,b,c_0)=\int_a^b(x-c_0)^2dx$ and $I_0(a,b)=\min_{c_0}J(a,b,c_0)$, we obtain by calculus $I_0(a,b)=\frac{1}{3}(b-a)^3$, and the optimal value of c_0 is $c^*_0=\frac{1}{2}(b+a)$. The solution to this minimization problem is shown in Fig. 12.17.

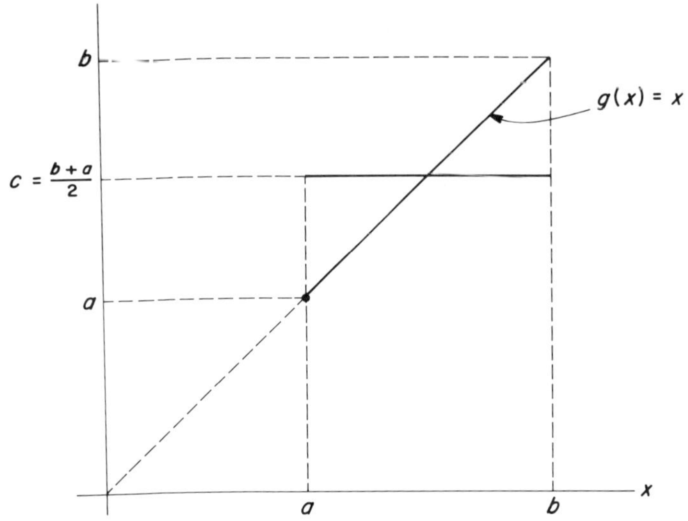

FIGURE 12.17

Next we consider the given problem of approximating $g(x)=x$ by two constant functions. We divide the interval $[0,1]$ into two subintervals $[0,u]$ and $[u,1]$, as shown in Fig. 12.18. The best approximation is obtained by using the recursive scheme of Problem 12.14, which is

$$I_1(0,1) = \min_{(c_1,u)} \left[\int_0^u (x-c_1)^2\,dx + I_0(u,1) \right]$$

$$= \min_{(c_1,u)} \left[\tfrac{1}{3}(u-c_1)^3 + \tfrac{1}{3}c_1^3 + \tfrac{1}{3}(1-u)^3 \right].$$

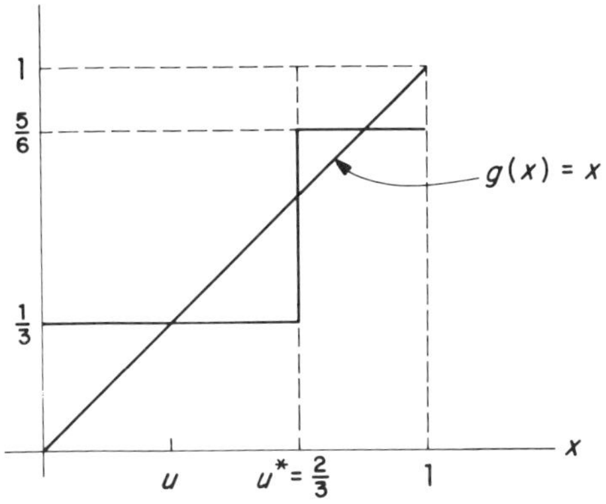

FIGURE 12.18

Solving the minimization problem by calculus, we have $u^* = \frac{2}{3}$ and $c_1^* = \frac{1}{3}$. Once u^* is known, we obtain $c_0^* = \frac{1}{2}(b + u^*) = \frac{5}{6}$. The optimal approximation is shown in Fig. 12.18.

Supplementary Problems

12.16 A research foundation has X dollars to allocate to N different research grants. Let x^i and $f_i(x^i)$ denote respectively the number of dollars allocated to the ith grant and the benefit obtained from the ith grant. Formulate this optimal allocation problem as a multistage decision process, approximate by an infinite-stage decision process, and obtain the optimum allocation policy. (*Hint:* See Examples 12.4 and 12.5.)

12.17 Suppose we want to divide $[0, 1]$ into two subintervals and approximate $g(x)$ by straight lines on these subintervals. Formulate this problem as a multistage decision process for $g(x) = x^2$ and obtain its solution. (*Hint:* See Problem 12.14.)

12.18 (Gold mining) Suppose we have two gold mines X and Y containing x and y tons of gold respectively. Let p_x denote the probability that a mining machine mines r_x tons of gold at mine X, $0 \leqslant r_x \leqslant x$, and remains in working order. Then $1 - p_x$ denotes the probability that the machine will be damaged before mining any gold at X. For mine Y, $0 \leqslant r_y \leqslant y$ and p_y are interpreted in the same way. Assuming that the machine is not damaged after the first use, it can be used again at either X or Y. However, after three jobs, the machine requires maintenance. Find the optimal policy for using the machine in order to maximize the expected amount of gold mined.

12.19 Consider the problem of finding the maximum value of $f(x) = x_1 x_2 \cdots x_N$ such that $x_i \geqslant 0$ and $\sum_{i=1}^{N} x_i = a$, where $a > 0$. Show that $I(a, 1) = a$ and

$$I(a, N) = \max_{0 \leqslant x \leqslant a} [xI(a - x, N - 1)].$$

12.20 Minimize $\sum_{i=0}^{N-1}[c(u^i)^2 + d(x^i - u^i)^2]$ with respect to u^i, $i = 0, \ldots, N-1$, where $c, d > 0$ and the state equation is given by $x^{i+1} = ax^i + bu^i$. (*Hint:* See Problem 12.12.)

12.21 Obtain the infinite-stage approximation to the multistage decision problem given in Problem 12.20 and solve the resulting functional equation. (*Hint:* See Problem 12.13.)

Chapter 13

Large-Scale Programming

In the first twelve chapters we have discussed various methods and algorithms for solving a variety of nonlinear programming problems. As in any nonlinear theory, each of these methods is useful only for a special class of problems. In addition to this limitation, each method also tends to break down as the problem dimensionality increases. In particular, convergence of the numerical algorithms cannot be obtained within reasonable computational costs when the number of problem variables $\mathbf{x}$ becomes large. Exactly what integer value corresponds to "large" is highly dependent on the problem structure. In general, the largest problems can be solved with linear programming, followed in order by quadratic programming, integer programming, certain nonlinear-programming direct algorithms, and dynamic programming. While linear programming problems having several thousand variables are routinely handled, dynamic programming problems containing more than two states may be prohibitive.

In order to extend any of these methods to solve problems of higher dimensionality, special techniques and modifications are needed. These methods have been studied extensively and now constitute the basis for the field of large-scale mathematical programming. In most cases, the techniques of large-scale mathematical programming exploit the structure of a particular class of problems to increase the dimensionality which can be handled. These methods of problem decomposition lead to the creation of several smaller subproblems which can be solved independently. The final step is to combine the subproblem solutions in such a way that they solve the original optimization problem. This step is referred to as the master problem or the second-level problem. When this approach is valid, the price which we pay for reduced problem dimensionality is that the individual subproblems must be solved many times before convergence of the second-level is obtained. Sometimes it is not practical to solve precisely for the optimum in very large problems. In these cases it may be sufficient to obtain a "good" feasible solution by terminating the higher level algorithm before complete convergence is obtained.

EXAMPLE 13.1 To illustrate the idea of decomposition, we consider the optimization of a typical industrial plant. This plant is composed of many individual unit processes, each of which is itself complex. Each process uses raw materials which are produced by some other process or purchased from an external supplier. In addition, each process may produce products which are further processed by other production units or sold externally. It is intuitively clear that the optimization of profit for the entire plant will require optimizing each unit process as well as optimizing the coordination or interaction between processes.

In order to decompose this problem to the solution of several independent subproblems, we will temporarily sever or ignore the interactions between processes. Then after optimizing each unit process independently, we will recombine the solutions in such a way that the overall optimum is obtained. However, since we severed the process interconnections initially, the first attempt at recombining the unit solutions may not be feasible, and hence several iterations of the second-level algorithm may be required.

To illustrate this decomposition process graphically, we show in Fig. 13.1 a diagram representing the product flow within a very simple plant where the variables $\mathbf{x}_i$ represent the amounts of products produced by the ith process. Figure 13.2 shows the optimization procedure for this plant, where the unit processes are optimized independently and then coordinated at a higher level.

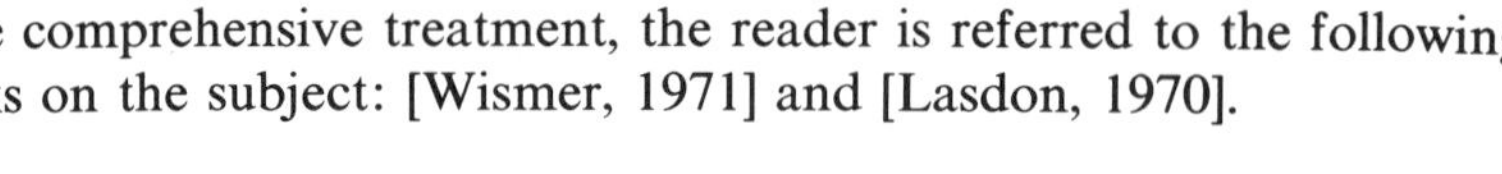

FIGURE 13.1

Since the literature in large-scale programming is very extensive, we will only be able to discuss a few of the many important topics here. For a more comprehensive treatment, the reader is referred to the following two books on the subject: [Wismer, 1971] and [Lasdon, 1970].

Lagrangian Decomposition

We consider first a problem of the form

$$\max \sum_{i=1}^{N} f_i(\mathbf{x}_i)$$

such that

$$\mathbf{g}_i(\mathbf{x}_1, \mathbf{x}_2, \ldots, \mathbf{x}_N) \geqslant 0, \qquad i = 1, 2, \ldots, N.$$

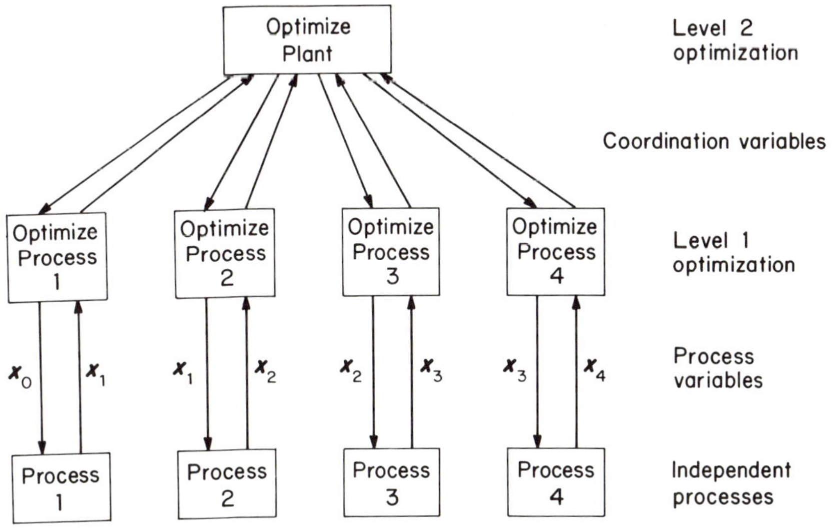

FIGURE 13.2

Forming the Lagrangian yields

$$L(\mathbf{x},\boldsymbol{\mu}) = \sum_{i=1}^{N} f_i(\mathbf{x}_i) + \sum_{i=1}^{N} \boldsymbol{\mu}_i^T \mathbf{g}_i(\mathbf{x}_1, \mathbf{x}_2, \ldots, \mathbf{x}_N),$$

or

$$L(\mathbf{x},\boldsymbol{\mu}) = \sum_{i=1}^{N} \left[f_i(\mathbf{x}_i) + \boldsymbol{\mu}_i^T \mathbf{g}_i(\mathbf{x}_1, \mathbf{x}_2, \ldots, \mathbf{x}_N) \right]. \tag{13.1}$$

In order to decompose the problem, we want to obtain a form for $L(\mathbf{x},\boldsymbol{\mu})$ such that the terms in brackets above are functions of the ith subsystem only. We begin by defining some additional variables $\mathbf{s}_j$, where

$$\mathbf{s}_j = \mathbf{x}_j, \qquad j = 1,2,\ldots,N. \tag{13.2}$$

This substitution can now be made in Eq. (13.1) if we also satisfy Eq. (13.2) by adding it to the Lagrangian. Thus

$$L(\mathbf{x},\boldsymbol{\lambda},\boldsymbol{\mu},\mathbf{s}) = \sum_{i=1}^{N} \left[f_i(\mathbf{x}_i) + \boldsymbol{\mu}_i^T \mathbf{g}_i(\mathbf{x}_i, \mathbf{s}_1, \mathbf{s}_2, \ldots, \mathbf{s}_j, \ldots, \mathbf{s}_N)_{j \neq i} + \boldsymbol{\lambda}_i^T (\mathbf{x}_i - \mathbf{s}_i) \right], \tag{13.3}$$

where $\boldsymbol{\lambda}_i = \mathbf{0}$ if $\mathbf{g}_j(\mathbf{x}_1, \ldots, \mathbf{x}_N)$ is not a function of $\mathbf{x}_i$ for all $j \neq i$.

The Lagrangian in (13.3) can now be decomposed in two different ways. The first way is to regard the $\mathbf{s}_i$ as fixed parameters. Then the Lagrangian for each subproblem can be written

$$L_i(\mathbf{x}_i, \boldsymbol{\lambda}_i, \boldsymbol{\mu}_i; \mathbf{s}_i) = f_i(\mathbf{x}_i) + \boldsymbol{\mu}_i^T \mathbf{g}_i(\mathbf{x}_i; \mathbf{s}_1, \mathbf{s}_2, \ldots, \mathbf{s}_j, \ldots, \mathbf{s}_N)_{j \neq i} + \boldsymbol{\lambda}_i^T(\mathbf{x}_i - \mathbf{s}_i)$$

$$i = 1, 2, \ldots, N, \tag{13.4}$$

where we have rearranged the terms so that the only variables not associated with the ith subsystem are the fixed parameters $\mathbf{s}_i$. The stationary points for each Lagrangian function $L_i(\mathbf{x}_i, \boldsymbol{\lambda}_i, \boldsymbol{\mu}_i; \mathbf{s}_i)$, $i = 1, 2, \ldots, N$, can now be determined independently as a function of $\mathbf{s}_i$. This task is referred to as a first-level optimization, and it is accomplished by applying the necessary conditions discussed in Chapter 4.

It is now necessary to determine a means of selecting the optimal values for the parameters $\mathbf{s}_i$. This task is referred to as a second-level optimization, and it is accomplished by methods to be discussed shortly. This method of Lagrangian decomposition is called *feasible decomposition*, since the interconnection constraints given by Eq. (13.2) are always satisfied. Thus even if the second-level optimization problem is terminated before complete convergence to the optimal values of $\mathbf{s}_i$ is obtained, the resulting solution to (13.4) is feasible.

Another way to decompose the Lagrangian in (13.3) is to regard the $\boldsymbol{\lambda}_i$ as fixed parameters. In this case we let

$$\mathbf{s}_{ij} = \mathbf{x}_j, \qquad i, j = 1, 2, \ldots, N, \quad i \neq j \tag{13.5}$$

where $\mathbf{s}_{ij}$ represents the presence of $\mathbf{x}_j$ in the ith subsystem.

Then, substituting in (13.1), and adding (13.5) with Lagrange multipliers $\boldsymbol{\lambda}_{ij}$, we have

$$L(\mathbf{x}, \boldsymbol{\lambda}, \boldsymbol{\mu}, \mathbf{s})$$

$$= \sum_{i=1}^{N} \left[f_i(\mathbf{x}_i) + \boldsymbol{\mu}_i^T \mathbf{g}_i(\mathbf{x}_i, \mathbf{s}_{i1}, \mathbf{s}_{i2}, \ldots, \mathbf{s}_{ij}, \ldots, \mathbf{s}_{iN})_{i \neq j} + \sum_{j=1}^{N} \boldsymbol{\lambda}_{ij}^T (\mathbf{x}_j - \mathbf{s}_{ij}) \right],$$

where $\boldsymbol{\lambda}_{ij} = \mathbf{0}$ if $g_i(\mathbf{x}_1, \ldots, \mathbf{x}_N)$ is not a function of $\mathbf{x}_j$.

Now the subsystem Lagrangians become

$$L_i(\mathbf{x}_i,\boldsymbol{\mu}_i,\mathbf{s}_{ij};\boldsymbol{\lambda}_{ij})=f_i(\mathbf{x}_i)+\boldsymbol{\mu}_i^T\mathbf{g}_i(\mathbf{x}_i,\mathbf{s}_{i1},\mathbf{s}_{i2},\ldots,\mathbf{s}_{ij},\ldots,\mathbf{s}_N)_{i\neq j}$$

$$+\sum_{j=1}^{N}\boldsymbol{\lambda}_{ji}^T\mathbf{x}_i-\sum_{j=1}^{N}\boldsymbol{\lambda}_{ij}^T\mathbf{s}_{ij}. \tag{13.6}$$

In (13.6) the stationary points with respect to $\mathbf{x}_i$, $\boldsymbol{\mu}_i$, and $\mathbf{s}_{ij}$ can be determined as functions of $\boldsymbol{\lambda}_{ij}$ by the Kuhn-Tucker theory. These values are determined in the first-level optimization, and the second-level optimization determines $\boldsymbol{\lambda}_{ij}$. This method of Lagrangian decomposition is called *dual feasible* because it can be shown that the second-level problem arising from (13.6) is the dual of the one arising from (13.4).

EXAMPLES

13.2 Use dual feasible decomposition to determine the subproblems for the large-scale linear programming problem

$$\min \sum_{i=1}^{N}\mathbf{c}_i^T\mathbf{x}_i \tag{13.7}$$

such that

$$\sum_{i=1}^{N}A_i\mathbf{x}_i=\mathbf{b},$$

$$B_i\mathbf{x}_i\geqslant\mathbf{b}_i, \tag{13.8}$$

$$\mathbf{x}_i\geqslant\mathbf{0}.$$

The only equations having any interconnections are (13.7) and (13.8). We can now write the Lagrangian as

$$L(\mathbf{x},\boldsymbol{\lambda})=\sum_{i=1}^{N}\left[\mathbf{c}_i^T\mathbf{x}_i+\boldsymbol{\lambda}_i^T(A_i\mathbf{x}_i-\mathbf{b})\right].$$

Thus the ith subproblem ($i=1,2,\ldots,N$) can be written as

$$\min_{\mathbf{x}_i}\,(\mathbf{c}_i^T+\boldsymbol{\lambda}_i^TA_i)\mathbf{x}_i-\boldsymbol{\lambda}_i^T\mathbf{b} \tag{13.9}$$

such that

$$B_i \mathbf{x}_i \geqslant \mathbf{b}_i,$$

$$\mathbf{x}_i \geqslant 0.$$

The optimal values of $\boldsymbol{\lambda}_i$ are determined by the second-level optimization, which in linear programming is called the *master problem*. Dual feasible decomposition is also called *goal coordination*, since the interconnection constraints (13.8) are satisfied by modifying the goals in the subproblems given by (13.9).

Most linear programming problems which can be solved by decomposition exhibit the block diagonal structure shown in Fig. 13.3.

x_1	x_2			x_N	
c_1	c_2	• • •		c_N	
A_1	A_2	• • •		A_N	b
B_1	0	• • •		0	b_1
	B_2			0	b_2
					⋮
	0	⋱	B_{N-1}	0	b_{N-1}
				B_N	b_N

FIGURE 13.3

13.3 Formulate two independent subproblems using feasible decomposition for a minimization problem where the objective function is

$$f(x_1, x_2) = x_1^2 - 12x_1 + \tfrac{1}{2}x_2^2 - x_1 x_2.$$

This objective function can be decomposed if we substitute $x_1 = s$ in the last term and write the Lagrangian

$$L(x_1, x_2, s, \lambda) = x_1^2 - 12x_1 + \tfrac{1}{2}x_2^2 - sx_2 + \lambda(x_1 - s).$$

For feasible decomposition s is treated as a known parameter. Thus the subproblem Lagrangians are

$$L_1(x_1, \lambda; s) = x_1^2 - 12x_1 + \lambda(x_1 - s),$$

$$L_2(x_2; s) = \tfrac{1}{2}x_2^2 - sx_2,$$

and the necessary conditions are

$$\frac{\partial L_1}{\partial x_1} = 2x_1 - 12 + \lambda = 0,$$

$$\frac{\partial L_1}{\partial \lambda} = x_1 - s = 0,$$

$$\frac{\partial L_2}{\partial x_2} = x_2 - s = 0.$$

Thus for subproblem 1,

$$x_1^*(s) = s, \qquad \lambda^*(s) = 12 - 2s,$$

and for subproblem 2,

$$x_2^*(s) = s.$$

13.4 For Example 13.3 sketch the two-level structure showing all information transfers.

The two-level structure is shown in Fig. 13.4.

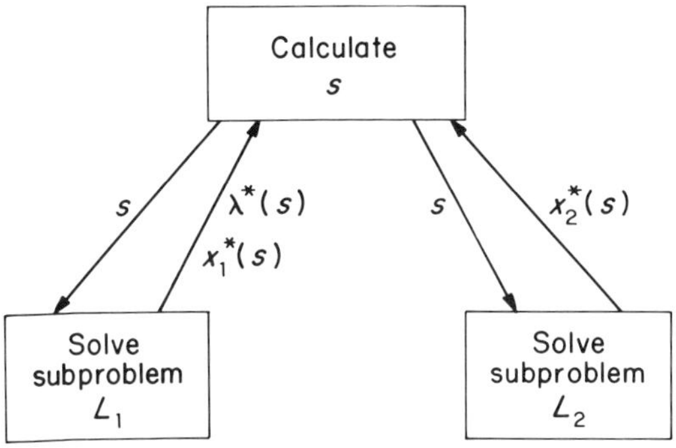

FIGURE 13.4

13.5 For Example 13.3, decompose the problem using the dual feasible method.

In this case λ is treated as a known parameter, the subproblem Lagrangians are

$$L_1(x_1;\lambda) = x_1^2 - 12x_1 + x_1,$$

$$L_2(x_2, s;\lambda) = \tfrac{1}{2}x_2^2 - sx_2 - \lambda s,$$

and the necessary conditions are

$$\frac{\partial L_1}{\partial x_1}=2x_1-12+\lambda=0,$$

$$\frac{\partial L_2}{\partial x_2}=x_2-s=0,$$

$$\frac{\partial L_2}{\partial s}=-x_2-\lambda=0.$$

Thus for subproblem 1,

$$x_1^*(\lambda)=6-\tfrac{1}{2}\lambda,$$

and for subproblem 2,

$$x_2^*(\lambda)=-\lambda, \qquad s^*(\lambda)=-\lambda.$$

13.6 For Example 13.5 sketch the two-level structure showing all information transfers.

The two-level structure is shown in Fig. 13.5.

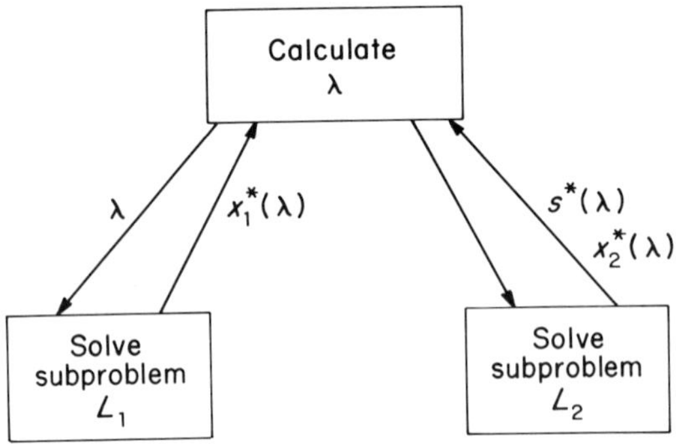

FIGURE 13.5

Feasible Methods

For the maximization problem

$$\max \sum_{i=1}^{N} f_i(\mathbf{x}_i)$$

such that

$$\mathbf{g}_i(\mathbf{x}_1, \mathbf{x}_2, \ldots, \mathbf{x}_N) \geqslant \mathbf{0}, \qquad i = 1, 2, \ldots, N,$$

the subproblem Lagrangians corresponding to feasible decomposition are given by Eq. (13.4). The Kuhn-Tucker necessary conditions for these subproblems are

$$\frac{\partial L_i}{\partial \mathbf{x}_i} = \frac{\partial f_i}{\partial \mathbf{x}_i} + \left(\frac{\partial \mathbf{g}_i}{\partial \mathbf{x}_i} \right)^T \boldsymbol{\mu}_i + \boldsymbol{\lambda}_i = \mathbf{0}, \tag{13.10a}$$

$$\frac{\partial L_i}{\partial \boldsymbol{\mu}_i} = \mathbf{g}_i(\mathbf{x}_i; \mathbf{s}_1, \mathbf{s}_2, \ldots, \mathbf{s}_j, \ldots, \mathbf{s}_N)_{i \neq j} \geqslant \mathbf{0}, \tag{13.10b}$$

$$\frac{\partial L_i}{\partial \boldsymbol{\lambda}_i} = \mathbf{x}_i - \mathbf{s}_i = \mathbf{0}, \tag{13.10c}$$

$$\boldsymbol{\mu}_i^T \mathbf{g}_i = 0, \qquad \boldsymbol{\mu}_i \geqslant \mathbf{0}. \tag{13.10d}$$

These necessary conditions are solved for $\mathbf{x}_i, \boldsymbol{\lambda}_i$, $i = 1, 2, \ldots, N$. The necessary condition which remains to be satisfied by the second level is

$$\frac{\partial L}{\partial \mathbf{s}_i} = \mathbf{0},$$

where

$$L = \sum_{i=1}^{N} L_i.$$

Thus

$$\frac{\partial L}{\partial \mathbf{s}_i} = \sum_{j=1}^{N} \left(\frac{\partial \mathbf{g}_i}{\partial \mathbf{s}_i} \right)^T \boldsymbol{\mu}_j - \boldsymbol{\lambda}_i = \mathbf{0}, \qquad i = 1, 2, \ldots, N. \tag{13.11}$$

This condition can be used to determine $\mathbf{s}_i$ by employing a gradient algorithm. If we rewrite Eq. (13.11) as

$$dL = \left[\sum_{j=1}^{N} \left(\frac{\partial \mathbf{g}_j}{\partial \mathbf{s}_i} \right)^T \boldsymbol{\mu}_j - \boldsymbol{\lambda}_i \right]^T d\mathbf{s}_i,$$

then we can choose

$$d\mathbf{s}_i = k\left[\sum_{j=1}^{N}\left(\frac{\partial \mathbf{g}_i}{\partial \mathbf{s}_i}\right)^T \boldsymbol{\mu}_j - \boldsymbol{\lambda}_i\right],$$

where $k>0$ to ensure that $dL>0$, which is necessary for maximization of L. Thus in discrete form the second-level gradient algorithm becomes

$$\mathbf{s}_i^{p+1} = \mathbf{s}_i^p + k\left[\sum_{j=1}^{N}\left(\frac{\partial \mathbf{g}_j}{\partial \mathbf{s}_i}\right)^T \boldsymbol{\mu}_j - \boldsymbol{\lambda}_i\right]_p, \tag{13.12}$$

where $k>0$. When this algorithm has converged, $\mathbf{s}_i^{p+1} \approx \mathbf{s}_i^p$ and (13.11) is satisfied. The only condition required for convergence of this two-level iterative procedure is the existence of feasible solutions to the subproblems.

EXAMPLE 13.7 Determine the second-level necessary condition and the gradient algorithm for Example 13.3.

The second-level necessary condition is

$$\frac{\partial L}{\partial s} = -x_2 - \lambda = 0,$$

or

$$x_2^* = -\lambda^*.$$

Thus (from Example 13.3)

$$s = 2s - 12, \qquad \text{or} \quad s^* = 12,$$

and

$$x_1^* = x_2 = 12, \qquad \lambda^* = -12.$$

The gradient algorithm is

$$s^{p+1} = s^p + k(-x_2 - \lambda)^p,$$

or

$$s^{p+1} = s^p + k(s-12)^p, \qquad k<0.$$

Thus

$$s^{p+1} = (1-k)s^p + 12k, \qquad k<0,$$

which converges toward the value $s^* = 12$, since for $0 < k < 1$ and $s^p = 0$, $s^{p+1} < 12$ and $1 - k > 0$; and for $k > 1$ and $s^p = 0$, $s^{p+1} > 12$ and $1 - k < 0$.

Dual Feasible Methods

For a dual feasible method, the subproblem Lagrangians were given by Eq. (13.6). The Kuhn-Tucker necessary conditions for these subproblems are

$$\frac{\partial L_i}{\partial \mathbf{x}_i} = \frac{\partial f_i}{\partial \mathbf{x}_i} + \left(\frac{\partial \mathbf{g}_i}{\partial \mathbf{x}_i} \right)^T \boldsymbol{\mu}_i + \sum_{j=1}^{N} \boldsymbol{\lambda}_{ji} = \mathbf{0}, \tag{13.13a}$$

$$\frac{\partial L_i}{\partial \boldsymbol{\mu}_i} = \mathbf{g}_i(\mathbf{x}_i, \mathbf{s}_{i1}, \mathbf{s}_{i2}, \ldots, \mathbf{s}_{ij}, \ldots, \mathbf{s}_{iN})_{i \neq j} \geqslant \mathbf{0}, \tag{13.13b}$$

$$\frac{\partial L_i}{\partial \mathbf{s}_{ij}} = \left(\frac{\partial \mathbf{g}_i}{\partial \mathbf{s}_{ij}} \right)^T \boldsymbol{\mu}_i - \boldsymbol{\lambda}_{ij} = \mathbf{0}, \qquad j = 1, 2, \ldots, N, \tag{13.13c}$$

$$\boldsymbol{\mu}_i^T \mathbf{g}_i = 0, \qquad \boldsymbol{\mu}_i \geqslant \mathbf{0}. \tag{13.13d}$$

These necessary conditions are solved for $\mathbf{x}_i$, $\boldsymbol{\mu}_i$, and $\mathbf{s}_{ij}$ as functions of $\boldsymbol{\lambda}_{ij}$. The necessary condition which remains to be satisfied by the second level is

$$\frac{\partial L}{\partial \boldsymbol{\lambda}_{ij}} = \mathbf{0},$$

where

$$L = \sum_{i=1}^{N} L_i.$$

Thus

$$\frac{\partial L}{\partial \boldsymbol{\lambda}_{ij}} = \mathbf{x}_j - \mathbf{s}_{ij} = \mathbf{0}, \qquad i,j = 1, 2, \ldots, N, \tag{13.14}$$

which can be solved for $\boldsymbol{\lambda}_{ij}$ using a gradient algorithm as explained previously. In this case,

$$d\boldsymbol{\lambda}_{ij} = k(\mathbf{x}_j - \mathbf{s}_{ij}) \tag{13.15}$$

or

$$\boldsymbol{\lambda}_{ij}^{p+1} = \boldsymbol{\lambda}_{ij}^{p} + k(\mathbf{x}_j - \mathbf{s}_{ij})^p,$$

where $k<0$, since the second-level problem is the dual of the first-level problem, and hence requires minimization because the first level was a maximization. This duality is proved in Problem 13.1, p. 339. The convergence condition for this two-level iterative procedure requires that subproblem maxima exist satisfying Eqs. (13.13). It is not sufficient that (13.13) be satisfied by minima or saddle points.

EXAMPLES

13.8 Determine the second-level necessary condition and the gradient algorithm for Example 13.5.

The second-level necessary condition for dual feasible decomposition is

$$\frac{\partial L}{\partial \lambda}=x_1-s=0,$$

and from Example 13.5,

$$x_1^*(\lambda)=6-\tfrac{1}{2}\lambda, \qquad s^*(\lambda)=-\lambda,$$

and

$$6-\tfrac{1}{2}\lambda=-\lambda.$$

So

$$\lambda^*=-12 \quad \text{and} \quad x_1^*=x_2^*=s^*=12.$$

The gradient algorithm is

$$\lambda^{p+1}=\lambda^p+k(x_1-s)^p, \qquad k<0,$$

or

$$\lambda^{p+1}=\lambda^p+k\left(6+\tfrac{1}{2}\lambda\right)^p=\left(1+\frac{k}{2}\right)\lambda^p+6k, \qquad k<0.$$

Thus if $\lambda^p=0$, then $\lambda^{p+1}=6k>0$ and the gradient algorithm moves away from the optimum $\lambda^*=-12$. We examine the reason in Example 13.9.

13.9 Determine why the second-level algorithm fails to converge in Example 13.8.

The convergence conditions require that both $L_1(x_1;\lambda)$ and $L_2(x_2,s;\lambda)$ have local minima with respect to x_1, x_2, and s for given values of λ.

From Example 13.8,

$$\frac{\partial L_1}{\partial x_1} = 2x_1 - 12 + \lambda,$$

so

$$\frac{\partial^2 L}{\partial x_1^2} = 2 > 0,$$

which implies that L_1 has a minimum with respect to x_1. For L_2 we have

$$\frac{\partial L_1}{\partial x_2} = x_2 - s,$$

$$\frac{\partial L_2}{\partial s} = -x_2 - \lambda,$$

so

$$\frac{\partial^2 L}{\partial x_2^2} = 1, \qquad \frac{\partial^2 L_2}{\partial s^2} = 0,$$

$$\frac{\partial^2 L_2}{\partial x_2 \partial s} = \frac{\partial^2 L_2}{\partial s\, \partial x_2} = -1.$$

Thus the Hessian of L_2 is

$$\begin{bmatrix} 1 & -1 \\ -1 & 0 \end{bmatrix},$$

and the principal minors are 1 and -1, so L_2 has a saddle point at $x_2^*(\lambda), s^*(\lambda)$ (see Chapter 2, p. 38). Thus L_2 does not have a local minimum with respect to x_2 and s, so the convergence conditions are not satisfied. This accounts for the behavior of the second-level gradient algorithm in Example 13.8.

Economic Interpretation

The dual feasible decomposition of the Lagrangian given by Eq. (13.6) has a very interesting economic interpretation which arises from the dual behavior of the first- and second-level problems.

Consider the behavior of a perfectly competitive economy, and let the ith subsystem represent a particular sector of the economy. Now assume:

$f_i(\mathbf{x}_i)$ is the "net profit" earned by producing finished goods $\mathbf{x}_i$ including (negative) cost of production. See Problem 13.4.

$\mathbf{P}_i = \sum_{j=1}^{N} \boldsymbol{\lambda}_{ji}$ is the selling price for each unit of $\mathbf{x}_i$.

$\boldsymbol{\lambda}_{ij}$ is the purchase price for each unit of raw material $\mathbf{s}_{ij}$ purchased from the jth sector of the economy.

Then

$$L_i = f_i + \mathbf{P}_i^T \mathbf{x}_i - \sum_{j=1}^{N} \boldsymbol{\lambda}_{ij}^T \mathbf{s}_{ij}$$

is the total net profit earned by the ith sector at current prices $\mathbf{P}_i$, $\boldsymbol{\lambda}_{ij}$ in all transactions with its environment. If resource constraints exist, then the terms

$$(\boldsymbol{\mu}_i)^T \mathbf{g}_i(\mathbf{x}_i, \mathbf{s}_{i1}, \mathbf{s}_{i2}, \ldots, \mathbf{s}_{ij}, \ldots, \mathbf{s}_{iN})_{i \neq j} = 0$$

in Eq. (13.6) may be included without loss of generality.

Now consider the job of the manager of the ith sector, which is to maximize profit L_i at given prices $\boldsymbol{\lambda}_{ij}$, $\mathbf{P}_i$ by choosing the production levels $\mathbf{x}_i$. Assuming that prices are set by a central agency, the task of this agency is to adjust prices until supply equals demand. Thus if

$$E_i(\boldsymbol{\lambda}_{ij}) = \mathbf{s}_{ij} - \mathbf{x}_i, \qquad j = 1, 2, \ldots, N, \tag{13.16}$$

represents the excess demand function for commodity $\mathbf{x}_i$, the central agency will adjust prices $\boldsymbol{\lambda}_{ij}$ until demand ($\mathbf{s}_{ji}$) equals supply ($\mathbf{x}_i$). At this point $E_i(\boldsymbol{\lambda}_{ij}) = 0$, $j = 1, 2, \ldots, N$, and prices cease to fluctuate. Since in Eq. (13.15)

$$d\boldsymbol{\lambda}_{ij} = kE_i(\boldsymbol{\lambda}_{ij}), \qquad k > 0, \quad j = 1, 2, \ldots, N,$$

when $E_i(\boldsymbol{\lambda}_{ij}) = 0$ for $j = 1, 2, \ldots, N$, prices do not fluctuate and the second-level algorithm has converged. On the other hand, when demand exceeds supply, then $E_i > \mathbf{0}$ and prices rise. Similarly, when supply exceeds demand, then $E_i < \mathbf{0}$ and prices fall.

To be precise, this economic analogy would have required us to write the interconnection constraint as

$$\mathbf{x}_i = \sum_{j=1}^{N} \mathbf{s}_{ij}$$

rather than $\mathbf{x}_i = \mathbf{s}_{ij}$, $j = 1, 2, \ldots, N$. Then

$$E_i(\boldsymbol{\lambda}_i) = \sum_{j=1}^{N} \mathbf{s}_{ij} - \mathbf{x}_i$$

in Eq. (13.16), and the above discussion is unchanged.

Solved Problems

13.1 Apply the dual feasible method to the problem

$$\min f(\mathbf{x})$$

such that

$$g_i(\mathbf{x}) = 0, \qquad i = 1, 2, \ldots, M,$$

and show that the first-level and the second-level problems are duals. In this case $g_i(\mathbf{x}) = 0$, $i = 1, 2, \ldots, M$, represent the interconnection constraints between subproblems.

Assume that a finite local minimum exists at $\mathbf{x}^\circ$, i.e.,

$$\min f(\mathbf{x}) = f(\mathbf{x}^\circ) \tag{13.17}$$

and

$$g_i(\mathbf{x}^\circ) = 0, \qquad i = 1, 2, \ldots, M.$$

Now form the Lagrangian

$$L(\mathbf{x}, \boldsymbol{\lambda}) = f(\mathbf{x}) + \boldsymbol{\lambda}^T \mathbf{g}(\mathbf{x})$$

and assume that there exists a $\boldsymbol{\lambda}^\circ$ associated with $\mathbf{x}^\circ$ and that within some region of $\boldsymbol{\lambda}^\circ$, a finite minimum of $L(\mathbf{x}, \boldsymbol{\lambda})$ exists.

Now define

$$L(\mathbf{x}^*, \boldsymbol{\lambda}) = \min_{\mathbf{x}} L(\mathbf{x}, \boldsymbol{\lambda}) = H(\boldsymbol{\lambda}), \tag{13.18}$$

where $\mathbf{x}^*$ corresponds to the minimum with respect to $\mathbf{x}$, given an arbitrary value of $\boldsymbol{\lambda}$. If $\boldsymbol{\lambda} = \boldsymbol{\lambda}^\circ$, then $\mathbf{x} = \mathbf{x}^*$. Let X be the set of all $\mathbf{x}$ such that $\mathbf{g}(\mathbf{x}) = \mathbf{0}$. Then from

(13.17) and (13.18),

$$\min_{\mathbf{x}\in X} L(\mathbf{x},\boldsymbol{\lambda}) = \min_{\mathbf{x}\in X} f(\mathbf{x}) = f(\mathbf{x}^\circ) = H(\boldsymbol{\lambda}^\circ).$$

But

$$H(\boldsymbol{\lambda}) = \min_{\text{all } \mathbf{x}} L(\mathbf{x},\boldsymbol{\lambda}) \leqslant \min_{x\in X} L(\mathbf{x},\boldsymbol{\lambda}) = H(\boldsymbol{\lambda}^\circ).$$

Thus

$$H(\boldsymbol{\lambda}) \leqslant H(\boldsymbol{\lambda}^\circ),$$

and $H(\boldsymbol{\lambda})$ takes on a maximum at $\boldsymbol{\lambda}^\circ$ while $f(\mathbf{x})$ takes on a minimum at $\mathbf{x}^*$. Hence the first-level problem and the second-level problem are duals.

In the dual feasible method, $\boldsymbol{\lambda}$ is the variable in the second level and

$$dH(\boldsymbol{\lambda}) = dL(\mathbf{x}^*(\boldsymbol{\lambda}),\boldsymbol{\lambda}) = \left(\frac{\partial L}{\partial \boldsymbol{\lambda}}\right)^T d\boldsymbol{\lambda} + \left(\frac{\partial L}{\partial \mathbf{x}}\right)^T \frac{d\mathbf{x}}{d\boldsymbol{\lambda}}\, d\boldsymbol{\lambda}$$

$$= \mathbf{g}(\mathbf{x}^*)^T d\boldsymbol{\lambda} + \left[\frac{\partial f}{\partial \mathbf{x}}(\mathbf{x}^*) + \boldsymbol{\lambda}^T \frac{\partial \mathbf{g}}{\partial \mathbf{x}}(\mathbf{x}^*)\right]^T \frac{d\mathbf{x}^*}{d\boldsymbol{\lambda}}\, d\boldsymbol{\lambda}.$$

But

$$\frac{\partial f}{\partial \mathbf{x}}(\mathbf{x}^*) + \boldsymbol{\lambda}^T \frac{\partial \mathbf{g}}{\partial \mathbf{x}}(\mathbf{x}^*) = \mathbf{0}$$

is ensured by the necessary conditions in the first level, so

$$\frac{dH}{d\boldsymbol{\lambda}}(\boldsymbol{\lambda}) = \mathbf{g}(\mathbf{x}^*).$$

Thus to maximize $H(\boldsymbol{\lambda})$, $dH(\mathbf{x}^*(\boldsymbol{\lambda})) > 0$, so we let

$$d\boldsymbol{\lambda} = k\mathbf{g}(\mathbf{x}^*),$$

where $k > 0$. In the form of a discrete algorithm, this expression can be written

$$\boldsymbol{\lambda}^{p+1} = \boldsymbol{\lambda}^p + k\mathbf{g}(\mathbf{x}^*)^p, \qquad k > 0.$$

13.2 The solution of the subsystem necessary conditions for both feasible and dual feasible decomposition is considerably simplified if we combine the methods and solve for both $\mathbf{s}_i$ and $\boldsymbol{\lambda}_i$ in the second level. For this approach, which we call *relaxation*, determine the first- and second-level necessary conditions.

Working with Eqs. (13.10) and (13.11), the first-level necessary conditions are

$$\frac{\partial L_i}{\partial \mathbf{x}_i}=\frac{\partial f_i}{\partial \mathbf{x}_i}+\left(\frac{\partial g_i}{\partial \mathbf{x}_i}\right)^T\boldsymbol{\mu}_i+\boldsymbol{\lambda}_i=\mathbf{0},$$

$$\frac{\partial L_i}{\partial \boldsymbol{\mu}_i}=\mathbf{g}_i(\mathbf{x}_i;\mathbf{s}_1,\mathbf{s}_2,\ldots,\mathbf{s}_j,\ldots,\mathbf{s}_N)_{i\neq j}\geqslant\mathbf{0},$$

$$\boldsymbol{\mu}_i{}^T\mathbf{g}_i=0,\qquad \boldsymbol{\mu}_i\geqslant 0,$$

which are solved for $\mathbf{x}_i(\boldsymbol{\lambda},\mathbf{s})$ and $\boldsymbol{\mu}_i(\boldsymbol{\lambda},\mathbf{s})$. The second-level necessary conditions are

$$\frac{\partial L}{\partial \mathbf{s}_i}=\mathbf{0},\qquad \frac{\partial L_i}{\partial \boldsymbol{\lambda}_i}=\mathbf{x}_i-\mathbf{s}_i=0,$$

or

$$\mathbf{s}_i=\mathbf{x}_i,$$

$$\boldsymbol{\lambda}_i=\sum_{j=1}^{N}\left(\frac{\partial \mathbf{g}_j}{\partial \mathbf{s}_i}\right)^T\boldsymbol{\mu}_j.$$

Thus using this method, both $\mathbf{s}_i$ and $\boldsymbol{\lambda}_i$ can be determined explicitly as functions of $\mathbf{x}$ and $\boldsymbol{\mu}$, and the gradient algorithm is not required. This method of decomposition is preferred in most problems because of its simplicity.

13.3 Explain clearly how to use the relaxation method of Problem 13.2 to solve a problem where $N=2$.

The necessary conditions to be solved by each first-level subproblem and also by the second level are shown in Fig. 13.6. The procedure is to guess $\boldsymbol{\lambda}_1^\circ$, $\boldsymbol{\lambda}_2^\circ$, $\mathbf{s}_1^\circ$, $\mathbf{s}_2^\circ$ and solve each subproblem independently for $\boldsymbol{\mu}_1(\boldsymbol{\lambda}_1^\circ,\mathbf{s}_2^\circ)$, $\mathbf{x}_1(\boldsymbol{\lambda}_1^\circ,\mathbf{s}_2^\circ)$, $\boldsymbol{\mu}_2(\boldsymbol{\lambda}_2,\mathbf{s}_1^\circ)$, and $\mathbf{x}_2(\boldsymbol{\lambda}_2^\circ,\mathbf{s}_1^\circ)$. Then calculate $\boldsymbol{\lambda}_1^{p+1}$, $\boldsymbol{\lambda}_2^{p+1}$, $\mathbf{s}_1^{p+1}$, and $\mathbf{s}_2^{p+1}$ by substituting the subproblem solutions into the second-level equations. Continue in this way until convergence is obtained.

13.4 Consider the operation of a decentralized plant where the ith sector is shown in Fig. 13.7. The variables associated with the ith sector are:

$\mathbf{X}_i$ = outputs going to other sectors for further processing,

$\mathbf{Y}_i$ = finished outputs requiring no further processing,

$\mathbf{M}_i$ = controls used to produce $\mathbf{Y}_i$ and $\mathbf{Z}_i$,

$\mathbf{S}_i$ = inputs from other sectors

Determine the objective function for each sector which will maximize profit for the entire plant. Use dual feasible decomposition, and assume that the sector models

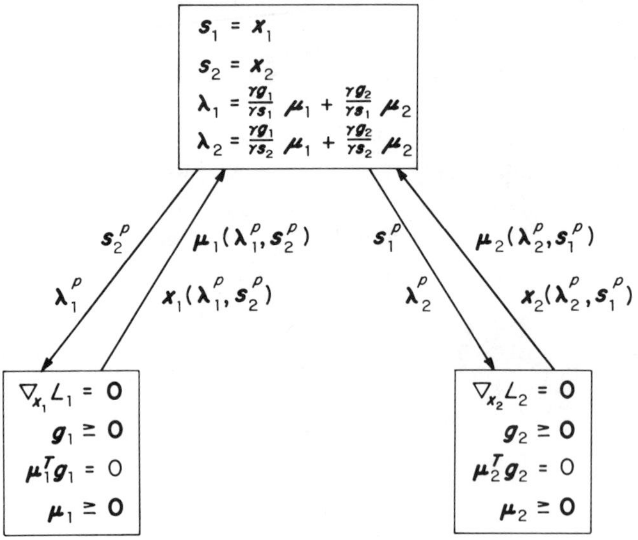

FIGURE 13.6

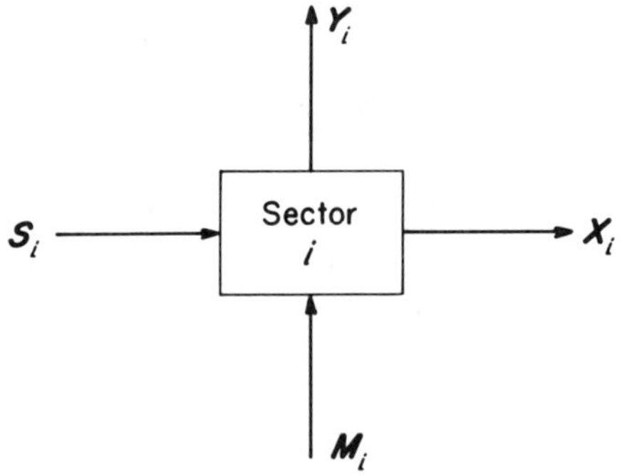

FIGURE 13.7

are given by

$$\mathbf{X}_i = \mathbf{T}_i(\mathbf{M}_i, \mathbf{S}_i)$$

$$\mathbf{Y}_i = \mathbf{U}_i(\mathbf{M}_i, \mathbf{S}_i)$$

over N sectors. The profit maximization is given by the objective function

$$\max \sum_{i=1}^{N} f_i(\mathbf{M}_i, \mathbf{Y}_i).$$

We assume that the interconnections between sectors can be modeled by

$$\mathbf{S}_i = \sum_{j=1}^{N} \mathbf{C}_{ij}\mathbf{X}_j,$$

where $\mathbf{C}_{ij} = [c_{ij}]$, a matrix of zeros and ones such that

$$c_{ij} = \begin{cases} 1 & \text{if } \mathbf{X}_j \to \mathbf{S}_i, \\ 0 & \text{otherwise.} \end{cases}$$

The Lagrangian is

$$L(\mathbf{M},\mathbf{X},\mathbf{Y},\mathbf{S},\boldsymbol{\lambda}) = \sum_{i=1}^{N} \left[f_i(\mathbf{M}_i, \mathbf{Y}_i) + \boldsymbol{\lambda}_i^T \left(\sum_{j=1}^{N} \mathbf{C}_{ij}\mathbf{X}_j - \mathbf{S}_i \right) \right]$$

or

$$L(\mathbf{M},\mathbf{S},\boldsymbol{\lambda}) = \sum_{i=1}^{N} \left[f_i(\mathbf{M}_i, \mathbf{U}_i(\mathbf{M}_i, \mathbf{S}_i)) + \boldsymbol{\lambda}_i^T \sum_{j=1}^{N} \mathbf{C}_{ij}\mathbf{T}_j(\mathbf{M}_j, \mathbf{S}_j) - \boldsymbol{\lambda}_i^T \mathbf{S}_i \right]$$

Now define

$$\mathbf{P}_i^T = \sum_{j=1}^{N} \boldsymbol{\lambda}_i^T \mathbf{C}_{ji}$$

and

$$f_i'(\mathbf{M}_i, \mathbf{S}_i) = f_i(\mathbf{M}_i, \mathbf{U}_i(\mathbf{M}_i, \mathbf{S}_i)).$$

Then

$$L(\mathbf{M},\mathbf{S};\boldsymbol{\lambda}) = \sum_{i=1}^{N} L_i(\mathbf{M}_i, \mathbf{S}_i; \boldsymbol{\lambda}_i),$$

where L_i is the objective function for the ith sector given by

$$L_i(\mathbf{M}_i, \mathbf{S}_i; \boldsymbol{\lambda}_i) = f_i'(\mathbf{M}_i, \mathbf{S}_i) + \mathbf{P}_i^T \mathbf{T}_i(\mathbf{M}_i, \mathbf{S}_i) - \boldsymbol{\lambda}_i^T \mathbf{S}_i, \qquad i = 1, 2, \ldots, N,$$

and P_i represents the selling price of products $\mathbf{X}_i$, while $\boldsymbol{\lambda}_i$ represents the cost of raw materials $\mathbf{S}_i$.

Because of duality, the sector problem is to minimize L_i by choosing $\mathbf{M}_i$ and $\mathbf{S}_i$. The prices $\boldsymbol{\lambda}_i$ and $\mathbf{P}_i$ are provided by some central agency responsible for coordinating all sectors (see Problem 13.5).

13.5 Derive the gradient algorithm which could be used by the central agency in Problem 13.4 to maximize profit for the entire plant.

The task of the central agency is to choose $\boldsymbol{\lambda}_i$, $i=1,2,\ldots,N$, so that

$$\mathbf{S}_i = \sum_{j=1}^{N} \mathbf{C}_{ij}\mathbf{X}_j.$$

Since $\mathbf{P}_i = \sum_{j=1}^{N} \mathbf{C}_{ij}^T\boldsymbol{\lambda}_j$ and $\mathbf{C}_{ij}$ is given, the price $\mathbf{P}_i$ is known once $\boldsymbol{\lambda}_j$ is specified for $j=1,2,\ldots,N$. Defining

$$\mathbf{E}_i(\boldsymbol{\lambda}) = \mathbf{S}_i - \sum_{j=1}^{N} \mathbf{C}_{ij}\mathbf{X}_j,$$

$$\mathbf{E}_i(\boldsymbol{\lambda}) = \mathbf{S}_i(\boldsymbol{\lambda}) - \sum_{j=1}^{N} \mathbf{C}_{ij}\mathbf{T}_j(\mathbf{M}_j(\boldsymbol{\lambda}), \mathbf{S}_i(\boldsymbol{\lambda}))$$

leads to the gradient algorithm

$$\boldsymbol{\lambda}_i^{p+1} = \boldsymbol{\lambda}_i^p + k\mathbf{E}_i(\boldsymbol{\lambda}^p), \qquad i=1,2,\ldots,N,$$

where $k<0$ since we are minimizing $L(\mathbf{M},\mathbf{S},\boldsymbol{\lambda})$ with respect to $\boldsymbol{\lambda}$. For $N=2$, this procedure leads to a decentralized plant operation as shown in Fig. 13.8

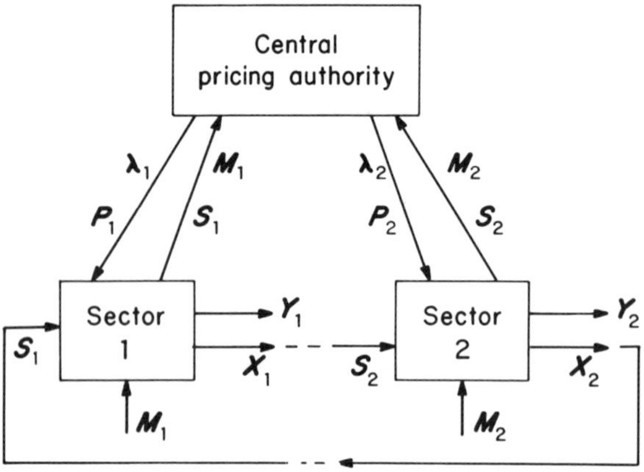

FIGURE 13.8

13.6 Use dual feasible decomposition to solve the problem

$$\min x_1^2 + x_2^2 + m^2$$

such that

$$x_1 = -x_2 + m,$$

$$x_2 = x_1 + k.$$

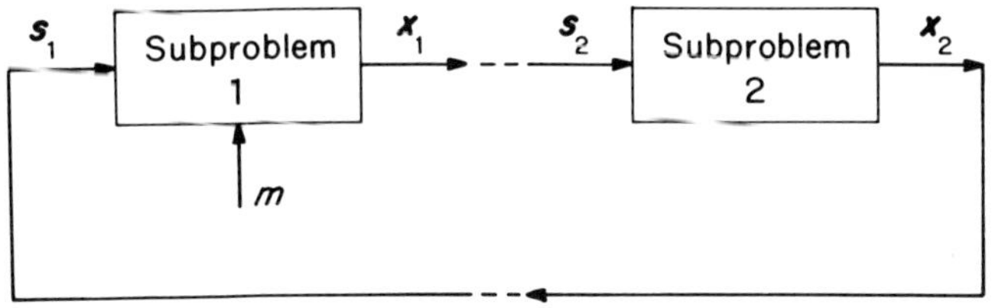

FIGURE 13.9

The problem is decomposed as shown in Fig. 13.9, which leads to the Lagrangian

$$L(\mathbf{x}, m, \mathbf{s}, \boldsymbol{\lambda}) = x_1^2 + x_2^2 + m^2 + \lambda_1(x_1 - s_2) + \lambda_2(x_2 - s_1).$$

Substituting the constraints to eliminate $\mathbf{x}$ yields

$$L(m, \mathbf{s}, \boldsymbol{\lambda}) = (-s_1 + m)^2 + (s_2 + k)^2 + m^2 + \lambda_1(-s_1 + m - s_2) + \lambda_2(s_2 + k - s_1).$$

Then

$$L_1(s_1, m; \lambda_1, \lambda_2) = (-s_1 + m)^2 + m^2 - \lambda_1 s_1 + \lambda_1 m - \lambda_2 s_1,$$

$$L_2(s_2; \lambda_1, \lambda_2) = (s_2 + k)^2 - \lambda_1 s_2 + \lambda_2 s_2 + k\lambda_2,$$

and

$$\frac{\partial L_1}{\partial s_1} = -2(-s_1 + m) - \lambda_1 - \lambda_2 = 0,$$

$$\frac{\partial L_1}{\partial m} = 2(-s_1 + m) + 2m + \lambda_1 = 0,$$

$$\frac{\partial L_2}{\partial s_2} = 2(s_2 + k) - \lambda_1 + \lambda_2 = 0,$$

which yields

$$m^*(\boldsymbol{\lambda}) = \tfrac{1}{2}\lambda_2,$$

$$s^*(\boldsymbol{\lambda}) = \lambda_2 + \tfrac{1}{2}\lambda_1,$$

$$s_2^*(\boldsymbol{\lambda}) = \tfrac{1}{2}(\lambda_1 - \lambda_2) - k.$$

The second-level gradient algorithm is given by

$$\lambda_1^{p+1} = \lambda_1^p + KE_1(\boldsymbol{\lambda}^p), \qquad K > 0,$$

$$\lambda_2^{p+1} = \lambda_2^p + KE_2(\boldsymbol{\lambda}^p), \qquad K > 0,$$

where

$$E_1(\boldsymbol{\lambda}) = x_1 - s_2 = -s_1 + m - s_2$$
$$= -\lambda_1 + k,$$
$$E_2(\boldsymbol{\lambda}) = x_2 - s_1 = s_2 + k - s_1$$
$$= -\tfrac{3}{2}\lambda_2.$$

When $E_1(\boldsymbol{\lambda}) = 0$, we have $\lambda_1 = k$ and $x_1^* = s_2$; and when $E_2(\boldsymbol{\lambda}) = 0$, we have $\lambda_2 = 0$ and $x_2^* = s_1$. Thus

$$m^*(\boldsymbol{\lambda}) = 0, \qquad s_1{}^*(\boldsymbol{\lambda}) = \frac{k}{2}, \qquad s_2{}^*(\boldsymbol{\lambda}) = -\frac{k}{2},$$

and

$$x_1^* = -\frac{k}{2}, \qquad x_2^* = \frac{k}{2}, \qquad m^* = 0.$$

13.7 For Problem 13.6 show that the convergence conditions for the second-level algorithm are satisfied.

From duality, the second-level problem requires maximization, since the first-level problems require minimization. The second-level gradient algorithm converges if the optimal solutions to each of the first-level subproblems correspond to local minima. For subproblem 1, the Hessian matrix is

$$\begin{bmatrix} \dfrac{\partial^2 L_1}{\partial s_1^2} & \dfrac{\partial^2 L_1}{\partial s_1 \partial m} \\ \dfrac{\partial^2 L_1}{\partial m \partial s_1} & \dfrac{\partial^2 L_1}{\partial m^2} \end{bmatrix} = \begin{bmatrix} 2 & -2 \\ -2 & 4 \end{bmatrix}$$

The principal minors, 2 and 4, are both positive, so the solution $s_1^*(\boldsymbol{\lambda}), m^*(\boldsymbol{\lambda})$ to this problem is a minimum of $L_1(s_1, m; \boldsymbol{\lambda})$. For subproblem 2, the Hessian is

$$\frac{\partial^2 L_2}{\partial s_2^2} = 2 > 0,$$

so $s_2^*(\boldsymbol{\lambda})$ is a minimum of $L_2(s_2; \boldsymbol{\lambda})$. Thus the convergence condition is satisfied and the gradient algorithm given in Problem 13.6 converges.

13.8 For Problem 13.6 state the computational procedure for dual feasible decomposition.

The computational steps are:

1. Guess λ_1^0, λ_2^0, and set $p=0$.
2. Minimize $L_1(s_1, m; \lambda_1^p, \lambda_2^p)$.
3. Minimize $L_2(s_2; \lambda_1^p, \lambda_2^p)$.
4. Compute $E_1(\boldsymbol{\lambda}^p)$ and $E_2(\boldsymbol{\lambda}^p)$.
5. If $\max_i |E_i(\boldsymbol{\lambda}^p) - E_i(\boldsymbol{\lambda}^{p-1})| < \varepsilon$, stop.
6. Otherwise compute λ_i^{p+1} and λ_2^{p+1}.
7. Increase p by 1 and go to step 2.

13.9 Determine the solution to Problem 13.6 by using feasible-decomposition methods.

Using feasible decomposition, the Lagrangians for each subproblem are given by

$$L_1(\lambda_1, m; s_1, s_2) = (-s_1 + m)^2 + m^2 + \lambda_1(-s_1 + m - s_2),$$

$$L_2(\lambda_2; s_1, s_2) = (s_2 + k)^2 + \lambda_2(s_2 + k - s_1),$$

and the necessary conditions are

$$\frac{\partial L_1}{\partial \lambda_1} = -s_1 + m - s_2 = 0,$$

$$\frac{\partial L_2}{\partial m} = 2(-s_1 + m) + 2m + \lambda_1 = 0,$$

$$\frac{\partial L_2}{\partial \lambda_2} = s_2 + k - s_1 = 0.$$

The solution to subproblem 1 is given by

$$m^*(\mathbf{s}) = s_1 + s_2,$$

$$\lambda_1{}^*(\mathbf{s}) = -2s_1 - 4s_2.$$

However, the necessary condition for subproblem 2 does not yield a solution $\lambda_2^*(\mathbf{s})$, and hence the method of feasible decomposition cannot be used to solve this problem.

13.10 For computational reasons, the decomposition methods discussed in this chapter are only practical when the subproblems can be solved easily and when the number of interconnecting variables is small. One class of optimization problems which has these properties involves the optimization of certain partial differential equations.

Consider the problem

$$\min_{m(x)} \int_0^1 \left[u^2(x) + m^2(x)\right] dx$$

such that

$$\frac{\partial^2 u}{\partial x^2} = m,$$

with boundary conditions

$$u(0)=0, \qquad u(1)=1.$$

If we consider discrete segments of length Δx, then the problem can be stated (approximately) as

$$\min_{m_i} \sum_{i=1}^{N} (u_i^2 + m_i^2), \qquad i=1,2,\ldots,N$$

such that

$$\frac{1}{\Delta x}\left[\frac{u_{i+1}-u_i}{\Delta x} - \frac{u_i - u_{i-1}}{\Delta x}\right] = m_i,$$

or

$$u_{i+1} + u_{i-1} - 2u_i = (\Lambda x)^2 m_i, \qquad i=1,2,\ldots,N,$$

where

$(u_{i+1} - u_i/\Delta x)$ is called the *forward difference approximation* of $\partial u/\partial x$, and $(u_i - u_{i-1}/\Delta x)$ is called the *backward difference approximation* of $\partial u/\partial x$.

In matrix form, the problem can be written

$$\min_{\mathbf{M}} (\mathbf{U}^T\mathbf{U} + \mathbf{M}^T\mathbf{M})$$

such that

$$\mathbf{AU} = (\Delta x)^2\mathbf{M} + \mathbf{F},$$

where

$$\mathbf{A} = \begin{bmatrix} -2 & 1 & & & & & & \\ 1 & -2 & 1 & & & & 0 & \\ & 1 & -2 & 1 & & & & \\ & & \cdot & \cdot & \cdot & & & \\ & & & \cdot & \cdot & \cdot & & \\ & 0 & & & \cdot & \cdot & \cdot & \\ & & & & & 1 & -2 & 1 \\ & & & & & & 1 & -2 \end{bmatrix}$$

and

$$\mathbf{F}^T = [0,0,0,\ldots,1].$$

Since the off-diagonal terms in the **A**-matrix represent the only interconnections between nodes, it is convenient to decompose the problem by partitioning the **A**-matrix. Thus

$$\mathbf{A}\left[\begin{array}{c|c|c} \mathbf{A}_{11} & & 0 \\ \hline & \mathbf{A}_{22} & \\ \hline 0 & & \mathbf{A}_{33} \end{array}\right],$$

where the interconnections between subproblems correspond to the nonzero terms outside the block-diagonal matrices. It has been found in practice [Wismer, 1971] that the subproblems of this type converge rapidly because of the strongly diagonal submatrices, and the multilevel problem also converges rapidly because of the relative sparsity of the **A**-matrix.

Supplementary Problems

13.11 Use dual feasible decompositon to solve the subproblems for

$$\min_{m}\,(x_2-k)^2+m^2$$

such that

$$x_1=\tfrac{1}{2}x_3^2+2m,$$

$$x_2=\tfrac{1}{2}x_1,$$

$$x_3=x_2.$$

13.12 Determine the second-level gradient algorithm for Problem 13.11. Are the convergence conditions satisfied?

13.13 Solve Problem 13.11 using feasible decomposition.

13.14 Solve Problem 13.11 using the relaxation method discussed in Problem 13.2.

13.15 Solve Problem 13.6 using the relaxation method discussed in Problem 13.2.

13.16 For Problem 13.15, state precisely the computational procedure.

13.17 Formulate Problem 13.10 using dual feasible decomposition.

13.18 Formulate Problem 13.10 using feasible decomposition.

13.19 Formulate Problem 13.10 using the relaxation method discussed in Problem 13.2.

13.20 Explain the significance of Problem 13.10 in terms of a physical example with all variables carefully defined.

13.21 Formulate the following two-dimensional optimization problem using dual feasible decomposition:

$$\min_{m} \int_0^1 \int_0^1 \left[u^2(x_1,x_2) + m^2(x_1,x_2) \right] dx_1\, dx_2$$

such that

$$\frac{\partial^2 u}{\partial x_1^2} + \frac{\partial^2 u}{\partial x_2^2} = m(x_1,x_2),$$

with boundary conditions

$$u(0,x_2) = u(1,x_2) = u(x_1,0) = u(x_2,0) = 1.$$

Appendix
Linear Programming

This appendix contains a brief discussion of linear programming problems and the simplex algorithm for their solution. The literature on linear programming is vast, and hence only a brief outline of the theory can be presented here. A knowledge of linear programming is very helpful in understanding nonlinear programming algorithms, since many such algorithms are based on extensions of the simplex method.

In linear programming, both the objective function and the constraint equations are linear in the independent variables. The objective function is

$$f(\mathbf{x}) = (\mathbf{c}, \mathbf{x}) = \mathbf{c}^T\mathbf{x}, \tag{A.1}$$

where $\mathbf{c}, \mathbf{x} \in E^n$. The constraint equations are

$$\begin{aligned} \mathbf{Ax} &\leqslant \mathbf{b}, \\ \mathbf{x} &\geqslant \mathbf{0}, \end{aligned} \tag{A.2}$$

where $\mathbf{b} \in E^m$ and $\mathbf{A}$ is an $m \times n$ matrix. The set of all feasible solutions is defined by $\mathbf{M} = \{\mathbf{x} | \mathbf{x} \in E^n, \ \mathbf{Ax} \leqslant \mathbf{b}, \ \mathbf{x} \geqslant \mathbf{0}\}$, and the linear programming problem consists of finding an extreme value of $f(\mathbf{x})$ on M. To be specific we shall consider $\max_{\mathbf{x} \in M} f(\mathbf{x})$ in the rest of this appendix.

The linear programming problem is said to be in the *normal form* if $\mathbf{b} \geqslant \mathbf{0}$ in (A.2), in which case $\mathbf{0} \in M$, ensuring that M is nonempty. Since an arbitrary linear programming problem can always be transformed into this form, we will restrict our discussion to linear programming in the normal form. If we transform constraint inequalities into equalities by introducing *slack variables* $\mathbf{u} = \mathbf{b} - \mathbf{Ax}$, the constraints (A.2) are equivalent to

$$\begin{aligned} \mathbf{Ax} + \mathbf{u} &= \mathbf{b}, \\ \mathbf{x}, \mathbf{u} &\geqslant \mathbf{0}. \end{aligned} \tag{A.3}$$

Letting $\mathbf{B} = [\mathbf{A}|\mathbf{I}]$ and $y = [\mathbf{x}^T, \mathbf{u}^T]^T$, we observe that every $\mathbf{y} \geqslant \mathbf{0}$ satisfying $\mathbf{By} = \mathbf{b}$ is a feasible solution of (A.3). The set of all feasible solutions can now be defined as $Y = \{\mathbf{y} | \mathbf{y} \in E^{m+n}, \ \mathbf{By} = \mathbf{b}, \ \mathbf{y} \geqslant \mathbf{0}\}$. If we augment the given vector $\mathbf{c}$ by attaching m zero components, the linear programming problem can be alternatively stated as $\max_{\mathbf{y} \in Y} (\mathbf{c}, \mathbf{y})$.

EXAMPLES

A.1 Consider an investment company that invests money in real estate and oil wells, and assume there are ceilings, denoted by b_1 and b_2 respectively, on the amount of capital that can be in invested in each venture. The total amount of capital available for investment is limited by b_3. The problem is to maximize the total profit of the investment company under the assumption that the profit is linearly related to the investment.

Let x_1 and x_2 denote respectively the investments in the real-estate and oil-well ventures. Then from the definitions, $x_1, x_2, b_1, b_2, b_3 \geqslant 0$, and the constraints on the investments are $0 \leqslant x_1 \leqslant b_1$, $0 \leqslant x_2 \leqslant b_2$, and $0 \leqslant x_1 + x_2 \leqslant b_3$. Letting $f(x_1,x_2) = c_1x_1 + c_2x_2$ denote the total profit from the investments x_1 and x_2, the problem is to maximize $f(x_1,x_2)$ with respect to x_1 and x_2 while satisfying the given constraints.

Letting

$$\mathbf{b} = [b_1, b_2, b_3]^T \quad \text{and} \quad \mathbf{A} = \begin{bmatrix} 1 & 0 \\ 0 & 1 \\ 1 & 1 \end{bmatrix},$$

the set of all feasible solutions becomes $M = \{\mathbf{x} | \mathbf{x} \in E^2, \mathbf{Ax} \leqslant \mathbf{b}, \mathbf{x} \geqslant \mathbf{0}\}$, as shown by the shaded region in Fig. A.1. The level curves of $f(x_1,x_2) =$ constant are also shown in this figure under the assumption of $c_1 = 2$ and $c_2 = 4$. As we move across these level curves in the direction of the arrow in Fig. A.1, the value of $f(x_1,x_2)$ increases until it reaches a maximum at $\mathbf{x}^* = [b_3 - b_2, b_2]^T$, where $f(x_1^*, x_2^*) = 2(b_3 - b_2) + 4b_2$. The optimum solution to the problem is given by $\mathbf{x}^*$. Since the oil wells produce twice as much profit as the real estate for the same amount of investment, it is obvious

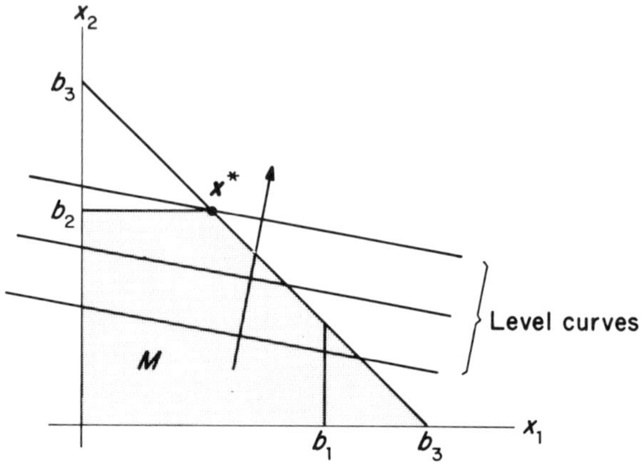

FIGURE A.1

that one should invest as much as allowable in the oil wells, and the rest in the real estate. The optimum solution $\mathbf{x}^*$ obtained graphically satisfies this distribution of capital between the two investments.

A.2 A farmer requires that the food ration for his cattle contain at least 4, 12, and 2 units of nutrients A, B, and C respectively. Two kinds of fodder are available, and the unit weights of each contain the following amounts of the three nutrients:

	Amount	
Nutrient	Fodder 1	Fodder 2
A	1	1
B	6	2
C	0	1

The cost per unit weight of each kind of fodder is 1 and 2 monetary units respectively, and the problem is to find a minimum-cost feed satisfying the nutrient requirements.

Letting x_1 and x_2 denote respectively the amounts of fodder 1 and fodder 2 in the feed, we obtain from physical considerations that $x_1, x_2 \geqslant 0$. To satisfy the need for nutrient A in the feed, we should require that $x_1 + x_2 \geqslant 4$. Similar considerations for nutrients B and C in the feed give $6x_1 + 2x_2 \geqslant 12$ and $0x_1 + x_2 \geqslant 2$. Letting

$$\mathbf{b} = [4, 12, 2]^T \quad \text{and} \quad \mathbf{A} = \begin{bmatrix} 1 & 1 \\ 6 & 2 \\ 0 & 1 \end{bmatrix},$$

the set of all feasible solutions becomes $M = \{\mathbf{x} | \mathbf{x} \in E^2,\ \mathbf{A}x \geqslant \mathbf{b},\ \mathbf{x} \geqslant \mathbf{0}\}$, as shown by the shaded region in Fig. A.2. The cost of the feed is given by $f(x_1, x_2) = x_1 + 2x_2$, and the level curves of $f(x_1, x_2)$ are also shown in Fig. A.2. As we move across these level curves in the direction of the arrow, the value of $f(x_1, x_2)$ decreases until it reaches a minimum at $\mathbf{x}^* = [2, 2]^T$. The optimum mix of feed is given by $x_1^* = 2$ units and $x_2^* = 2$ units.

A.3 Consider the set of all feasible solutions $M = \{\mathbf{x} | \mathbf{x} \in E^2,\ \mathbf{Ax} \leqslant \mathbf{b},\ \mathbf{x} \geqslant \mathbf{0}\}$, where

$$\mathbf{b} = [4, 4]^T \quad \text{and} \quad \mathbf{A} = \begin{bmatrix} -1 & 1 \\ 1 & -2 \end{bmatrix},$$

as shown by the shaded region in Fig. A.3. Find the maximum value of $f(x_1, x_2) = x_1 + x_2$ on this set M.

As shown in Fig. A.3, the set M is unbounded. As we move across the level curves $f(x_1, x_2) = \text{constant}$ in the direction of the arrow, the value of

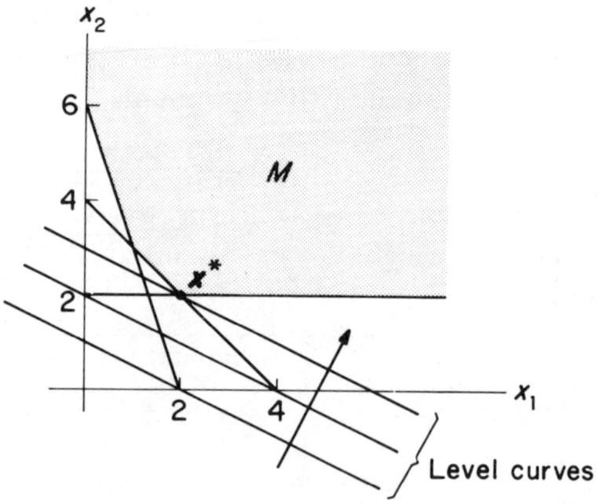

FIGURE A.2

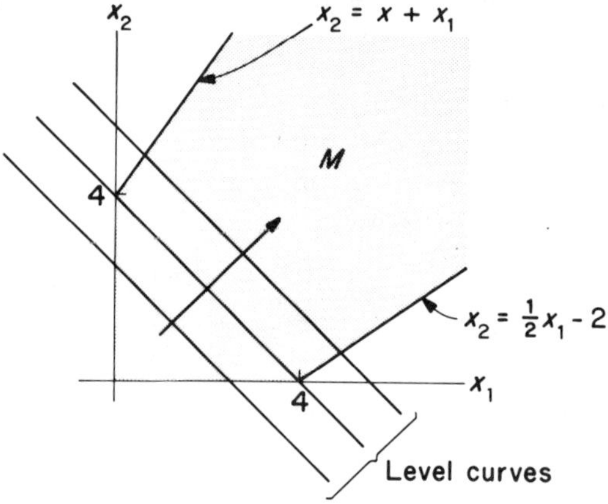

FIGURE A.3

$f(x_1, x_2)$ does not attain a maximum value for finite values of x_1 and x_2. Hence this problem does not have any finite optimum solutions.

A.4 Consider the problem of finding the maximum value of $f(x_1, x_2) = -x_1 + 4x_2$ on the set M shown by the shaded region of Fig. A.4. Solve this problem graphically.

The level curves of $f(x_1, x_2) = \text{constant}$ are also shown in Fig. A.4. As we move across these level curves in the direction of the arrow, the value of $f(x_1, x_2)$ increases and attains a maximum on M at $\mathbf{x}^* = [0, 2]^T$. Hence $\mathbf{x}^* = [0, 2]^T$ is the optimal solution of this problem. We note that although M is unbounded, we have a finite optimum solution.

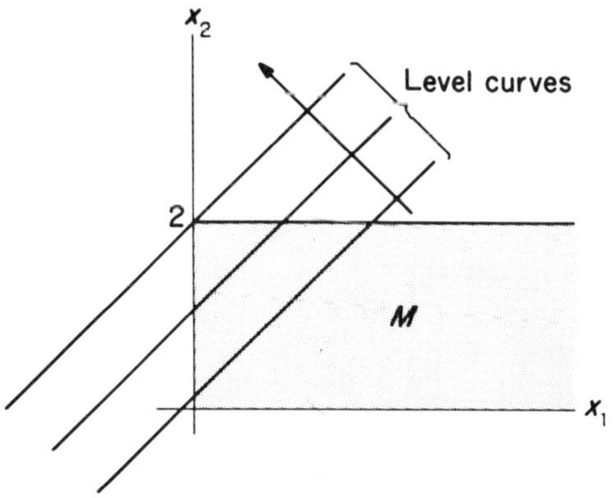

FIGURE A.4

Basic Feasible Solutions

We assume that the linear inequality constraints given by $\mathbf{Ax} \leqslant \mathbf{b}$ have been transformed into equality constraints $\mathbf{By} = \mathbf{b}$ by means of slack variables and that the linear programming problem is in the normal form, i.e., $\mathbf{b} \geqslant \mathbf{0}$. The equality constraints $\mathbf{By} = \mathbf{b}$ can be rewritten as

$$\begin{aligned} b_{11}y_1 + \cdots + b_{1,m+n}y_{m+n} &= b_1, \\ \vdots \qquad\qquad \vdots \qquad & \quad \vdots \\ b_{m1}y_1 + \cdots + b_{m,m+n}y_{m+n} &= b_m, \end{aligned} \tag{A.4}$$

where b_{ij} denotes the (i,j)th element of the $m \times (m+n)$ matrix $\mathbf{B}$, and y_j and b_j are the jth components of the vectors $\mathbf{y}$ and $\mathbf{b}$ respectively. In (A.4) we have m linear algebraic equations in the $m+n$ unknowns $y_1, \ldots, y_{m+n}$. If we set any of the n unknown variables equal to zero, then from (A.4) we obtain m equations in m unknowns. Without loss of generality we assume that $y_1 \neq 0, \ldots, y_m \neq 0$ and $y_{m+1} = \cdots = y_{m+n} = 0$. Then (A.4) can be solved to obtain unique values of $y_1, \ldots, y_m$ provided that the column vectors

$$\begin{bmatrix} b_{11} \\ \vdots \\ b_{m1} \end{bmatrix}, \ldots, \begin{bmatrix} b_{1m} \\ \vdots \\ b_{mm} \end{bmatrix} \tag{A.5}$$

are linearly independent and hence form a basis in E^m. Such a solution is called a *basic solution*, and the corresponding $y_1,\ldots,y_m$ are called *basic variables*. If $\mathbf{B}_1$ denotes the $m\times m$ matrix whose column vectors are the linearly independent vectors given in (A.5), then a basic solution can be written as $\mathbf{y}=[(\mathbf{B}_1^{-1}\mathbf{b})^T,(\mathbf{0})^{\mathrm{T}}]^{\mathrm{T}}$. Next we observe that a basic solution is not always a feasible solution, since we may have $y_j<0$, $1\leqslant j\leqslant m$, and so we call a basic solution a *basic feasible solution* if $y_j\geqslant 0$ for all j, $1\leqslant j\leqslant m$. If $y_j<0$ for all j, $1\leqslant j\leqslant m$, then we have a *nondegenerate* solution. If one or more of the basic variables are zero, we call the solution *degenerate*. Starting with a given basic solution, we can construct a new basic solution by setting any one of the m basic variables equal to zero and replacing it by a nonbasic variable. The nonbasic variable chosen for this purpose is called an *entering variable*, and the basic variable set equal to zero is called a *leaving variable*.

Assume that $\mathbf{y}^0$ is a basic feasible solution, which implies that $y_i^0>0$, $i=1,\ldots,m$, $y_i^0=0$, $i=m+1,\ldots,m+n$, and the vectors $\mathbf{b}^i$, $i=1,\ldots,m$, are linearly independent and form a basis in E^m. Then we can represent every $\mathbf{b}^j=[b_{1j},\ldots,b_{mj}]^T$ as $\mathbf{b}^j=\Sigma_{i=1}^m\lambda_{ij}\mathbf{b}^i$, where the vectors $\mathbf{b}^i$ are used as the basis vectors. Since $\mathbf{y}^0$ is a basic solution of (A.4), we have $\Sigma_{i=1}^m y_i^0\mathbf{b}^i=\mathbf{b}$; and using these representations of $\mathbf{b}$ and $\mathbf{b}^j$, we obtain

$$\sum_{i=1}^{m} y_i^0\mathbf{b}^i+\theta\left[\mathbf{b}^j-\sum_{i=1}^{m}\lambda_{ij}\mathbf{b}^i\right]=\mathbf{b},$$

or

$$\sum_{i=1}^{m}\left(y_i^0-\theta\lambda_{ij}\right)\mathbf{b}^i+\theta\mathbf{b}^j=\mathbf{b}. \tag{A.6}$$

Now if we slowly increase $\theta\geqslant 0$ while maintaining $y_i^0-\theta\lambda_{ij}\geqslant 0$ for $i=1,\ldots,m$, then the vector $\mathbf{b}^j$ becomes increasingly significant in the representation of $\mathbf{b}$.

Letting $y_k^0/\lambda_{kj}=\min(y_i^0/\lambda_{ij})$ for all $\lambda_{ij}>0$ and setting $\theta=y_k^0/\lambda_{kj}$ in (A.5), we observe that the coefficient of $\mathbf{b}^k$ becomes zero, i.e., $\mathbf{b}^k$ in (A.5) is replaced by $\mathbf{b}^j$. If $\mathbf{y}^1$ denotes the vector whose m nonnegative components are given by the nonzero coefficients in (A.5), then $\mathbf{y}^1$ is a new basic feasible solution obtained from $\mathbf{y}^0$. The following example provides a geometric interpretation of the algebraic procedure discussed above.

EXAMPLE A.5 Consider a linear programming problem where

$$\mathbf{B}=\begin{bmatrix}0 & 1 & 1 & 0 & 0\\ 1 & 1 & 0 & 1 & 0\\ 1 & -1 & 0 & 0 & 1\end{bmatrix}$$

and $\mathbf{b}=[6,10,6]^T$. Find the set M of feasible solutions, and starting with $\mathbf{y}^0=[4,6,0,0,8]^T$ construct all other basic feasible solutions obtainable from $\mathbf{y}^0$.

Following our convention of $\mathbf{B}=[\mathbf{A}|\mathbf{I}]$ and $\mathbf{y}=[\mathbf{x}^T,\mathbf{u}^T]^T$, we obtain in this case

$$\mathbf{A}=\begin{bmatrix} 0 & 1 \\ 1 & -1 \\ 1 & 1 \end{bmatrix}, \qquad \mathbf{x}=\left[x_1,x_2\right]^T, \qquad \mathbf{u}=\left[u_1,u_2,u_3\right]^T.$$

The set M containing all vectors $\mathbf{x}$ satisfying $\mathbf{Ax}\leqslant\mathbf{b}$ is shown by the shaded region in Fig. A.5.

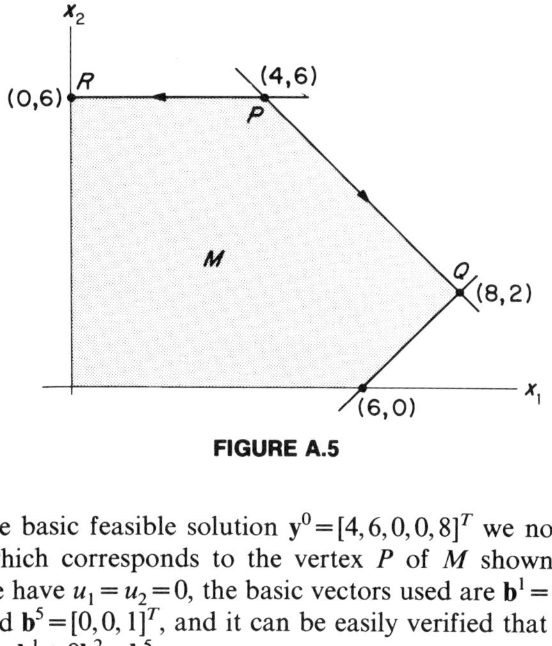

FIGURE A.5

Next, in the basic feasible solution $\mathbf{y}^0=[4,6,0,0,8]^T$ we note that $x_1=4$ and $x_2=6$, which corresponds to the vertex P of M shown in Fig. A.5. Since in $\mathbf{y}^0$ we have $u_1=u_2=0$, the basic vectors used are $\mathbf{b}^1=[0,1,1]^T$, $\mathbf{b}^2=[1,1,-1]^T$, and $\mathbf{b}^5=[0,0,1]^T$, and it can be easily verified that $\mathbf{b}^3=-\mathbf{b}^1+\mathbf{b}^2+2\mathbf{b}^5$ and $\mathbf{b}^4=\mathbf{b}^1+0\mathbf{b}^2-\mathbf{b}^5$.

Now suppose we introduce $\mathbf{b}^3$ into the set of basis vectors. This requires that we eliminate one of the original basis vectors $\mathbf{b}^1$, $\mathbf{b}^2$, and $\mathbf{b}^5$ from further consideration. To decide which one of $\mathbf{b}^1$, $\mathbf{b}^2$, and $\mathbf{b}^5$ we should eliminate, we use the expression for $\mathbf{b}^3$ given above and (A.5) to obtain

$$\left[4-\theta(-1)\right]\mathbf{b}^1+\left[6-\theta(1)\right]\mathbf{b}^2+\left[8-\theta(2)\right]\mathbf{b}^5+\theta\mathbf{b}^3=\mathbf{b}.$$

In this case by direct computation we have $\min(y_i^0/\lambda_{i3})=\min(6,8/2)=4$, and setting $\theta=4$, we obtain

$$8\mathbf{b}^1+2\mathbf{b}^2+0\mathbf{b}^5+4\mathbf{b}^3=\mathbf{b}.$$

Thus $\mathbf{b}^5$ is eliminated and we have a new basic feasible solution $\mathbf{y}^1=[8,2,4,0,0]^T$. In $\mathbf{y}^1$ we have $x_1=8$ and $x_2=2$, which corresponds to the vertex Q of M.

Next we introduce $\mathbf{b}^4$ into the set of basis vectors, and using the expression for $\mathbf{b}^4$ given above as well as (A.5), we obtain

$$[4-\theta(1)]\mathbf{b}^1+[6-\theta(0)]\mathbf{b}^2+[8-\theta(-1)]\mathbf{b}^5+\theta\mathbf{b}^4=\mathbf{b}.$$

In this case $\min(y_i^0/\lambda_{i4})=\min 4=4$, and setting $\theta=4$, we obtain

$$0\mathbf{b}^1+6\mathbf{b}^2+12\mathbf{b}^5+4\mathbf{b}^4=\mathbf{b}.$$

Thus $\mathbf{b}^1$ is eliminated and we have a new basic feasible solution $\mathbf{y}^2=[0,6,0,4,12]^T$. In $\mathbf{y}^2$ we have $x_1=0$ and $x_2=6$, which corresponds to the vertex of R of M. From Fig. A.5 we see clearly that the algebraic process of obtaining new basic feasible solutions corresponds to moving from a given vertex of M to neighboring vertices. Each vertex of M has an associated basic feasible solution, and every vertex of M can be obtained by computing the associated basic feasible solution.

Simplex Algorithm

The simplex algorithm was developed by Dantzig for solving linear programming problems. We note that in Examples A.1, A.2, and A.4, the optimum solutions were the vertices of the set M. When the objective function is linear, the set M is a convex polyhedron, and an optimum solution exists, then at least one of the vertices of M will correspond to the optimum solution. Therefore a linear programming problem can be solved by computing the values of the objective function at the vertices of M and comparing these values to locate the optimum solutions. The simplex algorithm is an efficient way of carrying out this program.

We have already discussed an algebraic procedure for moving from any vertex of M to a neighboring vertex via the associated basic feasible solutions. Part of the simplex algorithm uses this procedure to generate new vertices of M for test purposes. The simplex algorithm does not evaluate the objective function at every vertex of M, but rather economizes the search for the optimum solution by selecting certain promising vertices. To be specific, we assume that $f(\mathbf{x})$ is to be maximized over M. Then the simplex algorithm will generate a vertex of M only if the value of $f(\mathbf{x})$ at that vertex is expected to be at least as large as the value already computed.

We assume that the problem of linear programming is $\max(\mathbf{c},\mathbf{y})$ with $\mathbf{y}\in Y$, as stated in the introductory section. Let $\mathbf{y}^0$ denote the initial basic feasible solution, and let $\mathbf{y}$ be the new basic feasible solution obtained from $\mathbf{y}^0$ by using (A.5). Then evaluating the objective function at $\mathbf{y}$, we have

$$\begin{aligned}(\mathbf{c},\mathbf{y}) &= \sum_{i=1}^{m} \left(y_i^0 - \theta\lambda_{ij}\right)c_i + \theta c_j \\ &= (\mathbf{c},\mathbf{y}^0) - \theta\left[z_j - c_j\right], \end{aligned} \tag{A.6}$$

where $z_j = \sum_{i=1}^m \lambda_{ij} c_i$. The maximization of the objective function requires an evaluation of the following three cases.

Case I: There exists a j_0 between $m+1$ and $m+n$ such that $z_{j_0} - c_{j_0} < 0$ and $\lambda_{ij_0} > 0$ for some i. Letting $\theta = \min(y_i^0/\lambda_{ij_0})$, where the minimum is taken over all i for which $\lambda_{ij_0} > 0$, we observe that $\theta > 0$ and $y_i^0 - \theta\lambda_{ij_0} \geqslant 0$ for $i = 1,\ldots,m$, with equality holding for at least one value of i. The vector $\mathbf{y}^1$ determined by this choice of θ satisfies (A.6) and has at most m nonnegative components, and hence this is a basic feasible solution of (A.4). From Eq. (A.7), $(\mathbf{c},\mathbf{y}^1) > (\mathbf{c},\mathbf{y}^0)$, and thus we obtain a new basic feasible solution which provides a larger function value.

Case II: The inequality $z_j - c_j \geqslant 0$ holds for all j between $m+1$ and $m+n$. In this case $\mathbf{y}^0$ is an optimal basic feasible solution. To prove this statement we let $\mathbf{y}$ denote an arbitrary basic feasible solution of (A.4), which yields

$$\mathbf{b} = \sum_{j=1}^{n} y_j \mathbf{b}^j = \sum_{j=1}^{n} y_j \left[\sum_{i=1}^{m} \lambda_{ij} \mathbf{b}^i\right] = \sum_{i=1}^{m} \mathbf{b}^i \left[\sum_{j=1}^{n} \lambda_{ij} y_j\right] = \sum_{i=1}^{m} y_i^0 \mathbf{b}^i,$$

since $\mathbf{y}^0$ is also a basic feasible solution of (A.4). Now since the $\mathbf{b}^i$, $i = 1,\ldots,m$, are linearly independent, it follows that $y_i^0 = \sum_{j=1}^n \lambda_{ij} y_j$. Furthermore, the given condition is equivalent to $z_j \geqslant c_j$, $j = m+1,\ldots,m+n$, and $z_j = c_j$, $j = 1,\ldots,m$. Hence

$$\begin{aligned}\sum_{j=1}^{n} y_j c_j \leqslant \sum_{j=1}^{n} y_j z_j &= \sum_{j=1}^{n} y_j \left[\sum_{i=1}^{m} \lambda_{ij} c_i\right] \\ &= \sum_{i=1}^{m} c_i \left[\sum_{j=1}^{m} \lambda_{ij} y_i\right] \\ &= \sum_{i=1}^{m} y_i^0 c_i,\end{aligned}$$

i.e., for any arbitrary basic feasible solution $\mathbf{y}$ of (A.4), $(\mathbf{c},\mathbf{y}) \leqslant (\mathbf{c},\mathbf{y}^0)$ and hence $\mathbf{y}^0$ is an optimal solution.

Case III: There exists a j_0 between $m+1$ and $m+n$ such that $z_{j_0}-c_{j_0}<0$ and $\lambda_{ij_0}\leqslant 0$, $i=1,\ldots,m$. In this case $y_i^0-\theta\lambda_{ij}\geqslant 0$, and hence $\mathbf{y}^0$ is a basic feasible solution of (A.4) for all $\theta>0$. As $\theta\to+\infty$, it is clear from Eq. (A.7) that $(\mathbf{c},\mathbf{y})$ increases without any upper bound and hence the problem has no finite optimum solution.

It is now clear that the above scheme generates new basic feasible solutions of (A.4) such that the value of $(\mathbf{c},\mathbf{y})$ is successively increased. If the problem has finite optimal solutions, at least one of these will be found after a finite number of iterations. Otherwise Case III will be encountered, and the iterations can then be stopped.

Simplex Tableau

Suppose that we have a basic feasible solution $\mathbf{y}^0$ of (A.4) and we let $\mathbf{b}^j$, $j=1,\ldots,m$, denote the basis vectors associated with $\mathbf{y}^0$. We then compute the coefficients λ_{ij} in

$$\mathbf{b}^j=\sum_{i=1}^{m}\lambda_{ij}\mathbf{b}^i,\qquad j=1,\ldots,n,$$

where $\mathbf{b}^0=\mathbf{b}$, and enter them in Table A.1, which is called the *simplex tableau*. This tableau is used as a convenient method of keeping track of the various quantities involved in the simplex method.

Table A.1
Simplex Tableau

BASIS VECTORS	$\mathbf{b}^1$	$\cdots$	$\mathbf{b}^j$	$\cdots$	$\mathbf{b}^n$	$\mathbf{b}$
$\mathbf{b}^1$	λ_{11}	$\cdots$	λ_{1j}	$\cdots$	λ_{1n}	λ_{10}
$\vdots$	$\vdots$		$\vdots$		$\vdots$	$\vdots$
$\mathbf{b}^m$	λ_{m1}	$\cdots$	λ_{mj}	$\cdots$	λ_{mn}	λ_{m0}
z	z_1	$\cdots$	$z_j=\sum_{i=1}^{m}\lambda_{ij}c_i$	$\cdots$	z_n	
$z-c$	z_1-c_1	$\cdots$	z_j-c_j	$\cdots$	z_n-c_n	

We next check the elements of the last row of the simplex tableau; and if $z_j - c_j \geqslant 0$ for all j, then $\mathbf{y}^0$ is an optimal solution. However, if there exists a j_0 between $m+1$ and $m+n$ such that $z_{j_0} - c_{j_0} < 0$, we then check the elements λ_{ij_0} in the j_0th column, and if $\lambda_{ij_0} \leqslant 0$, $i = 1, \ldots, m$, then the problem has no finite solutions.

If $z_{j_0} - c_{j_0} < 0$ and $\lambda_{ij_0} > 0$ for some i, then we can find a new basic feasible solution of (A.4) with a higher function value. In practice it is advisable to choose the index j_0 for which $z_{j_0} - c_{j_0}$ has the largest negative value, since this choice has the effect of reducing the number of iterations required to find an optimal solution. The components of $\mathbf{y}^1$ are $y_i^1 = y_i^0 - \theta\lambda_{ij_0}$, $i = 1, \ldots, m$, $y_{j_0}^1 = \theta$, where

$$\theta = \min_{\lambda_{ij_0} > 0} \frac{y_i^0}{\lambda_{ij_0}},$$

with zeros elsewhere. Letting p_0 be an element such that $y_{p_0}^0 - \theta\lambda_{ij_0} = 0$, we observe that the basis vectors associated with $\mathbf{y}^1$ and $\mathbf{b}^i$, $i = 1, \ldots, m$ $(i \neq p_0)$, and $\mathbf{b}^{j_0}$. This is equivalent to replacing the vector $\mathbf{b}^{p_0}$ in the basis by $\mathbf{b}^{j_0}$. We can now compute the simplex tableau associated with $\mathbf{y}^1$, where to simplify the presentation we let $p_0 = 1$. In this case the new basis consists of the vectors $\mathbf{b}^2, \ldots, \mathbf{b}^m$ and $\mathbf{b}^{j_0}$. Any vector $\mathbf{b}^k$ can now be written as a linear combination of the new basis vectors. Thus

$$\mathbf{b}^k = \sum_{i=2}^{m} \hat{\lambda}_{ik}\mathbf{b}^i + \hat{\lambda}_{j_0 k}\mathbf{b}^{j_0}, \tag{A.8}$$

and we can rewrite $\mathbf{b}^{j_0} = \sum_{i=1}^{m} \lambda_{ij_0}\mathbf{b}^i$ as

$$\mathbf{b}^1 = -\sum_{i=2}^{m} \frac{\lambda_{ij_0}}{\lambda_{1j_0}}\mathbf{b}^i + \frac{1}{\lambda_{1j_0}} b^{j_0}.$$

Substituting this expression into $\mathbf{b}^k = \sum_{i=1}^{m} \lambda_{ik}\mathbf{b}^i$, we obtain

$$\mathbf{b}^k = -\lambda_{1k} \sum_{i=2}^{m} \frac{\lambda_{ij_0}}{\lambda_{1j_0}}\mathbf{b}^i + \frac{\lambda_{1k}}{\lambda_{1j_0}}\mathbf{b}^{j_0} + \sum_{i=2}^{m} \lambda_{ik}\mathbf{b}^i. \tag{A.9}$$

Comparing coefficients in (A.8) and (A.9) yields

$$\hat{\lambda}_{ik} = \begin{cases} \lambda_{ik} - \dfrac{\lambda_{1k}\lambda_{ij_0}}{\lambda_{1j_0}}, & i \neq j_0, \\[2ex] \dfrac{\lambda_{1k}}{\lambda_{1j_0}}, & i = j_0, \end{cases} \tag{A.10}$$

where we now use $\hat{\lambda}_{ik}$ given by (A.10) to form the simplex tableau associated with $\mathbf{y}^1$. The new tableau associated with $\mathbf{y}^1$ is next used to check whether $\mathbf{y}^1$ is an optimal solution or whether the problem has no finite optimal solutions. In either case the iterations are terminated. Otherwise this iteration process is continued until an optimal solution is found.

EXAMPLE A.6 For the linear programming problem stated in Example A.1, assume that $c_1=2$, $c_2=4$, $b_1>b_3>b_2$, and find the optimal soluton by the simplex method.

The set of all feasible solutions is shown by the shaded region in Fig. A.6 along with the level curves of the objective function $f(x_1,x_2)=2x_1+4x_2$. Using the graphical technique of Example A.1, we observe that the optimal solution is given by $\mathbf{x}^*=[b_3-b_2,b_2]^T$.

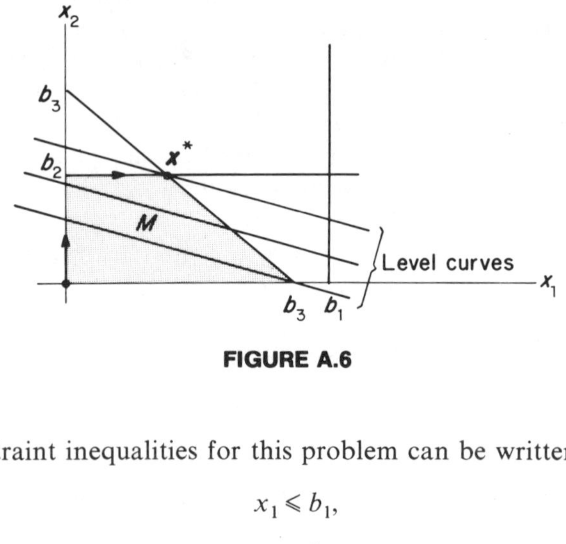

FIGURE A.6

The constraint inequalities for this problem can be written as

$$x_1 \leqslant b_1,$$

$$x_2 \leqslant b_2,$$

$$x_1+x_2 \leqslant b_3,$$

and using slack variables u_1, u_2, and u_3 these inequalities can be transformed into the equalities

$$x_1+u_1=b_1,$$

$$x_2+u_2=b_2,$$

$$x_1+x_2+u_3=b_3.$$

The matrix **B** is given by

$$\mathbf{B}=\begin{bmatrix} 1 & 0 & 1 & 0 & 0 \\ 0 & 1 & 0 & 1 & 0 \\ 1 & 1 & 0 & 0 & 1 \end{bmatrix},$$

and $\mathbf{y}=[x_1,x_2,u_1,u_2,u_3]^T$. To start the simplex method, we choose $\mathbf{y}^0=[0,0,b_1,b_2,b_3]^T$, which satisfies the equation $\mathbf{B}\mathbf{y}^0=\mathbf{b}$, where $\mathbf{b}=[b_1,b_2,b_3]^T$. The basic variables are u_1, u_2, and u_3, and the corresponding column vectors of the **B**-matrix ($\mathbf{b}^3=[1,0,0]^T$, $\mathbf{b}^4=[0,1,0]^T$, and $\mathbf{b}^5=[0,0,1]^T$) form a basis in E^3. Since $\mathbf{b}^3$, $\mathbf{b}^4$, and $\mathbf{b}^5$ are the coordinate vectors in E^3, the coefficients λ_{ij} can be obtained by inspection, and the resulting simplex tableau associated with $\mathbf{y}^0$ is given below:

Basis vectors	$\mathbf{b}^1$	$\mathbf{b}^2$	$\mathbf{b}^3$	$\mathbf{b}^4$	$\mathbf{b}^5$
$\mathbf{b}^3$	1	0	1	0	0
$\mathbf{b}^4$	0	1	0	1	0
$\mathbf{b}^5$	1	1	0	0	1
z	0	0	0	0	0
$z-c$	-2	$\boxed{-4}$	0	0	0

In the simplex tableau we observe that $z_1-c_1=-2<0$ and $z_2-c_2=-4<0$. Since $z_2-c_2<z_1-c_1$, we introduce $\mathbf{b}^2$ as the new vector in the basis. In this case x_2 becomes the entering variable, and we now proceed to determine the leaving variable. Since $b_3>b_2$, we have $\theta=\min(b_2,b_3)=b_2$ and $\mathbf{y}^1=[0,b_2,b_1-b_2,0,b_3-b_2]^T$. Since $u_2=0$ in $\mathbf{y}^1$, we conclude that u_2 is the leaving variable.

The basic variables corresponding to $\mathbf{y}^1$ are x_2, u_1, and u_3 respectively, and the new basis vectors are $\mathbf{b}^2=[0,1,1]^T$, $\mathbf{b}^3=[1,0,0]^T$, and $\mathbf{b}^5=[0,0,1]^T$. In this new basis we have $\mathbf{b}^1=0\mathbf{b}^2+\mathbf{b}^3+\mathbf{b}^5$ and $\mathbf{b}^4=\mathbf{b}^2+0\mathbf{b}^3-\mathbf{b}^5$, which gives us the values of the coefficients λ_{ij} resulting in the simplex tableau given below:

Basis vectors	$\mathbf{b}^1$	$\mathbf{b}^2$	$\mathbf{b}^3$	$\mathbf{b}^4$	$\mathbf{b}^5$
$\mathbf{b}^2$	0	1	0	1	0
$\mathbf{b}^3$	1	0	1	0	0
$\mathbf{b}^5$	1	0	0	-1	1
z	0	4	0	4	0
$z-c$	$\boxed{-2}$	0	0	4	0

In this simplex tableau we observe that $z_1 - c_1 = -2 < 0$, and so we introduce $\mathbf{b}^1$ as the new vector in the basis. Thus x_1 becomes the entering variable, and we proceed to determine the leaving variable as before. Since $b_1 > b_3 > b_2$, we have $\theta = \min(b_1 - b_2, b_3 - b_2) = b_3 - b_2$ and hence $\mathbf{y}^2 = [b_3 - b_2, b_2, b_1 - b_3, 0, 0]^T$, which shows that u_3 is the leaving variable.

Now for the basic feasible solution $\mathbf{y}^2 = [b_3 - b_2, b_2, b_1 - b_3, 0, 0]^T$, we have $x_1 = b_3 - b_2$, $x_2 = b_2$, and hence $x_1 = x_1^*$ and $x_2 = x_2^*$. The path of the simplex method is shown graphically in Fig. A.6.

Dual Problem

Every linear programming problem has associated with it another linear programming problem, and we call one of them the *primal* problem and the other one its *dual*. Let us take the maximization problem introduced earlier as our primal problem, i.e., $\max_{\mathbf{x} \in M}(\mathbf{c}, \mathbf{x})$, where $M = \{\mathbf{x} | \mathbf{x} \in E^n, \mathbf{A}\mathbf{x} \leqslant \mathbf{b}, \mathbf{x} \geqslant \mathbf{0}\}$. Then the associated dual problem is $\min_{\mathbf{z} \in N}(\mathbf{b}, \mathbf{z})$, where $N = \{\mathbf{z} | \mathbf{z} \in E^m, \mathbf{A}^T\mathbf{z} \geqslant \mathbf{c}, \mathbf{z} \geqslant \mathbf{0}\}$. We note that each constraint in one problem corresponds to a variable in the other problem. The constraint vector $\mathbf{b}$ of the primal problem generates the objective function of the dual problem, and vice versa. The direction of the constraint inequalities ($\leqslant$ in the primal problem) is reversed ($\geqslant$ in the dual problem).

The two problems and their optimal solutions are related, and the optimal solution of one problem yields complete information about the optimal solution of the other. In this section we explore the interrelations of these two problems by means of examples.

EXAMPLES

A.7 Show that $\max_{\mathbf{x} \in M}(\mathbf{c}, \mathbf{x}) \leqslant \min_{\mathbf{z} \in N}(\mathbf{b}, \mathbf{z})$.

Noting that $\mathbf{z} \geqslant \mathbf{0}$, we premultiply the constraints $\mathbf{A}\mathbf{x} \leqslant \mathbf{b}$ in the primal problem by $\mathbf{z}^T$ to obtain $\mathbf{z}^T\mathbf{A}\mathbf{x} \leqslant \mathbf{z}^T\mathbf{b} = (\mathbf{b}, \mathbf{z})$. Similarly noting that $\mathbf{x} \geqslant \mathbf{0}$ and premultiplying the constraints $\mathbf{A}^T\mathbf{z} \geqslant \mathbf{c}$ of the dual problem by $\mathbf{x}^T$, we obtain $\mathbf{x}^T\mathbf{A}\mathbf{z} \geqslant \mathbf{x}^T\mathbf{c} = (\mathbf{c}, \mathbf{x})$. Combining these two relations, we have $(\mathbf{c}, \mathbf{x}) \leqslant \mathbf{x}^T\mathbf{A}^T\mathbf{z} = \mathbf{z}^T\mathbf{A}\mathbf{x} \leqslant (\mathbf{b}, \mathbf{z})$ for all $\mathbf{x} \in M$, $\mathbf{z} \in N$, and the required result follows.

A.8 Show that for the optimal feasible solutions of the primal and dual problems, whenever inequality occurs in the kth constraint of either system (i.e., the corresponding slack variable is positive), the kth variable of its dual becomes zero.

Introducing slack variables $\mathbf{u} = \mathbf{b} - \mathbf{A}\mathbf{x} \geqslant \mathbf{0}$ and $\mathbf{v} = \mathbf{A}^T\mathbf{z} - \mathbf{c} \geqslant \mathbf{0}$, we rewrite the inequality constraints of the primal and dual problems as $\mathbf{A}\mathbf{x} + \mathbf{u} = \mathbf{b}$

and $\mathbf{A}^T\mathbf{z}-\mathbf{v}=\mathbf{c}$. Letting $g(\mathbf{z})=(\mathbf{b},\mathbf{z})$ and using $\mathbf{b}=\mathbf{Ax}+\mathbf{u}$, we have $g(\mathbf{z})=\mathbf{z}^T\mathbf{Ax}+\mathbf{z}^T\mathbf{u}$. Computing $g(\mathbf{z})-f(\mathbf{x})$, where $f(\mathbf{x})=(\mathbf{c},\mathbf{x})$, we obtain $g(\mathbf{z})-f(\mathbf{x})=\mathbf{z}^T\mathbf{u}+(\mathbf{A}^T\mathbf{z}-\mathbf{c})^T\mathbf{x}=\mathbf{z}^T\mathbf{u}+\mathbf{v}^T\mathbf{x}$. Next we consider the case when $\mathbf{x}=\mathbf{x}^*$ and $\mathbf{z}=\mathbf{z}^*$, the optimal solutions of the primal and dual problems respectively. Using Example A.7, we have $g(\mathbf{z}^*)-f(\mathbf{x}^*)=0$, i.e., $(\mathbf{z}^*)^T\mathbf{u}^*+(\mathbf{v}^*)^T\mathbf{x}^*=0$. Since $\mathbf{x}^*,\mathbf{z}^*,\mathbf{v}^*,\mathbf{u}^*\geqslant\mathbf{0}$, and if $u_k^*>0$ ($v_k^*>0$), then $z_k^*=0$ ($x_k^*=0$), which proves the given statement.

A linear programming problem can be formulated either in the primal form or in the dual form. The choice is made on the basis of computational efficiency. In general, a large number of constraints increases the complexity of computation, since the number of constraints determines the number of basis vectors needed to obtain a basic feasible solution. Therefore if the primal problem involves a large number of constraints and relatively few independent variables, it is computationally advantageous to consider the dual problem for solution. Problems A.7–A.9 show how the optimal solution of one problem can be computed from a knowledge of the optimum solution of its dual.

Solved Problems

A.1 Consider two electric power stations with power outputs denoted by x_1 and x_2 respectively. The maximum amounts of power available from these stations are denoted by p_1 and p_2 respectively. Letting $f(x_1,x_2)=x_1+2x_2$ denote the total cost of generating power, find the distribution of power between the two stations which minimizes the total cost of supplying a given load with L units of power.

The constraints on the power outputs of the two stations are $0\leqslant x_1\leqslant p_1$ and $0\leqslant x_2\leqslant p_2$ respectively. These two constraints define the shaded region in Fig. A.7,

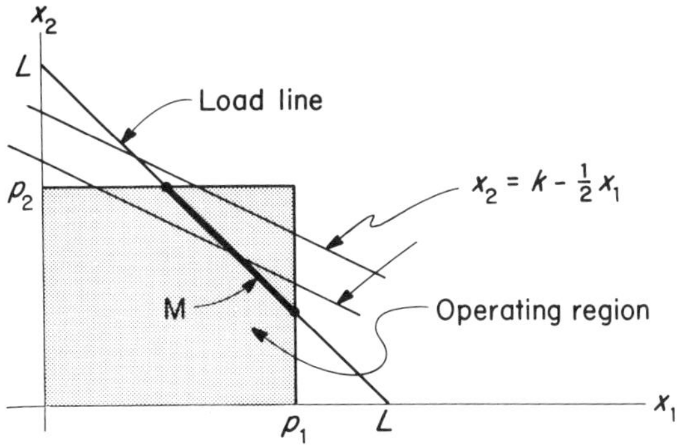

FIGURE A.7

called the *operating region*. Since the two stations are supplying a load of L units, we have $x_1+x_2=L$, which is called the *load line* and is shown in Fig. A.7. The feasible region M for this problem is that segment of the load line which lies inside the operating region.

As we move across the level curves $f(x_1,x_2)=\text{constant}$ in the direction of the arrow in Fig. A.7, the value of $f(x_1,x_2)$ decreases and attains a minimum at $\mathbf{x}^*=[p_1,L-p_1]^T$. Hence $\mathbf{x}^*=[p_1,L-p_1]^T$ represents the optimum operating point for the power stations when the load demand is L units.

A.2 In Problem A.1, find the optimum operating points as the load L changes.

The total cost of generating power is given by $f(x_1,x_2)=x_1+2x_2$. Since station 1 can produce the same amount of power as station 2 at half the cost, it is the more efficient one. Now if $L \geqslant p_1+p_2$, the power stations cannot supply the full power demand of the load, and hence the range of variation of L of interest is limited to $0 \leqslant L \leqslant p_1+p_2$.

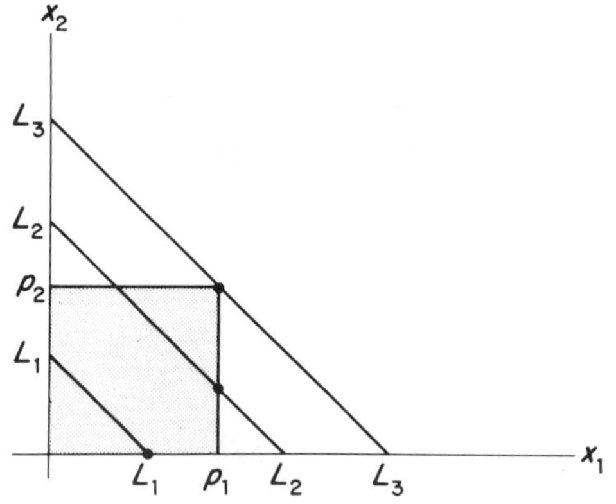

FIGURE A.8

Figure A.8 shows three load lines for three different values of L on the interval $[0,p_1+p_2]$, along with the corresponding optimal operating points, determined by the graphical method of Problem A.1. From this figure we obtain

$$\mathbf{x}^*=[p_1,0]^T \qquad \text{for} \quad 0 \leqslant L \leqslant p_1,$$

$$\mathbf{x}^*=[p_1,L-p_1]^T \qquad \text{for} \quad p_1 \leqslant L < p_1+p_2,$$

$$\mathbf{x}^*=[p_1,p_2]^T \qquad \text{for} \quad L=p_1+p_2,$$

which shows that the more efficient power station becomes operational first, followed by the other one when $L>p_1$.

A.3 An oil refinery uses two different blending processes whose input and output relations for each production cycle are given below:

	Input		Output	
PROCESS	CRUDE *A*	CRUDE *B*	GASOLINE *X*	GASOLINE *Y*
1	3	2	1	2
2	2	6	3	1

The maximum available amounts of crude *A* and *B* are 60 units and 90 units respectively. Market analysis shows that at least 30 units of gasoline *X* and 20 units of gasoline *Y* can be sold, and the profits for each production cycle from process 1 and process 2 are 1 and 2 units respectively. We wish to find an optimum mix of the two blending processes so as to supply the gasoline market at maximum profit.

Letting x_1 and x_2 denote the numbers of production runs of processes 1 and 2 respectively, the constraints on the availability of the crude oils can be written as

$$3x_1 + 2x_2 \leqslant 60,$$

$$2x_1 + 6x_2 \leqslant 90.$$

The constraints reflecting sales commitments of gasolines *X* and *Y* are

$$x_1 + 3x_2 \geqslant 30,$$

$$2x_1 + x_2 \geqslant 20.$$

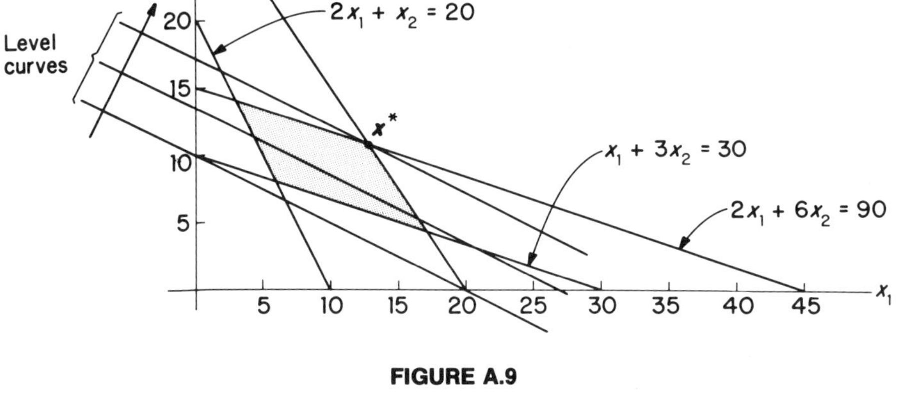

FIGURE A.9

The total profit from the sale of gasoline is given by $f(x_1,x_2)=x_1+2x_2$, which is to be maximized with respect to x_1 and x_2 while satisfying the above constraints.

The set of all feasible solutions is shown by the shaded region in Fig. A.9. As we move across the level curves $f(x_1,x_2)=$ constant in the direction of the arrow shown in Fig. A.9, the value of $f(x_1,x_2)$ increases, attaining a maximum at $\mathbf{x}^*=[12.80, 10.7]^T$; and the corresponding optimum number of production cycles is given by $x_1^*=12$ and $x_2^*=10$.

A.4 Consider the set of all feasible solutions defined as $M=\{\mathbf{x}|\mathbf{x}\in E^n,\ \mathbf{Ax}\leqslant\mathbf{b},\ \mathbf{x}\geqslant\mathbf{0}\}$, and show that M is a convex set.

Let $\mathbf{x}^1$ and $\mathbf{x}^2$ denote any two arbitrary points of M, and define $\mathbf{x}^3=\lambda\mathbf{x}^1+(1-\lambda)\mathbf{x}^2$, where $0<\lambda<1$. Then $\mathbf{x}^3\geqslant\mathbf{0}$ by construction, and $\mathbf{Ax}^3=\lambda\mathbf{Ax}^1+(1-\lambda)\mathbf{Ax}^2\leqslant\lambda\mathbf{b}+(1-\lambda)\mathbf{b}=\mathbf{b}$. Thus $\mathbf{x}^3\in M$. Since the line segment joining any two points in M is contained in M, it follows that M is a convex set.

A.5 An extreme point of a convex set is any point in it that cannot be represented as a convex combination of two other points in the set. Letting M be defined as in Problem A.4, show that $\mathbf{0}$ is an extreme point of M.

Let $\mathbf{x}^1$ and $\mathbf{x}^2$ be any two points in M, and define $\mathbf{x}^3=\lambda\mathbf{x}^1+(1-\lambda)\mathbf{x}^2$, $0<\lambda<1$. Since $\mathbf{x}^1,\mathbf{x}^2\geqslant\mathbf{0}$, it follows that $\mathbf{x}^3\geqslant\mathbf{0}$ and the ith component of $\mathbf{x}^3$ can be written as $x_i^3=\lambda x_i^1+(1-\lambda)x_i^2$. Clearly $x_i^3=0$ if and only if $x_i^1=x_i^2=0$, implying $\mathbf{x}^1=\mathbf{x}^2=\mathbf{0}$, and hence $\mathbf{0}$ is an extreme point of M.

A.6 Find the maximum of $f(x_1,x_2)=8x_1-4x_2$ on the set M defined by $x_1,x_2\geqslant 0$, $x_1-x_2\leqslant 2$, and $2x_1-x_2\leqslant 6$.

The set of all feasible solutions M is shown by the shaded region in Fig. A.10 along with the level curves $f(x_1,x_2)=$ constant. These level curves are parallel to the constraint hyperplane given by $2x_1-x_2=6$, and from graphical considerations it is clear that all points on the line $2x_1-x_2=6$ beyond the point of intersection with the line $x_1-x_2=2$ are optimal points. Note that the values of x_1^* and x_2^* in M can be increased indefinitely without increasing the optimal value of the objective function, $f(x_1^*,x_2^*)=6$.

A.7 Find the dual of the linear programming problem solved in Example A.2.

In Example A.2, the constraint inequalities were written as $\mathbf{Ax}\geqslant\mathbf{b}$. In order to conform to standard convention we rewrite these constraints in the form $\mathbf{Ax}\leqslant\mathbf{b}$, where

$$\mathbf{A}=\begin{bmatrix} 1 & -1 \\ -6 & -2 \\ 0 & -1 \end{bmatrix} \quad \text{and} \quad \mathbf{b}=\begin{bmatrix} -4 \\ -12 \\ -2 \end{bmatrix}.$$

The objective function in Example A.2 is $f(\mathbf{x})=x_1+2x_2$, so that $\mathbf{c}=[1,2]^T$. Hence the dual problem becomes $\max_{\mathbf{z}\geqslant\mathbf{0}} g(\mathbf{z})=-4z_1-12z_2-2z_3$, where the constraints

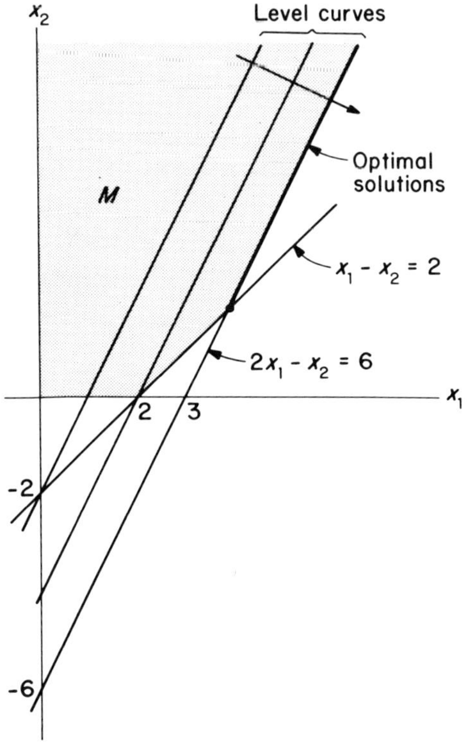

FIGURE A.10

are $-z_1-6z_2+0z_3 \geqslant 1$ and $-z_1-2z_2-z_3 \geqslant 2$, which can also be written in the form $\mathbf{A}^T\mathbf{z} \geqslant \mathbf{c}$.

A.8 Using the optimal solution $\mathbf{x}^*$ obtained in Example A.2 and the results of Example A.8, find the optimal solution $\mathbf{z}^*$ of the dual problem given in Problem A.7.

Using slack variables $u_1, u_2, u_3 \geqslant 0$ and $v_1, v_2 \geqslant 0$, the constraints of the primal and dual problems can be written as

$$-x_1 - x_2 + u_1 + 0u_2 + 0u_3 = -4,$$

$$-6x_1 - 2x_2 + 0u_1 + u_2 + 0u_3 = -12,$$

$$0x_1 - x_2 + 0u_1 + 0u_2 + u_3 = -2,$$

and

$$-z_1 - 6z_2 + 0z_3 + v_1 + 0v_2 = 1,$$

$$-z_1 - 2z_2 - z_3 + 0v_1 + v_2 = 2.$$

Substituting $x_1^*=2$, $x_2^*=2$ from Example A.2 in the constraint equations of the primal problem, we have $u_1^*=u_3^*=0$ and $u_2^*=2$. Since $u_2^*>0$, by the results of Example A.8 we have $z_2^*=0$. Using $x_1^*,x_2^*>0$ and the results of Example A.8, we also have $v_1^*=0$ and $v_2^*=0$. Substituting these values of z_2^*, v_1^*, and v_2^* in the constraint equations of the dual problem and solving for z_1 and z_3, we have $z_1^*=-1$ and $z_3^*=-1$, and therefore $z^*=[-1,0,-1]^T$.

As a check we evaluate $f(\mathbf{x}^*)=(\mathbf{c},\mathbf{x}^*)=x_1^*+2x_2^*=6$ and $g(\mathbf{z}^*)=(\mathbf{b},\mathbf{z}^*)=-4z_1^*-12z_2^*-2z_3^*=6$. As expected from Example A.7, we have $f(\mathbf{x}^*)=g(\mathbf{z}^*)$.

A.9 Find the optimal solution of the dual of the linear programming problem solved in Problem A.3.

Using slack variables $u_1,u_2,u_3,u_4\geqslant 0$, we write the constraints of the primal problem as

$$3x_1+2x_2+u_1=60,$$

$$2x_1+6x_2+u_2=90,$$

$$-x_1-3x_2+u_3=-30,$$

$$-2x_1-x_2+u_4=-20.$$

The optimal solution $\mathbf{x}^*$ of the primal problem is given in Problem A.3, and we have $x_1^*=12.8$ and $x_2^*=10.7$. Substituting these values of x_1^* and x_2^* in the constraint equation of the primal problem, we have $u_1^*=0.2$, $u_2^*=0.2$, $u_3^*=14.9$, and $u_4^*=16.3$. Since $u_1^*,u_2^*,u_3^*,u_4^*>0$ by the results of Example A.8, we have $z_1^*=z_2^*=z_3^*=z_4^*=0$.

Next we check this answer by using the results of Example A.7. From Problem A.3 we have $f(\mathbf{x}^*)=32$, and evidently in this case $g(\mathbf{z}^*)=0$. Since $f(\mathbf{x}^*)\neq g(\mathbf{z}^*)$, we conclude that the value of $\mathbf{z}^*$ derived earlier is incorrect. Looking back, we see that u_1^* and u_2^* are almost zero; suspecting computational errors, we set $u_1^*=u_2^*=0$, which means that z_1^* and z_2^* need no longer be zero. Since $x_1^*,x_2^*>0$, we know from Example A.8 that $v_1^*=v_2^*=0$ in the dual problem. Substituting these values of z_3^*, z_4^*, v_1^*, and v_2^* in the dual constraint equations

$$3z_1+2z_2-z_3-z_4+v_1=1,$$

$$2z_1+6z_2-3z_3-z_4+v_2=2,$$

and solving for z_1^*,z_2^*, we have $z_1^*=\frac{1}{7}$ and $z_2^*=\frac{2}{7}$, and $g(\mathbf{z}^*)=34.2$. Since $g(\mathbf{z}^*)$ is now reasonably close to $f(\mathbf{x}^*)$, we accept $\mathbf{z}^*=[\frac{1}{7},\frac{2}{7},0,0]^T$ as the optimal solution of the dual problem within the limits of computational error.

Supplementary Problems

A.10 Consider the problem of preparing a minimum-cost daily diet consisting of two types of food, where the diet is required to satisfy an average adult's daily need for proteins, fats, and carbohydrates. The relevant information is given in the table below:

Food type	Yield per unit			Cost
	Fats	Carbohydrates	Proteins	per unit
1	2	1	6	1
2	4	5	1	2
Minimum daily requirement	8	5	12	

Formulate this as a linear programming problem and find the optimal solution.

A.11 Find the maximum value of $f(x_1,x_2)=\min[3x_1-12, -4x_1+5]$ subject to $0 \leqslant x_1 \leqslant 5$.

A.12 Find the maximum value of $f(x_1,x_2,x_3)=2x_1+2x_2+4x_3$ subject to $x_1,x_2,x_3 \geqslant 0$, $x_1+2x_2+x_3 \leqslant 400$, $3x_1+2x_2 \leqslant 420$, and $x_1+4x_2 \leqslant 450$, by the simplex method.

A.13 Find the maximum value of $f(x_1,x_2)=3x_1+4x_2$ subject to $x_1,x_2 \geqslant 0$, $5x_1+4x_2 \geqslant 100$, $3x_1+5x_2 \leqslant 150$, and $5x_1+4x_2 \leqslant 200$.

References

Chapter 1 The Optimization Problem

ARROW, KENNETH J., L. HURWICZ, and H. UZAWA. *Studies in Linear and Non-Linear Programming*. Stanford University Press, 1958.

BEVERIDGE, G. S., and R. S. SCHECHTER. *Optimization Theory and Practice*. New York: McGraw-Hill, 1970.

BIRKHOFF, G., and S. MACLANE. *A Survey of Modern Algebra*. New York: Macmillan, 1941.

DORN, W. S. Non-linear programming—a survey. *Management Science*, 9 (1963): 171–208.

HADLEY, G. *Linear Algebra*. Reading: Addison-Wesley, 1961, Chap. 6.

HOHN, FRANZ, E. *Elementary Matrix Algebra* (2nd ed.). New York: Macmillan, 1964.

VALENTINE, F. A. *Convex Sets*. New York: McGraw-Hill, 1964.

ZADEH, L. A. *Linear System Theory: The State Space Approach*. New York: McGraw-Hill, 1963, Appendix C.

ZOUTENDIJK, G. Nonlinear programming: a numerical survey. *SIAM J. on Control*, 4 (1966): 194–210.

Chapter 2 Classical Optimization

BUCK, R. C. *Advanced Calculus*. New York: McGraw-Hill, 1956.

COOPER, L., and D. STEINBERG. *Introduction to Methods of Optimization*. Philadelphia: Saunders, 1970.

COXETER, H. S. M. The golden section, phyllotaxis, and Wythoff's game. *Scripta Mathematica* (1954): 135–43.

HANCOCK, H. *Theory of Maxima and Minima*. New York: Ginn and Company, 1917.

SMAIL, L. L. *Calculus*. New York: Appleton-Century, 1949.

WILDE, D. J. Differential calculus in nonlinear programming. *Opns. Res.*, 10 (1962): 764–73.

WILDE, D. J. Jacobians in constrained nonlinear optimization. *Opns. Res.*, 13 (1965): 848–56.

Chapter 3 Constraints and Lagrange Multipliers

ADBY, P. R., and M. A. DEMPSTER. *Introduction to Optimization Methods*. New York: Halsted Press, 1974.
BROOKS, R., and A. M. GEOFFRION. Finding Everetts' Lagrange multipliers by linear programming. *Operations Research*, 14 (1966): 1149–1153.
EVERETT, HIGH, III. Generalized Lagrange multiplier method for solving problems of optimum allocation of resources. *Operations Research*, II (1963): 399–417.
GOTTFRIED, B. S., and J. WEISMAN. *Introduction to Optimization Theory*. Englewood Cliffs: Prentice-Hall, 1973.
INTRILIGATOR, M. D. *Mathematical Optimization and Economic Theory*. Englewood Cliffs: Prentice-Hall, 1971.
JOHN, F. Extremum problems with inequalities as subsidiary conditions. In *Studies and Essays (Courant Anniversary Volume)*. New York: Interscience, 1948, 187–204.
KUNZI, H. P., et. al. *Numerical Methods of Mathematical Optimization with ALGOL and FORTRAN Programs*. New York; Academic Press, 1968.
PIERRE, D. A. *Optimization Theory with Applications*. New York: Wiley, 1969.
RUSSELL, D. L. *Optimization Theory*. Reading: Addison-Wesley, 1970.
WILDE, D. J., and C. S. BEIGHTLER. *Foundations of Optimization*. Englewood Cliffs: Prentice-Hall, 1967.

Chapter 4 Optimization with Inequality Constraints

BERNHOLTZ, B. A new derivation of the Kuhn-Tucker conditions. *Operations Research*, 12 (1964): 295–299.
BURLEY, D. M. *Studies in Optimization*. New York: Halsted Press, 1974.
COTTLE, RICHARD W. Nonlinear programs with positively bounded Jacobians. *SIAM J. on Applied Mathematics*, 14 (1966): 147–158.
HADLEY, G. *Nonlinear and Dynamic Programming*. Reading: Addison-Wesley, 1964.
HUARD, PIERRE. Dual programs. In *Recent Advances in Mathematical Programming* (Robert L. Graves and Philip Wolfe, eds.). New York: McGraw-Hill, 1963, pp. 55–62.
KUHN, H. W., and A. W. TUCKER. *Linear Inequalities and Related Systems*. Princeton University Press, 1956.
KUHN, H. W., and A. W. TUCKER. Nonlinear programming. In *Proceedings of the Second Berkeley Symposium on Mathematical Statistics and Probability* (J. Neyman, ed.). University of California Press, 1951, pp. 481–492.
MANGASARIAN, O. L. *Nonlinear Programming*. New York: McGraw-Hill, 1969.
TUCKER, A. W. Linear and nonlinear programming. *Operations Research*, 5 (1957): 244–257.
WHITTLE, P. *Optimization under Constraint: Theory and Applications of Nonlinear Programming*. New York: Wiley, 1971.
WOLFE, PHILIP. A duality theorem for nonlinear programming. *Quarterly of Applied Mathematics*, 19 (1961): 239–244.

Chapter 5 One-Dimensional Search Techniques

FIBONACCI, LEONARDO. *Algebra et almuchabala (liber abbaci)* (1202). Cited in Bell, E. T.. *The Development of Mathematics*. New York: McGraw-Hill, 1940, p. 160.

JOHNSON, S. M. *Optimal Search for a Maximum in Fibonaccian*, RAND Corporation Report, 1956, p. 856.

MURRAY, W. (ed.). *Numerical Methods for Unconstrained Optimization*. New York: Academic Press, 1972.

NEWMAN, D. J. Location of the maximum on unimodal surfaces. *J. Assn. Comp. Mach.*, 12 (1965): 395–98.

OLIVER, L. T., and D. J. WILDE. Symmetric sequential minimax search for a maximum. *Fibonacci Quart.*, (1964): 169–75.

PUN, L. *Introduction to Optimization Practice*. New York: Wiley, 1969.

VEINOTT, ARTHUR F., JR. The supporting hyperplane method for unimodal programming. *Operations Research*, 15 (1967): 147–152.

WILDE, D. J. *Optimum Seeking Methods*. Englewood Cliffs: Prentice-Hall, 1964.

Chapter 6 Unconstrained Gradient Techniques

ADABIE, J. (ed.). *Nonlinear Programming*. Amsterdam: North-Holland, 1967.

BOX, G. E. P., and K. B. WILSON. On the experimental attainment of optimum conditions. *J. Roy. Stat. Soc.*, B13 (1951): 1.

BOX, G. E. P. The exploration and exploitation of response surfaces. *Biometrics*, 10 (1954): 16.

BUEHLER, R. J., B. M. SHAH, and O. KEMPTHORNE. Some properties of steepest ascent and related procedures for finding optimum conditions. Iowa State University Statistical Lab., Ames, Iowa, April 1961.

GRAVES, ROBERT L., and PHILIP WOLFE. *Recent Advances in Mathematical Programming*. New York: McGraw-Hill, 1963.

KOWALIK, J., and M. R. OSBORNE. *Methods for Unconstrained Optimization Problems*. New York: Elsevier North-Holland, 1969.

MCCORMICK, GARTH P., Anti-zig-zagging by bending. *Management Science*, 15 (1969): 315–320.

WOLFE, PHILIP, Accelerating the cutting plane method for nonlinear programming, *SIAM J. on Applied Mathematics*, 9 (1961): 481–488.

Chapter 7 Constrained Gradient Techniques

APOSTOL, T. M. *Mathematical Analysis*. Reading: Addison-Wesley, 1957.

BOX, M. M. *Nonlinear Optimization Techniques*. New York: Longman, 1969.

FLETCHER, R., and M. J. D. POWELL. A rapidly convergent descent method for minimization. *Comp. J.*, 6 (1963): 163–68.

GILL, P. E., and W. MURRAY. *Numerical Methods for Constrained Optimization*. New York: Academic Press, 1974.

HASDORFF, L. *Gradient Optimization and Nonlinear Control*. New York: Wiley, 1976.

JACOBY, S. L. *Iterative Methods for Nonlinear Optimization Problems*. Englewood Cliffs: Prentice-Hall, 1972.

KELLEY, J. E., JR. The cutting plane method for solving convex programs. *J. Soc. Ind. Appl. Math.*, 8 (1960): 703–712.

POWELL, M. J. D. Method for minimizing a sum of squares of nonlinear functions without calculating derivatives. *Comp. J.*, 7 (1965): 303–307.

POWELL, M. J. D. An efficient method for finding the minimum of a function of several variables without calculating derivatives. *Comp. J.*, 7 (1964): 155–162.

ROSEN, J. B. The gradient projection method for nonlinear programming, Part I, Linear constraints. *J. Soc. Ind. Appl. Math.*, 8 (1960): 181–217.

ROSEN, J. B The gradient projection method for nonlinear programming, Part II, Nonlinear constraints. *J. Soc. Ind. & Appl. Math.*, 9 (1961): 514–532.

ROSENBROCK, H. H. An automatic method for finding the greatest or least value of a function. *Comp. J.*, 3 (1960): 175–184.

SHAH, B. V., R. J. BUEHLER, and O. KEMPTHORNE. Some algorithms for minimizing a function of several variables. *J. Soc. Ind. Appl. Math.*, 12 (1964): 74–92.

WAGNER, HARVEY M. *Principles of Operations Research*. Englewood Cliffs: Prentice-Hall, 1969.

ZOUTENDIJK, G. *Methods of Feasible Directions*. Amsterdam: Elsevier, 1960.

Chapter 8 Penalty-Function Methods

BELTRAMI, E. J. *An Algorithmic Approach to Nonlinear Analysis and Optimization*. New York: Academic Press, 1970.

FIACCO, A. F., and G. P. MCCORMICK. Computational algorithm for the sequential unconstrained minimization technique for nonlinear programming. *Man. Sci.*, 10 (1964): 601–617.

FIACCO, A. F., and G. P. MCCORMICK. *Nonlinear Programming, Sequential Unconstrained Minimization Techniques*. New York: Wiley, 1968.

HARTLEY, H. O. Non-linear programming by the simplex method. *Econometrica*, 29 (1961): 223–237.

HARTLEY, H. O., and R. R. HOCKING. Convex programming by tangential approximation. *Management Science*, 9 (1963): 600–612.

KIEFER, J. Optimum sequential search and approximation methods under minimum regularity assumptions. *SIAM J. on Applied Mathematics*, 5 (1957): 105–136.

LAVI, A., and T. P. VOGL (eds.). *Recent Advances in Optimization Techniques*. New York: Wiley, 1966.

MARTOS, BELA. Hyperbolic programming. *Naval Research Logistics Quarterly*, 11 (1964): 135–155.

PIERSKALLA, W. P. Mathematical programming with increasing constraint functions. *Management Science*, 15 (1969): 416–425.

SAATY, T. L., and J. BRAM. *Nonlinear Mathematics*. New York: McGraw-Hill, 1964, pp. 168–170.

Chapter 9 Quadratically Convergent Minimization Algorithms

ARROW, KENNETH J., and A. C. ENTHOVEN. Quasi-concave programming. *Econometrica*, 29 (1961): 779–800.

DAVIDON, W. C. Variable metric method for minimization. *A. E. C. Research and Development Report*, ANL–5590, 1959.

DAVIDON, W. C. Variable metric method for minimization, *A. E. C. Research and Development Report*, ANL–5990 (Rev.), 1959.

FLETCHER, R., and M. J. D. POWELL. A rapidly convergent descent method for minimization. *Computer Journal*, 6 (1963): 163–68.

FLETCHER, R., and C. M. REEVES. Function minimization by conjugate gradients. *Computer Journal*, 7 (1964): 149–154.

HESTENES, M. R., and E. STIEFEL. Methods of conjugate gradients for solving linear systems. *J. Res. Nat. Bur. Standards*, 49 (1952).

HOOKE, R., and T. A. JEEVES. Direct search solution of numerical and statistical problems. *J. of Assn. Computing Mach.*, 8 (1961).

POWELL, M. J. D. An efficient method for finding the minimum of a function of several variables without calculating derivatives. *Computer Journal*, 7 (1964): 152–162.

TOPKIS, DONALD M., and ARTHUR F. VEINOTT, JR. On the convergence of some feasible direction algorithms for nonlinear programming. *SIAM J. on Control*, 5 (1967): 268–279.

WHINSTON, A. Conjugate functions and dual programs. *Naval Research Logistics Quarterly*, 21 (1965): 315–322.

ZANGWILL, W. I. *Nonlinear Programming*. Englewood Cliffs: Prentice-Hall, 1969.

Chapter 10 Quadratic Programming

BEALE, E. M. L. On quadratic programming. *Naval Research Logistics Quarterly*, 6 (1969): 227–243.

BEALE, E. M. L. Note on a comparison of two methods of quadratic programming. *Operations Research*, 14 (1966).

BOOT, J. C. G. On sensitivity analysis in convex quadratic programming problems. *Operations Research*, 11 (1963): 771–786.

BRACKEN, J., and G. MCCORMICK. *Selected Applications of Nonlinear Programming*. New York: Wiley, 1968.

D'ESPO, D. A convex programming procedure. *Naval Research Logistics Quarterly*, 6 (1959): 33–42.

DORN, W. S. Self dual quadratic programs. *SIAM J. of Applied Mathematics*, 9 (1961): 51–54.

FRANK, MARGUERITE, and PHILIP WOLFE. An algorithm for quadratic programming. *Naval Research Logistics Quarterly*, 3 (1956): 95–110.

HILDRETH, C. G. A quadratic programming procedure, *Naval Research Logistics Quarterly*, 4 (1957): 79–85.

HOUTHAKKER, H. S. The capacity method of quadratic programming. *Econometrica*, 28 (1960): 62–87.

KRIEBEL, CHARLES H. Coefficient estimation in quadratic programming models. *Management Science*, 13 (1967): B-473–B-486.

KUNZE, H. P., and W. KRELLE *Nonlinear Programming*. Waltham, Mass.: Blaisdell, 1966.

LEMKE, C. E. A method of solution for quadratic programs. *Management Science*, 8 (1962): 442–453.

THEIL, H., and C. VAN DE PANNE. Quadratic programming as an extension of conventional quadratic maximization. *Journal of the Institute of Management Science*, 7 (1961): 1–20.

WOLFE, P. The simplex method for quadratic programming. *Econometrica*, 27 (1959): 382–398.

Chapter 11 Integer Programming

ABADIE, J. (ed.). *Integer and Nonlinear Programming*. Amsterdam: North-Holland, 1970.

ALCALY, R. E., and A. K. KLEVORICK. A note on the dual prices of integer programs. *Econometrica*, 34 (1966): 206–214.

BALAS, E. An additive algorithm for solving linear programs with zero-one variables. *Operations Research*, 13 (1965): 517–544.

BALINSKI, M. L. Integer programming: methods, uses, computation. *Management Sciences*, 12 (1965): 253–313.

BELLMORE, M., and G. L. NEMHAUSER. The traveling salesman problem: a survey. *Operations Research*, 16 (1968): 538–558.

CABOT, A. VICTOR, and ARTHUR P. HURTER, JR. An approach to zero-one integer programming. *Operations Research*, 16 (1968): 1206–1211.

COOPER, L., and C. DREBES. An approximate solution method for the fixed charge problem. *Naval Research Logistics Quarterly*, 14 (1967): 101–113.

EFROYMSON, M. A., and T. L. RAY. A branch-bound algorithm for plant locations. *Operations Research*, 14 (1966): 361–368.

GAVETT, F. W., and N. V. PLYTER. The optimal assignments of facilities to locations by branch and bound. *Operations Research*, 14 (1966): 210–232.

GEOFFRION, ARTHUR M. Integer programming by implicit enumeration and Balas' method. *SIAM Review*, 9 (1967): 178–190.

GLOVER, FRED. Surrogate constraints. *Operations Research*, 16 (1968): 741–749.

GLOVER, FRED. A multiphase-dual algorithm for zero-one integer programming problems. *Operations Research*, 13 (1965): 879–919.

GLOVER, FRED. A new foundation for a simplified primal integer programming algorithm. *Operations Research*, 16 (1968): 727–740.

GOMORY, R. E. An algorithm for integer solutions to linear programs. In *Recent Advances in Mathematical Programming* (F. L. Graves and P. Wolfe, eds.). New York: McGraw-Hill, 1963, pp. 269–302.

GOMORY, R. E. Outline of an algorithm for integer solutions to linear programs. *Bull. Amer. Math. Soc.*, 64 (1958): 275–278.

GOMORY, R. E. On the relation between integer and noninteger solutions to linear programs. *National Academy of Sciences*, 53 (1965): 260–265.

GRAVES, G. W., and A. WHINSTON. A new approach to discrete mathematical problems. *Management Science*, 11 (1964): 308–315.

HU, T. C. *Integer Programming and Network Flows*. Reading: Addison-Wesley, 1969.

MURTY, K. G. *Linear and Combinatorial Programming*. New York: Wiley, 1976.

MUTH, J. F., and G. L. THOMPSON. *Industrial Scheduling*. Englewood Cliffs: Prentice-Hall, 1963.

SAATY, T. L. *Optimization in Integers and Related External Problems*. New York: McGraw-Hill, 1970.

Chapter 12 Dynamic Programming

AIRS, R. *Discrete Dynamic Programming*. New York: Blaisdell, 1964.

BELLMAN, R. E., and S. DREYFUS. *Applied Dynamic Programming*. Princeton University Press, 1967.

BELLMAN, R. E. *Dynamic Programming*. Princeton University Press, 1957.

BREIMAN, LEO Stopping-rule problems. In *Applied Combinatorial Mathematics* (Edwin F. Beckenbach, ed.). New York: Wiley, 1964, pp. 284–319.

DAN, S. *Nonlinear and Dynamic Programming: An Introduction*. New York: Springer Verlag, 1975.

DERMAN, CYRUS, and G. J. LIEBERMAN. A Markovian decision model for a joint replacement and stocking problem. *Management Science*, 13 (1967): 609–617.

GREENBERG, H. J. The use of branching in dynamic programming for parametric analysis. *Operations Research*, 15 (1967): 976–977.

HOWARD, RONALD A. *Dynamic Programming and Markov Processes*, M. I. T. Press, 1960.

KARP, RICHARD M., and M. HELD Finite-state processes and dynamic programming. *SIAM J. on Applied Mathematics*, 15 (1967): 693–718.

KARUSH, WILLIAM. On a class of minimum-cost problems. *Management Science*, 4 (1958): 136–153.

KAUFMANN, A. *Graphs, Dynamic Programming and Finite Games* New York: Academic Press, 1967.

LARSON, R. E. *State Increment Dynamic Programming*. New York: Elsevier North-Holland, 1968.

MARTIN, J. J. *Bayesian Decision Problems and Markov Chains*. New York: Wiley, 1967.

NEWHAUSER, G. L. *Introduction to Dynamic Programming*. New York: Wiley, 1966.

VEINOTT, ARTHUR F., JR., and HARVEY M. WAGNER. Optimal capacity scheduling —I. *Operations Research*, 10 (1962): 518–532.

VEINOTT, ARTHUR F. JR., and HARVEY M. WAGNER. Optimal capacity scheduling—II. *Operations Research*, 10 (1962): 533–546.

WHITE, D. J. *Dynamic Programming*, San Francisco: Holden-Day, 1969.

WONG, PETER J., and DAVID G. LUENBERGER. Reducing the memory requirements of dynamic programming. *Operations Research*, 16 (1968): 1115–1125.

Chapter 13 Large-Scale Programming

DANTZIG, G. B., and PHILIP WOLFE. Decomposition principle for linear programs. *Operations Research*, 8 (1960): 101–111.

GEOFFRION, ARTHUR M. Reducing concave programs with some linear constraints. *SIAM J. on Applied Mathematics*, 15 (1967): 653–664.

HIMMELBLAU, D. M. *Decomposition of Large-Scale Problems*. New York: Elsevier North-Holland, 1973.

LASDON, L. S. *Optimization Theory for Large Systems*. New York: Macmillan, 1970.

ROSEN, J. B., and J. C. ORNEA. Solution of nonlinear programming problems by partitioning. *Management Science*, 10 (1963): 160–173.

SANDERS, J. L. A nonlinear decomposition principle. *Operations Research*, 13 (1965): 266–271.

WISMER, D. A. (ed.). *Optimization Methods for Large Scale Systems*. New York: McGraw-Hill, 1971.

Appendix Linear Programming

BEALE, E. M. L. *Mathematical Programming in Practice*. London: Pitman, 1968.

Dantzig, G. B. *Linear Programming and Extensions*. Princeton University Press, 1963.

DANTZIG, G. B., L. R. FORD, and D. R. FULKERSON. A primal-dual algorithm for linear programming. In (H. W. Kuhn and A. W. Tucker, eds.). *Linear Inequalities and Related Systems* Princeton University Press, 1956.

FRAZER, J. R. *Applied Linear Programming*. Englewood Cliffs: Prentice-Hall, 1968.

GARVIN, W. W. *Introduction to Linear Programming*. New York: McGraw-Hill, 1960.

GASS, S. I. *Linear Programming Methods and Applications* (3rd ed.). New York: McGraw-Hill, 1969.

HADLEY, G. *Linear Programming*. Reading: Addison-Wesley, 1962.

LEMKE, C. E. The dual method of solving the linear programming problem. *Naval Research Logistics Quarterly*, 1 (1954): 36–47.

ORCHARD-HAYS, W. *Advanced Linear Programming Computing Techniques*. New York: McGraw-Hill, 1968.

VAJDA, S. *Mathematical Programming*. Reading: Addison-Wesley, 1961.

Answers to Supplementary Problems

Chapter 1

1.18 (a) $[x_1, x_2]\begin{bmatrix} 1 & -1 \\ -1 & 4 \end{bmatrix}\begin{bmatrix} x_1 \\ x_2 \end{bmatrix}$

(b) $\begin{bmatrix} 2 & -2 \\ -2 & 8 \end{bmatrix}$

(d) Yes

(e) $z = 12x_2 - 12$

1.21 (a) and (b)

1.22 (a) $\begin{bmatrix} \dfrac{3}{x_2} + \dfrac{4}{x_2} & e^{x_1/x_2} \\ -\dfrac{3x_1}{x_2^2} - \dfrac{4x_1}{x_2^2} & e^{x_1/x_2} \end{bmatrix}$

(b) $\begin{bmatrix} -x_2^{-1} & x_1^{-2} \\ -x_1^{-1} & x_2^{-2} \end{bmatrix}$

1.23 (a) $\begin{bmatrix} \dfrac{4}{x_2^2} e^{x_1/x_2} & -\dfrac{3}{x_2^2} - \dfrac{4}{x_2^2}\left(1 + \dfrac{x_1}{x_2}\right)e^{x_1/x_2} \\ -\dfrac{3}{x_2^2} - \dfrac{4}{x_2^2}\left(1 + \dfrac{x_1}{x_2}\right)e^{x_1/x_2} & \dfrac{6x_1}{x_2^3} + \dfrac{4x_1}{x_2^3}\left(2 + \dfrac{x_1}{x_2}\right)e^{x_1/x_2} \end{bmatrix}$

(b) $\begin{bmatrix} 2x_1^{-3}x_2^{-1} & (x_1x_2)^{-2} \\ (x_1x_2)^{-2} & 2x_1^{-1}x_2^{-3} \end{bmatrix}$

1.24 (a) Indefinite

(b) Indefinite

(c) Indefinite

1.25 (a) $\mathbf{H} = \begin{bmatrix} 2 & \frac{1}{2} & \frac{9}{2} \\ \frac{1}{2} & 3 & \frac{1}{2} \\ \frac{9}{2} & \frac{1}{2} & 0 \end{bmatrix} \mathbf{q} = \begin{bmatrix} 0 \\ 2 \\ 0 \end{bmatrix}$

(b) $\mathbf{H}=\begin{bmatrix} 1 & 0 & -3 \\ 0 & 0 & 0 \\ -3 & 0 & 1 \end{bmatrix}$ $\mathbf{q}=\begin{bmatrix} 0 \\ 0 \\ 9 \end{bmatrix}$

1.26 $\mathbf{x}^*=\begin{bmatrix} 9/2 \\ 3/2 \end{bmatrix}$

1.27 $\mathbf{x}^*=\begin{bmatrix} 3 \\ 1 \end{bmatrix}$

1.28 $\mathbf{x}^*=\begin{bmatrix} 3/2 \\ 9/4 \end{bmatrix}$

1.29 $\mathbf{x}^*=\begin{bmatrix} 10/3 \\ 5/3 \end{bmatrix}$

1.30 $\mathbf{x}^*=\begin{bmatrix} 3 \\ 2 \end{bmatrix}$

Chapter 2

2.12 (a) $\mathbf{x}^*=\begin{bmatrix} -23.5 \\ 5.5 \\ 8 \end{bmatrix}$

(b) $\mathbf{x}^*=\begin{bmatrix} 11 \\ 95 \\ 78 \end{bmatrix}$

(c) $\mathbf{x}^*=\begin{bmatrix} 0 \\ 0 \\ 0 \end{bmatrix}$

2.13 (a) Minimum
(b) Minimum
(c) Saddle point

2.14 $[0,0]^T$; $\left[\dfrac{a_1^4}{a_2^2}, \dfrac{a_1^2}{a_2^2}\right]^T$; $\left[\dfrac{3a_1^4}{8a_2^2}, \dfrac{a_1^2}{2a_2^2}\right]^T$

2.15 $h_{11}<0$
$h_{11}h_{22}-h_{12}h_{21}>0$

2.16 (a) Inflection point at 0
Maximum at 27/12
(b) Minimum at 2

2.17 Does not exist

2.18 $x^*=0$

2.19 No finite maximum or minimum

2.20 $\dfrac{\text{Circle perimeter}}{\text{Square perimeter}}=\dfrac{\pi}{16}$

2.21 $[6,6,3]^T$

2.22 $\begin{bmatrix} -5 \\ 7/2 \end{bmatrix}$

2.23 $\sqrt{1.2} - 1.1$

2.24 $\sqrt{1.2} - 1.095$

2.25 $\ln(1.05) - 0.04875$

2.26 $x_{n+1} = x_n - \dfrac{x_n e^n - 2}{(x_n + 1)e^{x_n}}$

2.27 $x_{n+1} = x_n - \dfrac{\sin x_n - \dfrac{x_n + 1}{x_n - 1}}{\cos x_n + \dfrac{2}{(x_n - 1)^2}}$

2.28 $x_{n+1} = x_n - \dfrac{x_n^3 - 5x_n^2 - 6x_n + 4}{3x_n^2 - 10x_n - 6}$

2.29 $\begin{bmatrix} x_{n+1} \\ y_{n+1} \end{bmatrix} = \begin{bmatrix} x_n \\ y_n \end{bmatrix} - \left(\dfrac{x_n y_n}{36 \times 10^6} \right)^4 \begin{bmatrix} 1 - 3000 x_n^{-2} y_n^{-1} \\ 2 - 4000 x_n^{-1} y_n^{-2} \end{bmatrix}$

Chapter 3

3.14 $T_{n+1} = T_n - \dfrac{e^{T_n}(e^{T_n} - 2) + T_n}{e^{T_n}(2e^{T_n} - 2) + 1}$

3.15 $\left[\dfrac{-1 + \sqrt{5}}{2}, \sqrt{\dfrac{-1 + \sqrt{5}}{2}} \right]$

3.16 $[\sqrt{1/2}, \sqrt{1/2}]^T$

3.17 $[2, 2]^T$

3.20 $[\sqrt{1/2}, \sqrt{1/2}]^T$

3.21 $[-16, 12]^T$

3.22 $\left[\dfrac{\pi}{4}, \dfrac{\pi}{4} \right]^T$

3.23 $c_1 \geqslant 4/3$

Chapter 4

4.18 Primal solution: $[1/2, 3/2]^T$
Dual solution: $[0, 1]^T$

4.19 $2\sqrt{2}$

4.22 No

4.24 No

4.25 Yes

4.26 $[5, 1]^T$

4.28 $a_1 \geqslant 4/3$

Chapter 5

5.17

TRIAL	a_i^i	$f(a^i)$	b^i	$f(b^i)$
1	$\frac{21}{13}$	-3.048	$\frac{18}{13}$	-2.93
2	$\frac{23}{13}$	-2.981	$\frac{20}{13}$	-3.0353
3	$\frac{22}{13}$	-3.0305		
$L_6 =$	$\frac{1}{13}$			

5.18 (a) 6
(b) 5

5.19 $\hat{\lambda} = 0.3$
$\lambda^* = 0.5$

Chapter 6

6.14

ITERATION	$[\mathbf{x}^p]^T$	K^*_p	$[\mathbf{x}^{p+1}]^T$
1	$[2, -1, 1]$	$3/8$	$[-1/2, -1/2, -1/2]$
2	$[-1/2, -1/2, -1/2]$	$3/10$	$[-1/5, -1/5, 1/10]$
3	$[-1/5, -1/5, 1/10]$	$3/8$	$[-1/20, -1/20, -1/20]$
4	$[-1/20, -1/20, -1/20]$	$3/10$	$[-1/50, -1/50, 1/100]$

6.15

ITERATION	$[\mathbf{x}^p]^T$	K^*_p	$[\mathbf{x}^{p+1}]^T$
1	$[2, -2, 1, -1]$	$1/2$	$[0, 0, 0, 0]$

Chapter 7

7.22 $\mathbf{P} = I$
rank $\mathbf{P}$ = number of active constraints

7.23 $\mathbf{A}$ has rank m

Chapter 8

8.9 $\mathbf{x}^* = [0,0]^T$

8.10 $\mathbf{x}^* = [1/2, 3/2]^T$

8.11 $\mathbf{x}^* = [3,2]^T$

Chapter 9

9.11 $\mathbf{x}^* = [0,1]^T$

Chapter 11

11.9 (1,3,4,2,1)

11.10 Item 3, Item 1, Item 2

11.11 $\mathbf{x}^* = [0,3]^T$

11.12 $\mathbf{x}^* = [3,1]^T$

Chapter 12

12.16 *Hint*: See Examples 12.4 and 12.5.

12.17 *Hint*: See Problem 12.14.

12.20 *Hint*: See Problem 12.12.

12.21 *Hint*: See Problem 12.13.

Index

A

Absolute minimum, 33
Absolute global minimum, 33
Active constraints, 248
Algorithms, 139
  second-level, 326, 338
  quadratically convergent, 219
Allocation problem, 297, 308
Approximation, 307
  infinite stage, 307, 320
  polynomial, 130
Approximations, successive, 307, 308, 309
Ascent, steepest, 140, 219
Assignment problem, 271, 280
Asymptotic convergence rate, 137
Automatic stopping rule, 166

B

Backward difference, 348
Basic feasible solution, 277, 355, 356, 357, 360
Basic solution, 261, 356
Basic variables, 277, 279, 356
Basic vectors, 360
Basis, 3, 223, 363
Beltrami, E., 200
Block diagonal matrix, 330, 349
Bottleneck problem, 315
Bound, 272
Boundary following, 173
  equality constraints, 173
  methods, 163
Bound
  greatest lower, 92
  lower, 91, 134, 287
Branch, 272
Branch-and-bound-method, 272, 274, 276, 280

C

Candidate solutions, 213
Capacity constraint, 258
Capacity method, 258
  of Houthakker, 258
Central agency, 343
Classical optimization, 33
Closed half space, 5
Colinear vectors, 166
Colinearity, 166
Complementary slackness, 178
Complementary slackness condition, 85
Computational considerations, 44, 51
Computational cost, 325
Computational methods, 67
Computational requirements, with two state variables, 304
Concave function, 7, 25
Concave, strictly, 23
Condition for optimal descent, 224
Conditions, sufficient, 42, 44, 52, 90
Conjugate-direction algorithm, 226
Conjugate directions, 223, 224, 229
Conjugate gradient algorithm, 226

Conjugate-gradient method, 227
Constrained gradient techniques, 163
Constrained optimization, 2, 163
Constrained optimization problems, 197
Constraint elimination, 173
Constraint hyperplane, 276, 368
Constraint, inactive, 178, 248
Constraint inequalities, 364, 368
Constraint matrix, 189
Constraint qualification, 86, 88, 90, 103, 104, 200, 201
  geometrical properties, 88
Constraints, 55, 248
  active, 248
  capacity, 258
  inactive, 248
  inequality, 2, 14, 65, 75, 81
  integer, 274
Continuous gradient path, 151
Continuous variables, 301
Convergence, 47, 68, 142, 148, 154, 200, 219, 307
Convergence condition, 336
Convergence criterion, 205
Convergence
  Finite, 154, 219
  Monotonic, 142, 307
  Of gradient algorithm, 148
  Of Newton-Raphson algorithm, 67
  Order of, 219
  Quadratic, 219
Convergence rate, 45, 129, 137
  asymptotic, 137
Convex, 4
Convex functions, 7, 20, 25
Convexity, 4, 12, 20, 56, 166
  of Lagrangian, 56
Convex polyhedron, 358
Convex sets, 12
Convex, strictly, 7, 23, 24, 48
Convex subset, 93
Coordination, 326
  goal, 330
Cramer's rule, 135
Cubic polynomials, 130
Curve fitting, 321
Curve-fitting problem, 100
Cutting-plane method, 278
Cutting-plane methods, 276
Cutting planes, 276

D

Davidon, W. C., 234
Decentralized plant, 341
Decision process
  multistage, 291, 293, 297, 304, 307, 310, 317
  two stage, 297, 302
Decision variable, 291
Decision vector, 291
Decomposition, 326
  dual feasible, 329, 336, 337, 340, 344, 346
  feasible, 328, 330, 341, 347
  Lagrangian, 91, 326, 328
Definite
  negative, 24
  positive, 24
Degenerate solution, 356
Degrees of freedom, 14
Demand function, 338
Descent methods, 219
Descent, steepest, 140, 219
D'Espo, D.A., 254
Determinants, 12
Diagonal matrices, 154
Dichotomous search, 119, 120, 133
  measure of effectiveness, 133
Differentiable function, 139, 219, 233
  gradient, 139
Differential functions, 2
Dimensionality, 55, 67
Direct computation, 255
Direct enumeration, 292, 302
Directional derivative, 9
Direct methods, 139, 197
Direct optimizing methods, 45
Direct search, 199
Discrete variables, 301
Discretization, 301
Discriminant, 38, 51, 70, 221
Dispatching, 74

Dispatching problem, 74
Dot product, 3
Dual feasible decomposition, 329, 336, 337, 340, 344, 346
Dual feasible method, 339
Dual feasible methods, 335
Dual function, 91, 93, 110
Dual problem, 92, 93, 114, 253, 267, 364, 369
Duality, 11, 253, 343, 364
Duality conditions, 93
Duality theorem, 92, 111
Dynamic optimization, 1
Dynamic programming, 291, 296, 300
  computational requirements, 300

E

Economical dispatch rule, 75
Economic growth problem, 317
Economic interpretation, 337
Economic interpretation of Lagrange mutlipliers, 60
Economy, perfectly competitive, 338
Eigenvalue, 154, 223
  positive, 154
Eigenvector, 223
Elimination, 173, 174
Entering variable, 356
Equal-interval search, 120
  three point, 120
  two point, 120
Equality constraints, 2, 14, 55
Errors, round-off, 232
Euclidean norm, 3
Euclidean *n*-space, 2
Euclidean space, 223
Exhaustive search, 132
Existence of Lagrange multipliers, 63, 210
Experiments, 117
Exterior methods, 167, 200

F

Feasible decomposition, 328, 330, 341, 347
Feasible direction method, 169
Feasible directions, 169
Feasible directions algorithm, 168, 182, 185, 186
Feasible-directions method, 163
Feasible directions, method of, 169
Feasible methods, 332
Feasible point, initial, 173, 179, 182, 189
Feasible points, 14, 28, 170
Feasible region, 170, 173, 200
Feasible set, 14
Feasible solutions, 245, 274, 277
Feasible, strictly, 182
Fiacco and McCormick, 204
Fibonacci, 124
Fibonacci search, 122, 125, 126, 129, 135, 137
  advantages, 126
  algorithm, 125
  disadvantages, 126
  minimal measurement separation, 125
Fibonacci sequence, 124
Finite convergence, 154, 219
Finite optimum, 354
First-level subproblem, 341
First-order Taylor series, 46
Fletcher and Powell, 234
Fletcher-Powell algorithm, 234, 235, 241
Fletcher, Powell, and Davidon, 234
Fletcher-Reeves algorithms, 226
Forms
  indefinite, 24, 49, 50
  quadratic, 10, 12, 21, 27, 50, 71
Forward difference approximation, 348
Functional equation, 294, 307, 318
Functional equation of dynamic programming, 295
Function evaluation, 135, 301
Functions
  positive definite quadratic, 220
  quadratic, 200, 219, 233, 238, 239
  twice differentiable, 12, 38
Fundamental inequality, 94

G

General nonlinear programming problem, 65
Geometric considerations, 256
Geometric properties, 247
GLB, 92
Global minimization, 90
Global minimum, 33, 48, 53, 233, 238, 257
 absolute, 33
 proper, 34
Global optimization, 33
Goal coordination, 330
Golden section, 129
Golden-section algorithm, 127
Golden-section search, 127, 137
Gomory, R., 276, 278
Gomory cut, 278
Gradient, 8, 22, 23, 130, 139, 168
Gradient algorithm, 139, 145, 148, 150, 152, 155, 159, 205 333, 335, 336, 343
 convergence, 148, 159
 discrete, 152
 rate of functional improvement, 150
 step size, 152
Gradient direction, 139, 153
Gradient
 normalized, 146, 168
 optimal, 130
Gradient path, 140, 149, 152
 continuous, 152
 disadvantage, 149
Gradient, projection method, 163, 174, 178, 189, 190, 192
Gradients, 179
Gradient techniques, 163
 constrained, 163
Gradient vector, 221
Graph, 272
Greatest lower bound, 92
Grid points, 301, 302

H

Half space
 closed, 5
 open, 5
Hemstitching, 164
Hemstitching algorithm, 179
Hessian matrix, 9, 23, 36, 41, 45, 69, 94, 153, 216, 233, 235, 247, 346
Hildreth and D'Espo, 254
Houthakker, H. S., 258
Hyperellipsoid, 245
Hyperplane, 4, 276, 368
Hyperplane of support, 6, 21, 178

I

Identically zero, 49
Identity matrix, 176
Imbedded, iteration, 205
Inactive constraints, 178, 248
Incremental cost, 292
Indefinite, 11, 24, 49, 50
Independent variables, 351
Indeterminate forms, 22
Indirect method for optimization, 44
Indirect methods, 139, 197
Induction, 241, 242, 309
Inequalities, 247, 362
Inequality constraints, 2, 14, 65, 75, 81, 364, 368
Inequality, Schwarz, 3
Infinite
 stage approximation, 320
 stage process, 307
Inflection point, 39
Initial feasible point, 173, 179, 182, 189
Initial points, 150
Inner product, 3
Integer constraints, 274
Integer linear programming, 274, 276
Integer program, 269
Integer program, linear, 269
Integer programming, 15, 269
Integer programming, nonlinear, 269
Integer-valued functions, 276
Interaction, 326
Interconnections, 349
Interconnecting variables, 347
Interconnections between sectors, 343
Interior methods, 203
Interpolation, 303
 linear, 303
 quadratic, 129, 135, 151

Interval of uncertainty, 117, 119, 125, 127
Interval of uncertainty, constant, 127
Inverse penalty function, 213
Iterative numerical procedures, 139
Iterative procedure, 205

J

Jacobian, 88

K

Knapsack problem, 269, 283
Kuhn-Tucker Conditions, 82, 83, 84, 86, 94, 97, 101, 103, 178, 212, 247, 253, 260, 333
  necessary, 86
  sufficient, 86
Kuhn-Tucker, constraint qualification, 103
Kuhn-Tucker saddle-point conditions, 82
Kuhn-Tucker stationary conditions, 82, 90, 97, 98
Kuhn-Tucker theorem, 94, 247, 263
Kuhn-Tucker theory, 201

L

Lagrange, 55
Lagrange multipliers, 55, 63, 69, 73, 141, 197, 210, 247, 264, 304
  economic interpretation of, 60
  existence of, 63, 73, 210
Lagrangian, 55, 67, 72, 76, 92, 176, 200, 212, 252, 327, 328, 329, 337, 339, 343
  computational solution of, 67
  decomposition, 91, 326, 328
  saddle point of, 92
Lagrangians, subproblem, 330, 335
Large-scale mathematical programming, 325
Large-scale programming, 325
Lasdon, L., 326
Leaving variable, 356
Leonard of Pisa, 124
Level curves, 15, 27, 256, 368
l'Hospital's rule, 22
Line, 4
Line segment, 4
Linear algebra, 223
Linear equations, 254
Linear independence, 224
Linear integer programs, 269
Linear interpolation, 303
Linearizing, 177
Linearly independent, 224
Linear objective function, 351
Linear program, integer, 274, 276, 269
Linear programming, 1, 19, 30, 91, 170, 245, 325, 351
  mixed-integer, 269, 276
Linear programming problem, 171, 183
Linear subspace, 178
Linear vector space, 3
Load lines, 366
Local maximum, strong, 49
Local minimum, 33, 47, 48, 49, 149
  strong, 40
  weak, 34
Local optimum, 140
Logarithmic penalty function, 208
Lower bound, 91, 134, 287
  greatest, 92

M

Marginal cost, 63
Marginal return, 63
Master problem, 330
Mathematical preliminaries, 2
Mathematical programming, large-scale, 325
Matrix, 64, 154
  block diagonal, 330, 349
  negative definite, 11, 49
  negative semidefinite, 11, 49
  nonsingular, 21, 59, 64, 176
  positive definite, 11, 22, 24, 25, 27, 42, 49, 50, 154, 223, 239, 240
  positive semidefinite, 11, 49
  projection, 177, 187, 191
  rank of, 64, 86
  square, 223
  symmetric, 9, 222, 264

Maxima, 33
Maximal rank, 88
  of Jacobian, 88
Maximization problem, 17
Maximum, 53
  relative, 52
  strong local, 49
McCormick, Fiacco and, 204
Method of feasible directions, 169
Method of Hildreth and D'Espo, 254
Method of parallel tangents
Method of Theil and Van de Panne, 250, 266
Metric, 234
Minima, 33
Minimization algorithms, quadratically convergent, 219
Minimization, one-dimensional, 117, 147, 225, 229, 234, 258
Minimization problem, 17
Minimizing, 91
Minimizing down the line, 130
Minimizing nonconvex functions, 91
Minimizing nondifferentiable functions, 91
Minimum, 33, 53
  global, 33, 48, 53, 90, 233, 238, 257
  local, 33, 47, 48, 49, 140, 149
  proper global, 34
  relative, 52
  strong local, 40
  weak local, 34
Minors, principal, 12, 22, 27
Mixed integer linear program, 276
Mixed integer program, 269
Modified gradient direction, 242
Modified simplex methods, 260
Monotonic convergence, 142, 307
Multilevel problem, 349
Multimodal, 37
Multiple gradient summation, 167, 180, 181
Multiplication, scalar, 2
Multipliers
  Lagrange, existence of, 73, 210
  Lagrange, 55, 63, 69, 73, 197, 210, 247, 264, 304
Multistage decision process, 291, 293, 297, 304, 307, 317, 319
Mutual conjugacy, 224
Mutually conjugate, 223
Mutually conjugate directions, 239
Mutually orthogonal, 223

N

National economy, 317
Necessary conditions, 42, 44, 55, 90, 94, 99, 335
  for a minimum, 42
  Kuhn-Tucker, 86, 101, 212, 247, 253, 260, 333
  second-level, 334, 336
Negative definite matrix, 11, 49
Negative gradient path, 182
Negative semidefinite matrix, 11, 49
Net profit, 338
Newton-Raphson algorithm, 67, 69, 102
Newton-Raphson, convergence, 67
Newton-Raphson method, 45, 51, 67
Newton's algorithm, 46
Newton's method, 45
Node, 272
Nonconvexity, 166
Nondegenerate solution, 356
Nonlinear integer programs, 269
Nonlinear programming, 1, 14, 27, 90, 325
  primal, 11
  primal problem, 90
Nonlinear programming problem, 65
Nonnegative definite, 36
Nonnegativity, 178
Nonnegativity constraint, 96
Nonsingular matrices, 21
Nonsingular matrix, 59, 64, 176
Nonsingular submatrix, 64
Norm, 2, 234
Normal, 5
Normal form, 351
Normalization, 183
Normalized gradient, 146, 168
Normalized projection matrix, 177
Normalizing, 10, 133, 184
Normalizing factor, 142
Numbers, real, 220

Numerical algorithms, 325
Numcrical methods, 45, 67
Numerical procedures, iterative, 139, 205
Numerical results, 45

O

Objective function, 1, 351
  linear, 351
Objective functions, 51
One-dimensional minimization, 147
One-dimensional search, 117, 225, 229, 234, 258
Open half space, 5
Optimal allocation of a scarce resource, 61
Optimal descent, 224, 225
Optimal-descent method, 239
Optimal gradient, 130, 149, 155, 225
  algorithm, 155, 219
  orthogonal steps, 155
Optimality, principal of, 294
Optimal step size, 147
Optimal tour, 270, 299
Optimization, 1, 33
  constrained, 2, 163, 197
  dynamic, 1
  second-level, 328, 330
  static, 1
Order of convergence, 219
Orthogonal, 3
Orthogonality, 3, 144
Overconstrained, 14

P

Parallelogram rule, 2
Parallel tangents, 158
  method of, 150
Parameters, 328
PARTAN, 150, 158, 235
Partial Derivatives, 8
Partitioning, 349
Penalties, 198
Penalty function, 209, 213
  inverse, 213
  logarithm, 208
Penalty function methods, 163, 197, 200, 212
Penalty function, reciprocal, 213
Penalty functions, 197
Perfectly competitive economy, 338
Point of tangency, 16
Polynomial approximation, 130
Polynomials:
  cubic, 130
  quadratic, 130
Positive definite, 11, 22, 24, 25, 27, 42, 49, 50
Positive definite matrix, 223, 229, 240
Positive definite quadratic form, 227
Positive definite quadratic function, 220, 242, 264
Positive eigenvalues, 154
Positive semidefinite matrix, 11, 49
Powell, M.J.D., 229
Powell's algorithm, 235
Powell's method, 229
Prices, shadow, 61
Primal nonlinear programming, 90
Primal problem, 11, 90, 92, 108, 364
Principal minors, 12, 22, 27
Principal of optimality, 294
Process, multistage decision, 291, 293, 297, 304, 307, 317, 319
Production levels, 338
Programming, quadratic, 245, 251, 260
Projection matrix, 177, 187, 191
  normalized, 177
Proper global minimum, 34
Purchase price, 338

Q

Quadratic convergence, 225
Quadratic convergence rates, 219
Quadratic equation, 221
Quadratic form, 10
Quadratic forms, 12, 21, 27, 50, 71
Quadratic function, 219, 220, 233, 238, 239
  positive definite, 220, 242, 264
Quadratic interpolation, 129, 135, 151
Quadratic polynomials, 130
Quadratic programming, 245, 260

Quadratic programming problem, 251
Quadratically convergent algorithms, 241
Quadratically convergent minimization algorithms, 219

R

Rank of a matrix, 64, 86
Reactor, 76
Reactor optimization, 76
Real numbers, 220
Reciprocal penalty function, 213
Recurrence relation, 45
Recursion equation, 310
Recursive scheme, 300
Relative maximum, 52
Relative minima, 33
Relative minimum, 33
Relaxation, 340
Reliability problem, 312
Resource allocation, 297, 308
Rounding, 271
Round-off errors, 232
Round-trip tour, 270, 285

S

Saddle point, 33, 38, 48, 49, 50, 51, 70, 107, 336
Saddle-point conditions, 90, 93. 106
Saddle point
  necessary conditions, 107
  sufficient conditions, 107
Scalar, 4
Scalar multiplication, 2
Scarce resource, allocation of, 61
Schwarz inequality, 3
Search, one-dimensional, 117, 147, 225, 229, 234, 258
Second derivatiries, 9
Second-level algorithm, 326, 338
Second-level necessary condition, 334, 336
Second-level optimization, 328, 330
Second-level problem, 339
Selling price, 338
Sensitivity coefficients, 61
  Lagrange multipliers as, 61
Sequence of points, 45
Sequential unconstrained minimization techniques (SUMT), 203
Shadow prices, 61
Shortest route, 299, 309
Sign definiteness, 13, 27
Simplex algorithm, 358
Simplex method, 260, 261
Simplex tableau, 262, 360, 363
Simultaneous equations, 46
Singular point, 22
Slack variables, 65, 170, 247, 263, 351, 362, 364
Solutions
  candidate, 213
  tree of, 272, 284
Solved problems, 47
Space, open half, 5
Span, 3, 223
Sparsity, 349
Square matrix, 223
Stage variable, 291
Standard form, 17
Starting rules, 150
State equation, 319
State variable, 291
State vector, 291
Static optimization, 1
Stationarity, 55
Stationarity conditions, 83, 85, 93, 98, 113
Stationary points, 43, 50, 52, 328
Steepest ascent, 140
Steepest descent, 140
Steepest descent method, 219
Step function, 200
Step length, 234
Step size, 130, 147, 170
  optimal, 147
Stirling's formula, 300
Stopping rule, automatic, 166
Stopping rules, 150, 166
Storage requirements, 300
Strictly concave, 23
Strictly convex, 7, 23, 24, 48
Strictly feasible, 182

Strong local maximum, 49
Strong local minimum, 40, 49
Strong relative minimum, 33
Strong relative local minimum, 33
Strong relative maximum, 42
Structure, two-level, 331
Subproblem Lagrangians, 330, 335
Subproblems, 334, 349
  first-level, 341
Successive-approximation, 307, 309
Successive-approximation technique, 307, 308
Sufficient condition, 52
Sufficient conditions, 42, 44, 90
Sufficient conditions for a strong relative minimum, 42
Sufficient conditions, Kuhn-Tucker, 86
SUMT, 203, 205
Supporting hyperplane, 21, 178
Sylvester's theorem, 12, 22, 37
Symmetric matrices, 9
Symmetric matrix, 222, 264
Symmetric positive definite matrix, 234
Symmetric trails, 127
Symmetry, 9

T

Tangent hyperplane, 170
Tangent, point of, 16
Taylor series, 40, 41, 46, 47, 94, 153, 169, 220, 234
Theil & Van de Panne, 250
Theorem of Kuhn and Tucker, 81
Theorem, Sylvester's, 12, 22, 37
Three-point equal interval search, 120, 134
Total cost, 292
Tour, optimal, 270, 299
Transportation problem, 281
Traveling salesman problem, 270, 285, 286, 299
Tree of solutions, 272, 284
Trials, 117, 217
  symmetric, 127
Truncation, 271
Twice differentiable functions, 12, 38
Two-level structure, 331
Two-point equal-interval search, 120, 134
Two-stage decision process, 297, 302

U

Unbounded solution, 353
Uncertainty, interval of, 117, 119, 125, 127
Unconstrained gradient techniques, 139
Unconstrained optimization, 2, 33, 40, 49, 197
Unimodal, 37, 140
Unimodal functions, 117
Unimodality, 119
Upper bound, 130, 281

V

Van de Panne, C.R., 250
Variable-metric algorithm, 232, 237
Variables
  slack, 65, 170, 247, 263, 351, 362, 364
  stage, 291
  state, 291
Vector normal, 5
Vectors, 224
Vector, state, 291
Vector summation, 167
Vector summation of gradients, 167
Vertex, 358

W

Weak local maximum, 49
Weak local minimum, 34, 49
Wismer, David A., 326, 349
Wolfe, Philip, 261
Wolfe's method, 261, 263
Wolfe's modified simplex method, 262

Z

Zigzag algorithm, 164